The Chimpanzees of the Budongo Forest

The Chimpanzees of the Budongo Forest

Ecology, Behaviour, and Conservation

Vernon Reynolds

*Professor Emeritus of Biological Anthropology,
Oxford University*

OXFORD
UNIVERSITY PRESS

OXFORD

UNIVERSITY PRESS

Great Clarendon Street, Oxford OX2 6DP

Oxford University Press is a department of the University of Oxford.
It furthers the University's objective of excellence in research, scholarship,
and education by publishing worldwide in

Oxford New York

Auckland Cape Town Dar es Salaam Hong Kong   Karachi
Kuala Lumpur Madrid Melbourne Mexico City Nairobi
New Delhi Shanghai Taipei Toronto

With offices in

Argentina Austria Brazil Chile Czech Republic France Greece
Guatemala Hungary Italy Japan Poland Portugal Singapore
South Korea Switzerland Thailand Turkey Ukraine Vietnam

Oxford is a registered trade mark of Oxford University Press
in the UK and in certain other countries

Published in the United States
by Oxford University Press Inc., New York

© Oxford University Press 2005

British Library Cataloguing in Publication Data

(Data available)

Library of Congress Cataloging-in-Publication Data

Reynolds, Vernon.
The chimpanzees of the Budongo Forest : ecology, behaviour, and conservation /
Vernon Reynolds.
      p. cm.
Includes bibliographical references and index.
ISBN 0–19–851545–6 (alk. paper) — ISBN 0–19–851546–4 (alk. paper)
1. Chimpanzees—Uganda—Budongo Forest.   I. Title.
QL737.P96R46 2005
599.885′096761—dc22

2004030381

Typeset by Newgen Imaging Systems (P) Ltd., Chennai, India
Printed in Great Britain
on acid-free paper by
Biddles Ltd., King's Lynn

ISBN 0–19–851545–6 (Hbk)   978–0–19–851545–6
ISBN 0–19–851546–4 (Pbk)   978–0–19–851546–3

10 9 8 7 6 5 4 3 2 1

To the people — past, present and future — who have helped, do help now, and will in future years help make the Budongo Forest Project all the many things it is and aspires to be. And to the chimpanzees themselves; long may they live and prosper!

Contents

Acknowledgements

Support for the Budongo Forest Project (BFP), which has made this study of the Sonso chimpanzees possible, has come from very many sources. I especially want to acknowledge the help of Jane Goodall, Andy Johns and Derek Pomeroy in the very early stages when it was sink or swim for BFP, and to Jeff Burley, Howard Wright and Per Wegge a little later when their help made all the difference. I am indebted to the late Peter Miller for his support and encouragement in the early days, and also to Moses Okua, John Okedi and Lawrence Kiwanuka for permission to study the chimpanzees living in the Budongo Forest Reserve.

My family has been a constant source of encouragement and help. My wife Frankie was with me from the beginning, when we lived in Budongo Forest for a year, cutting and creeping through the forest to watch the shy and frightened chimpanzees. She has been a constant source of help and optimism for which I am deeply grateful. My son Jake underestimates the huge contribution he made during his year as a volunteer organizing the building of our wonderful camp out of ruined sawmill buildings in the heart of the forest. My daughter Janie likewise came out in the early years and made a video about the work of the Project which has been shown at conferences and meetings and has helped put us on the map.

I thank my good friend David Dickins for allowing me to accompany and assist him on his fieldwork visits to Lundy Island to study kittiwakes; during the years we did this in the 1980s I realized that I too wanted to get more involved in fieldwork and this got me started on the idea of going back to Budongo. For the final irresistible stimulus to go back I thank Shirley McGreal (see Appendix F).

For their support over the years I would like to thank Chuck Southwick, Doug Candland, Duane Quiatt and Janette Wallis. All have been tremendous assets at times when the going got tough.

I thank my collaborators at Sonso (we are still collaborating!), in particular Janette Wallis, Linda Vigilant, Leslie Knapp and Jenny Greenham. Their inputs into the work reported in the book have been considerable.

Without the support of our Directors and Assistant Directors the BFP would have been a pale shadow of itself: Chris Bakuneeta, Andy Plumptre, Jeremy Lindsell, Fred Babweteera, Mark Attwater, Hugh Notman, Lucy Beresford-Stooke, Michael Mbogga, Sean O'Hara, Gerald Eilu and Nick Newton-Fisher — all have made mighty contributions to the Project's success.

I want to thank Rhiannon Meredith, Susie Whiten, Robin May, Wilma van Riel, Juliet Craig, Richard Gregory, Rory Lynch, Michelle Higgins, Grant Joseph, Samantha Ralston, Nicky Jenner and Ilona Johnston for their voluntary work which has contributed to the Project in many ways: they have renovated the herbarium, created the laboratory, museum and library at Sonso, helped with our extension projects in surrounding villages, entered data on computers and applied paint to walls that badly needed it.

David Bowes-Lyon first mapped our trail system; he used a compass, tape measure, paper and pencil and his map has been useful to all of us down the years. Glenn Bush later mapped the new western trails, and Emily Bethell mapped the new eastern and southeastern sections. The original maps in this book were made by Leela Hazzah and Mary Reuling, International Cartography and Wildlife Conservation (ICWC). To all I am deeply grateful.

Gladys Kalema came from Kampala while she was the Uganda Wildlife Authority vet and taught our staff the importance of strict hygiene, also what to do in the event of finding a sick or injured or dead chimpanzee. Thanks to Gladys for making us conscious at an early stage of the importance of doing all we could to avoid disease transmission to our chimpanzees.

Fred Babweteera, Janette Wallis and Chris Fairgrieve were tremendously helpful in writing the Reports of our conferences and the workshop held at Nyabyeya Forestry College. Zoe Wales deserves a special mention for the superb job she did in writing the document *What We Know About Budongo Forest*.

Without the work of the many students and senior researchers who have worked at Budongo this book could not have been written. Their names are mostly included in Appendix F. Their contributions will be abundantly evident to all who read the book. I wish to thank most sincerely all those who have given permission to use the results of their studies or to use their figures. They are individually acknowledged in the section 'Permissions'.

For statistical help in writing papers and in the preparation of this book I should like to thank Francis Marriott.

Thanks to all who have commented on chapters or sections of text: Bob Plumptre, Tim Synnott, Nick Newton-Fisher, Janette Wallis, Richard Wrangham, Toshisada Nishida, Melissa Emery Thompson, Andy Plumptre, Linda Vigilant, Fred Babweteera, Duane Quiatt, Tweheyo Mnason, Jo Thompson, Phyllis Lee, Alan Dixson, Lucy Bates, Sean and Catherine O'Hara, Katie Fawcett, Kate Hill, Julie Munn, Hugh Notman, Lori Oliver, Emma Stokes, Dick Byrne, Emily Bethell, Kate Arnold, Andy Whiten, Ella Chase, Jenny Greenham, Zephyr Kiwede, Geresomu Muhumuza, Katie Slocombe and Klaus Zuberbühler.

The Boise Fund and the Jane Goodall Institute (JGI) via Richard Wrangham provided start-up funds for the BFP, enabling us to get off the ground, and for that I am forever grateful. For help with core funding for the Project and its staff, without which we could not have survived down the years, I am very indebted to the Overseas Development Administration (now the Department for International Development) through its Forest Research Programme for funding from 1991 to 1997 and to the Norwegian Agency for

Development through its support to Makerere University Faculty of Forestry and Nature Conservation for funding from 1997 to 2005. The National Geographic Society, through its Committee for Research and Exploration, has assisted with the funding of our chimpanzee studies down the years, and for that I am hugely grateful; it has helped to fund demographic monitoring of the Sonso community, as well as research in the fields of reproduction, phytochemistry of food choices and genetics. The Rainforest Action Fund, the International Primate Protection League, the British Council, the Bristol and West of England Zoological Society, the Percy Sladen Trust, the British Ecological Society and the Leakey Foundation have all provided much needed grant support. The US Agency for International Development helped us in the early days with funds to enable us to renovate and, where necessary, rebuild our camp. Conservation International, through its Margot Marsh Foundation, and the WildiZe Foundation, have assisted with much needed funding. The American Society of Primatologists provided funds to Janette Wallis to enable her to begin the Kasokwa project. The International Fund for Animal Welfare, Cleveland Zoological Society and Oakland Zoo have given us funds for staff development, snare removal, two forest guards at Kasokwa, student scholarships and community education. St Andrew's Church in Alfriston (and in particular the rector, Frank Fox-Wilson) have helped raise funds for our staff housing project, Alfriston Women's Institute has provided a bursary for further education for girl students and Alfriston Primary School has helped the primary school at Nyabyeya. Oxford University gave me a secure base to work from. Geoffrey Harrison, Head of the Institute of Biological Anthropology, my department at Oxford, was extremely generous at all times in support of the Project; I am deeply grateful to him for his unstinting help and encouragement.

The Budongo Forest Project, at the time of writing, looks forward to closer cooperation with the Royal Zoological Society of Scotland, and with the Scottish Primate Research Group, in particular the Department of Psychology at St Andrews University.

Individuals who have given to BFP, emotionally and in other ways, include Jo Thompson, Siddhartha Singh, Liz Rogers, Jim and Sandy Paterson, Beatrice Hahn, Chris Fairgrieve, and all the others who have provided items small and large and support of so many kinds for the BFP and its staff.

A big thankyou also to Andrew Brownlow and Nick Newton-Fisher who have created and kept up to date the truly excellent website for the Budongo Forest Project, www.budongo.org.

And finally I thank the readers of my initial proposal for this book, who made valuable comments, my editors at Oxford University Press, Ian Sherman, Abbie Headon and Anita Petrie for their help and patience, and Alison Jolly who wisely suggested that I change the title of Chapter 10 from 'The Human Backdrop' to 'The Human Foreground' — an insightful suggestion if ever there was one.

Permissions

I thank the following for permission to use their text, their figures or their work in the preparation and writing of this book:

Kate Arnold, Fred Babweteera, Lucy Bannon, Lucy Bates, Emily Bethell, Wayne Boardman, Ella Chase, Benedict Dempsey, Melissa Emery Thompson, Chris Fairgrieve, Katie Fawcett, Jenny Greenham, David Harris, Kate Hill, Kirstin Johnson, Helen Jones, Zephyr Kiwede, Mikala Lauridsen, Jeremy Lindsell, Heidi Marriott, Tweheyo Mnason, Geresomu Muhumuza, Julie Munn, Nick Newton-Fisher, Hugh Notman, Sean and Catherine O'Hara, Lori Oliver, Andy Plumptre, Graham Preece, Duane Quiatt, Donna Sheppard, Katie Slocombe, Bob Smith, Emma Stokes, John Waller, Janette Wallis, Cristy Watkins, Andy Whiten and Klaus Zuberbühler.

Figures and tables taken or modified from the work of the above and others are acknowledged in the appropriate place.

Maps

Introduction by Jane Goodall

I have known Vernon Reynolds since we were both students. I had been working at Gombe for about three years and Vernon and his wife Frankie had spent a year studying chimpanzees in Uganda's Budongo Forest. We discussed our plans. I was on my way back to Tanzania to continue my study of the chimpanzees. Vernon was going to find a teaching job at a university for a while and then return to the Budongo Forest. However, while my plan worked out fine, the political unrest in Uganda that followed the seizing of power by Idi Amin forced Vernon to put off his return to Budongo. Uganda became increasingly unsafe for foreign researchers, and Vernon immersed himself in primate academia at Oxford University.

His return to the field was triggered by a newspaper cutting sent to him by the chairwoman of the International Primate Protection League, Dr Shirley McGreal. It told the story of two infant chimpanzees who had been smuggled out of Uganda and sent to Dubai — where they were discovered, confiscated and sent back again. They had almost certainly been captured in the Budongo Forest, the place that had captured Vernon's heart almost three decades earlier. Two infant chimpanzees — infants who should have been safe with their mothers, learning chimpanzee lore in the safety of their social group. It reawakened his all but forgotten plan to return to Uganda — this time not only to learn about but also to try to protect the chimpanzees.

First, though, he had to get funding. Eventually, after many disappointments, he managed to get enough money for a return fare to Uganda and, in 1990, 27 years after leaving, he was back in Budongo. I can imagine his excitement when he actually glimpsed chimpanzees in the forest. But they were scared and moved off hastily through the trees. Hardly surprising as poachers had been roaming the forest, shooting females and stealing their infants. And setting snares. It was clear that something had to be done, and urgently.

Back in the UK he wrote to me, begging for help. And amazingly, although we were a tiny organization, the Jane Goodall Institute-UK (JGI-UK) was able to find a small amount of money, enough for Vernon to return to Budongo and employ a keen and talented young Ugandan, Chris Bakuneeta. Chris set up a camp, hired and trained some field staff and marked out a study site in the centre of the forest. Once things had got going on the ground it was easier to apply for grants, and within a year a basic research camp was firmly established. It was at about that time that I visited the project. I found

Chris to be enthusiastic about field work, and wanting to get further education so that he could do more to help study and protect the chimpanzees he had come to know. Each night we talked about the Budongo chimpanzees, compared their behaviour with those at Gombe and discussed the many dangers that beset them. Chris was full of questions — so was I!

The original camp has now become a fine research station in the heart of the Budongo, with accommodation for staff, students and visitors. Vernon realized from the outset the importance of attracting good students, for he himself had a full-time job at Oxford and could only spend a couple of months a year in the field. He encouraged students to get grants to work with the chimpanzees, and this book describes, among other things, the results of the many studies they have made of the Sonso community. As well, some students researched forest ecology and others were encouraged to study the impact on the habitat of the people who lived around and made use of the forest.

All these studies have added to our growing understanding of chimpanzees, their ecology and behaviour. The structure of the DNA of chimpanzees and humans differs by only about 1% and there are clear-cut biological similarities in blood composition, immune responses and the anatomy of the brain. It has become increasingly clear that we humans are not the only beings on this planet with distinct personalities, minds capable of rational thought, and emotions such as happiness and sadness, fear and despair, mental as well as physical suffering. Chimpanzees, like humans, are capable of compassion and altruism. And like us they are capable of brutal behaviour, even a kind of primitive warfare. There are long-term supportive and affectionate bonds between family members and a long period of childhood during which learning, not only resulting from trial and error and social facilitation, but also from observation and imitation, plays an important role in the acquisition of adult behaviour. In all places where chimpanzees have been studied there are different traditions, in tool-using and tool-making, feeding and a whole variety of social behaviours. Many scientists (including myself) believe that these are cultural differences. The tragedy is that these primitive cultures, along with the chimpanzee communities that practise them, are disappearing before we have been able to document the extent of behavioural flexibility across the chimpanzees' range.

It became clear to Vernon, as it had to me in Tanzania, that the future of the chimpanzees depends to a large extent on the attitude of the local people. It is very obvious that many of their activities in the Budongo have a direct and negative impact on the forest and its fauna. Most horrible is the setting of snares. These are designed to catch small antelopes and wild pigs for food, but chimpanzees sometimes get the wire nooses around their hands or feet. Typically they are able to break the wire but this pulls the noose tightly around the hand or foot. They suffer excruciating pain as the blood supply is cut off: some die of gangrene; many eventually lose the hand or foot. In the Sonso community as many as one-third are maimed.

The problems are grim — and it is the same for chimpanzees across most of their range in Africa. The mushrooming human population needs ever more land for growing crops and grazing cattle and is gradually invading more of the chimpanzees' forests. The bushmeat trade, the *commercial* hunting of wild animals for their meat, is decimating chimpanzees and everything else, especially in West and Central Africa. This has been made possible by the logging companies that build roads deep into the heart of the last rain forests, opening the way for hunters from the towns. And still, in some parts of their range in Africa, chimpanzee mothers are shot simply so that the hunter can steal the infants and sell them — as pets, or to the international entertainment industry.

Chimpanzees are recognized as an endangered species — their numbers have plummeted from about two million a hundred years ago to some 150 000 today, spread over 21 African nations. It is illegal to hunt, kill and sell endangered species in all countries that have signed onto the Convention on International Trade in Endangered Species (CITES) — and that includes all the chimpanzee range countries. But even in areas officially protected as reserves, such as the Budongo Forest, there is little incentive for government officials to enforce the law.

Thus the future for chimpanzees in Africa is grim. How shocking if we allow them to become all but extinct in the wild, surviving only in small heavily protected areas and in captivity. They need all our help. This is why the work in the Budongo Forest is so important, for research projects focus attention on specific areas and animals, attract tourism and provide jobs. How fortunate that Vernon did not give up on his dream of returning to the forest. His is one of the growing number of voices speaking out on behalf of an extraordinary species, our closest living relative. Thank you Vernon for caring, for persisting, and for writing this book.

1. *The Budongo Forest*

Up early, fried mashed spuds from last night + fried egg, all done in oil and v. tasty ... Geresomu takes the temp (max and min) & rainfall readings at 7 a.m. People are getting busy. Kingfishers, doves and hornbills with a background of colobus make up the morning sounds, plus a few human sounds — chopping, children's voices, sound of gum boots walking across the wet grass (22 March 1999, 7 a.m.).

The Budongo Forest lies like a sleeping giant at the top of the Albertine Rift, part of the great Rift Valley that cuts through Africa from North to South (Fig 1.1). It is a large forest by Ugandan standards, though by comparison with the great Congo forest across Lake Albert to the west it is tiny. Altogether it comprises 435 km^2 of continuous forest cover. The land undulates gently, with an overall downward slope from southeast to northwest, so that its rivers flow towards the northwest, leaving the forest to run down the escarpment that leads to the Butiaba Flats along the eastern side of the Rift Valley. These rivers include one, the River Sonso, that runs past the site of our Budongo Forest Project (BFP) and provides us with water in the dry season.

A good place to see the whole of Budongo Forest is from the fire tower on Nyabyeya Hill, above the Nyabyeya Forestry College situated about 5 km from our Sonso camp. When you climb up to the tower and look out over the forest stretching away to the west, north and east, you cannot help being impressed by the sea of trees as far as the horizon. The forest looks quiet, undisturbed and timeless. Towards the east, at the forest edge, you may see a tiny vehicle making its way along a road or a wisp of smoke where someone has a fire going, and further away to the east you can perhaps see the Kinyara Sugar Works factory with its permanent pall of steam. But the general impression is one of dark green, luxuriant vegetation (Synnott 1985 records 725 species of trees, epiphytes, lianas, herbs and shrubs in Budongo Forest).

This forest, so quiet and peaceful from afar, is in fact teeming with animal as well as plant life, and inside the forest there is constant activity and noise. It is home for an estimated 584 chimpanzees (Plumptre *et al.* 2003), together with thousands of redtail monkeys, blue monkeys, black and white colobus monkeys and forest baboons. Among its 226 bird species (128 of them true forest birds, according to Friedmann and Williams, 1973) it has the rare Nahan's francolin, three species of black and white casqued hornbills, turacos, grey parrots, brilliantly coloured forest kingfishers and forest

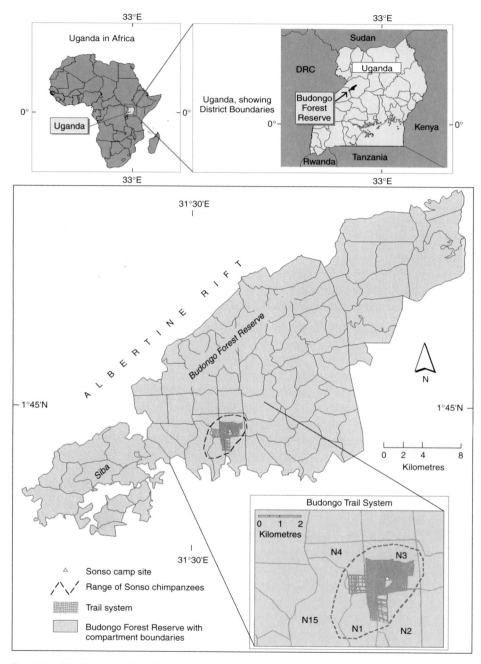

Fig. 1.1:   The Budongo Forest Reserve showing Sonso camp site and trail system (copyright Hazzah and Reuling).

guinea fowl. It has blue duikers, red duikers, bushbuck, elephant shrews, bushpigs, red river hogs, porcupines and honey badgers. Two thousand elephants used to come to Budongo in the course of their annual migrations from the Butiaba flats along the shore of Lake Albert, up the escarpment, into the forest and then back down to Lake Albert again, but today there are no elephants in the forest, all having been killed for food by soldiers and others during the 1971–1986 civil war that ruined Uganda. This situation will probably change slowly as the elephant population in Murchison Falls National Park to the north of Budongo recovers, and elephants begin to return to their old patterns of movement.

History of the forest

Nor is the forest timeless. Every forest has its own particular history. Looked at in the long term, forests come and go as the climate becomes wetter or drier. At times Budongo has been joined to its sister forests along the western margin of Uganda. Today the closest of these forests is Bugoma Forest, separated from Budongo by a distance of 40 km. Beyond Bugoma are Uganda's other western forests: Kibale, Ruwenzori, Maramagambo, Kasyoka-Kitomi, Semliki, Kalinzu, Itwara and Bwindi. Many centuries ago these now separate forests formed a continuum of trees emerging from the Congo forest just south of the Ruwenzori mountains, where the Semliki Forest now lies, and then stretching up along the eastern side of the mountains where Kibale Forest now lies, and on to Bugoma and finally Budongo, the most northerly part of this ancient forest. With the coming of cooler and, in Africa, drier conditions the continuous forest became divided into the series of forests we see today. Indeed, at various times these forests were smaller than they are now. Archaeological investigations in Bugoma Forest have uncovered traces of human settlement (pottery, evidence of fire, remains of cattle) deep within what is now forest; it was woodland or grassland then.

Jolly *et al.* (1997) report on a study of the montane forests of western Uganda from the end of the Pleistocene or late glacial times to the present. Although their study is of montane forests we can expect that drier and wetter periods there would coincide with those lower down. At the time of the last glacial maximum (ice age) around 18 000 years ago, because water was locked up as ice in the northern and southern hemispheres, sea levels globally were lower than they are today and this in turn led to drier conditions in the tropics than we see today. We can thus envisage a time some 18 000 years ago when the forest cover in western Uganda (as well as in the region as a whole) was appreciably less than it is today. It may be that the Budongo forest did not exist at that time, conditions having become too dry for its survival. Since then conditions have become wetter again. During the Early Holocene, from around 11 000 years ago, lower montane forest was on the increase, with a rich forest flora by 10 000 years ago. The same may well be true of semi-deciduous forests such as Budongo. In the Mid-Holocene around 3000–4000 years ago, there were clear signs of seasonality, with wet and dry periods such as we have in Budongo today. In the late Holocene, from around 2300 years ago, there is evidence of forest disturbance, presumably resulting from human activity, i.e. cultivation. This may mark the onset of the Early Iron Age in the region.

With the onset of human activity we have two sets of forces shaping the forests. On the one hand we have the climate, with forests expanding during wetter periods and contracting during drier ones. On the other we have the human factor, with cultivation and cattle herding increasing and decreasing as human populations grew in number and then, because of droughts and diseases, fell back again.

Forest expansion at lower altitudes may have reached a maximum around 7000–5000 years ago, at which time Budongo may have been the northernmost part of the long continuous forest already described, running from the Semliki junction with the Congo forest in a northeasterly direction along the eastern side of the Albertine Rift. After this time, slow drying became the dominant process of climatic change, with a particularly dry period from 1000 to 700 years ago. This may have reduced forest cover to what are now riverine areas, with drought-induced human migrations bringing people into close proximity with the remnant forests at that time, such as the earthworks found in the Bugoma Forest to the south of Budongo already referred to (Shiel 1996).

Thus the western Ugandan forests waxed and waned over the centuries, and the wildlife within them increased during periods of forest growth and declined as the forests retreated. These periods of growth and decline were occurring along an axis from south-west to northeast, on the eastern side of the Rift Valley. The point of contact with the great forest of Congo was the Semliki Forest region to the south of the Ruwenzori mountains. This area was the entry point for forest species into Uganda during wetter periods. Species entered at those times and survived in the newly formed forests to the east of the Rift Valley, then as conditions grew drier or as humans cut down the forest they died out. With wetter conditions again or with human decline due to disease, species entered again and re-established themselves. We don't know how often this may have happened. But we can see the effects of all this activity and they have been quite dramatic. Because the entry point has always been to the southwest, more species have always been able to enter and re-enter from there and so today more species are found in the more southerly forests than in the more northerly ones. More monkey species are found as you travel southwest from Budongo. For example, the red colobus monkey (*Procolobus badius*) and grey-cheeked mangabey monkey (*Cercocebus albigena*) are found in Kibale forest but not in Bugoma or Budongo, both of which provide perfectly acceptable habitats for them. Budongo Forest, with its four forest monkey species (blues, redtails, black and white colobus and forest baboons) and its single common prosimian species (pottos, *Periodicticus potto*), is poor in terms of numbers of primate species and this is because it is almost at the end of the migration or recolonization line, with only small forests such as Zoka, Mt. Kei and Otzi further north. It is understocked with primate species. It has very high densities of monkeys: the forest is full of monkeys, but it is species-poor. And when it comes to the prosimians, Budongo Forest has almost none of the bushbabies (*Galago* spp.) found further south; it just has pottos.

This brings us to a puzzle: Budongo, at or near the end of the line, does have a goodly population of chimpanzees. So does Bugoma, as well as the forests farther south. In other words, whereas some monkey species and some prosimian species failed to reach Budongo in the most recent period of forest expansion, chimpanzees did. Why? The answer may have something to do with the fact that chimpanzees are very good at moving between forest patches, as we know from our own studies at Budongo and from studies elsewhere. They are not confined to continuous forest by any means, and so would be quick to explore northwards along forest strips and patches. But we need to remember that blue monkeys, redtail monkeys and black and white colobus monkeys, all true denizens of the forest which do not normally cross open country for any great distances, also made it to Budongo. This leaves the question open as to whether and when the Budongo Forest was last reduced to a fragment of its present size, or disappeared altogether. If it disappeared during the last glaciation 18 000 years ago and did not reappear until around 10 000–12 000 years ago, then we can assume that the three forest monkey species and the potto made it to Budongo at that time or later. As regards the chimpanzees, perhaps they never became locally extinct in the way we think the monkeys would have done. Depending how dry it became, they may just have dispersed into the woodlands or bushy terrain where the forest once was. We know from studies in West Africa by Nissen in Guinea and by McGrew in Senegal that chimpanzees are able to survive in very dry conditions, albeit at very low densities. This would mean that their response to environmental change may be less dramatic than that of the monkeys and prosimians. Genetic data on East Africa's chimpanzees indicate a high degree of relatedness between all the chimpanzees of western Uganda (Goldberg 1997; Goldberg and Ruvolo 1997) but this cannot tell us whether they dispersed at the time of forest decline or disappeared to be replaced by new arrivals from the southwest. It is also not clear why chimpanzee distribution is confined to the western forests of Uganda and does not extend farther east, despite the existence of forest patches and some fairly continuous forest between the Rift Valley and Lake Victoria, e.g. the 'Mubende corridor' (D. Pomeroy, pers. comm.).

The forest in more detail

Today Budongo Forest is a medium altitude, moist, semi-deciduous tropical rain forest. It is situated between 1° 37′N–2° 03′N and 31° 22′–31° 46′E. The average altitude is 1100 m. The whole forest slopes down gently in a southeast–northwest direction, towards the Albertine Rift. Four rivers run through the forest: from east to west they are the Waisoke, the Sonso, the Kamirambwa and the Siba. The mass of the forest comprises some 35 200 ha (352 km^2) to which we must add the Siba Forest, which is an extension of Budongo to the southeast and comprises some 8300 ha (83 km^2) making a total of 435 km^2 of continuous forest cover. Outside these two main areas there are numerous lengthy strips of riverine forest to the south and southwest of Budongo, perhaps some 100 km of riverine forest in all, forming arms of forest that stretch out into the surrounding areas, grassland in the past but today mainly cropland (much of it sugar cane) or under human occupation.

Climate

Rainfall data were collected at Busingiro and at Nyabyeya in the past but both these sites are on the forest edge. The BFP has now been collecting rainfall and temperature data

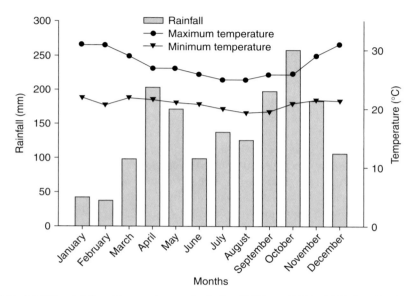

Fig. 1.2:   Monthly rainfall and temperatures collected by BFP from 1993 to 2000 (from Tweheyo *et al.* 2003).

daily at our Sonso camp in the middle of the forest for over 10 years. Rainfall varies between 1240 mm and 2187 mm per annum with a mean of 1600 mm. Its distribution over the year has a bimodal pattern with most rain falling between March and May and again between September and November (Fig. 1.2). There is one major dry season, between mid-December and mid-February when rainfall normally drops below 50 mm (and in some years below 10 mm) a month. At the height of this dry season, in January, the forest becomes hot and dry inside, leaves crunch underfoot and the trees begin to look weary and lifeless. At this time too the wildlife suffers from the daytime heat and chimpanzees, for example, spend much of the middle of the day down on the forest floor in areas of deep shade; their food supply is also depleted at this time. The minor dry season is around June–July when again the food supply is diminished. The combination of two dry seasons alternating with two wet seasons gives the Budongo Forest its characteristic bimodal pattern (Fawcett 2000).

Temperatures are rather even during the year, with an upturn in the dry season. Monthly maximum and minimum temperatures vary between 32°C and 19°C.

Disease and the human population

During the twentieth century Budongo Forest was expanding along its southern border, colonizing the surrounding grassland with a mixture of tree species, notably *Maesopsis*. In the twenty-first it is being cut back by encroaching settlers and refugees from the west, north and south. Besides these victims of political upheaval there are a number of other reasons for population increase around Budongo. First, health has improved in the area after an epidemic of trypanosomiasis (sleeping sickness) at the beginning of the twentieth century (described, among others, by Winston Churchill in his 1908 book *My African Journey*). Paterson (1991) has summarized the devastating series of diseases that afflicted this area (formerly Bunyoro, today Masindi District): trypanosomiasis entered Bunyoro from north of the Nile in 1904 and in a few years surrounded Budongo Forest on its western, northern and eastern sides. The cause of this disease, which spread across Uganda and decimated the human and cattle populations, may have been linked to the colonial administrators' efforts to stop the practice of grassland burning which was a tradition among the local cattle herders as it provided fresh grazing and suppressed the bush land that harboured the tsetse fly (Langlands 1967). In response the colonial administration evacuated the entire population of these areas in 1909–1912, and shot most of the large wild animals because it was feared they harboured the tsetse fly. This included buffaloes, wild pigs, bushbuck and Uganda kob (Harris 1934; Buechner *et al.* 1963). Elephants were spared and their numbers increased accordingly. Large elephant herds moved to and from the Budongo Forest in the early part of the twentieth century until the Game Department initiated elephant control measures in the 1950s in order to protect valuable timber saplings in Budongo Forest. Together with or following the outbreak of trypanosomiasis there was an outbreak of rinderpest, a highly contagious, fatal disease of cattle, which effectively ended the pastoral way of life of the Bunyoro people of this area, and they were obliged to take up the more agricultural way of life we see today (for a summary of these events, see Shiel 1996).

Thus what had been a populated area became an empty one, and only after 1936 did the population begin to build up again, albeit slowly. In the 1950s and 1960s the area attracted people to come and work at the Budongo Sawmill and the other three mills. When my wife and I were in Budongo in 1962 there was a local concentration of people around the mill, and there were a number of traditional villages around Budongo, but most of the land was covered in *Pennisetum* elephant grass. After the Amin and Obote civil wars that devastated the country from 1971 to 1986, there was economic growth, once again bringing job opportunities to the area. Growth has been particularly strong around the renovated Kinyara Sugar Works, and there has been a resurgence of commercial optimism leading to the establishment once again of tobacco growing. Today, the land around Budongo Forest is mostly under cultivation with houses, villages, schools and markets everywhere.

As elsewhere in Africa, the pace of change for the Budongo Forest has accelerated in the last few decades. Few parts of Africa have escaped the pressures of human population increase, of human conflicts and the damage they have caused, of technological progress, of large-scale urbanization and of continuing poverty. Standing on the fire tower we could be forgiven for forgetting for a moment these pressures that are squeezing the life out of Africa. But it is as with the myriad species that are there, almost unseen from a distance except maybe for the white flash of a hornbill flying at the forest edge. From a distance the human pressures are almost unseen too. But in the space of the last 50 years the whole of the area around Budongo Forest has come to be inhabited by people. We shall look at the details of the human population in Chapter 10, but we should note that today, as never before, any discussion or description of Budongo Forest needs to consider the people living around it. They are chipping away at its borders, turning forest edge into cropland. They are putting snares and traps on their fields to catch baboons, forest pigs and porcupines that come to eat their crops. They are going into the forest to cut poles for their houses and collect medicinal plants. They are setting snares in the forest to catch duikers for their cooking pots (and in the course of doing so they are inadvertently catching chimpanzees which suffer serious limb injuries as a result). And during the last century, with ever-increasing speed, they have felled and sold the fine mahogany trees that made the Budongo Forest estate the most valuable in Uganda, so that today its commercial value is small and will take half a century at the very least, and perhaps much longer, to recover its former glory (Reynolds 1993).

Dynamics of Budongo Forest

Two early descriptions of Budongo Forest are those by Harris (1934) and Brasnett (1946). Harris was the first to discuss the idea that the mahoganies and other species of valuable timber trees that occurred in Budongo might be subject to eventual exclusion by a more powerful species, *Cynometra alexandri*, or the ironwood tree. This idea provided the basis for a major silvicultural programme run by the then British-led Uganda Forest Department in later years. Brasnett concurred. He described Budongo as a relatively young forest, possibly some 300–500 years old. He noted a number of features of the forest, in particular that it contains both young (Colonizing), intermediate (Mixed species) and mature (Climax) forest.

   This introduces the idea of a succession of stages in the life of a forest. Succession is a fundamental idea in the understanding of how forests grow. Initially, when grassland or bush is being taken over by trees, whether due to a gradual sustained increase in rainfall from one decade to the next, or to the disappearance of pastoralists, the trees that manage to flourish are pioneer species, that is species whose seeds, seedlings and saplings thrive in conditions of bright hot sunshine such as *Acanthus* spp. Once one or more such species have established themselves and grown to maturity, they inevitably change the habitat, primarily in two ways. First, the colonizing species remove some of the moisture and nutrients from the soil. And second they provide a degree of shade which prevents evaporation to some extent. In such conditions, other species that are better adapted to these new conditions can gain an advantage. If birds or other seed dispersers bring in seeds of species that use less moisture from the soil, or have deeper roots, or can use other soil nutrients, and in particular if the newly arrived species can flourish in a degree of shade, i.e. in slightly cooler conditions with less solar radiation reaching the leaves of the growing seedlings and saplings, then such species will out-perform the colonizers and the next successional stage of the forest will be reached. This process will be continued again and again during the history of the forest. Finally a climax phase may be reached in which a particular monodominant species (i.e. species that can exclude all other species) becomes established, with highly efficient roots and whose seedlings and saplings thrive in conditions of extreme shade. (I say 'may be reached' because this successional process does not occur everywhere, e.g. in many of the forests of South America where the successional process is absent.)

Brasnett, Philip and Eggeling all shared the view that in Budongo Forest there were three successional stages: Colonizing, Intermediate and Climax. If that is right, then the part of Budongo that has today reached the Climax phase would be the part that was first colonized and therefore the oldest part, while the youngest forest would be the part that still shows the Colonizing species such as *Maesopsis eminii*. As shown by Plumptre (1996), when looked at as a whole, Budongo Forest shows a gradual change of tree species composition from the southwest to the northeast. The southwest has more species associated with Colonizing and Mixed Forest, while only in the northeast do we find heavy concentrations of *Cynometra* Forest. From this we can deduce that the initial colonization of grassland or bush during the present phase of growth of Budongo Forest occurred in what is now the northeast part, and from there the forest spread southwards and westwards.

We can now return to the issue raised by Brasnett in stating that the Budongo Forest is a young forest, only 300–500 years old. Because the forest has all three forest types he made the assumption that it had only reached Climax conditions in the areas where *Cynometra* was dominant, and elsewhere had not yet reached the Climax phase. In his view, it would take some 300–500 years for the present tree distribution to come about, from an original state of grassland or bush.

A major study of the ecology of the Budongo Forest was published after those of Harris (1934) and Brasnett (1946). This was a lengthy and detailed study by Eggeling (1947). His descriptions of the three major forest types in Budongo were based on extensive tree enumerations and surveys in and around Budongo. His classification was as follows: the forest is surrounded by grassland. Closest to the grassland is Colonizing Forest. This has two types, each of which is maintained for one generation and then develops into Mixed Forest. The two types are as follows: *Maesopsis* Forest, which develops on better soils, and Woodland Forest which develops on poorer soils. Colonizing Forest is succeeded by Mixed Forest. This is rich in species, with almost 50 species reaching the canopy. It contains the mahoganies (*Khaya anthotheca* and three species of *Entandrophragma*). As the forest ages, more shade-tolerant species are found in Mixed Forest, including *Cynometra* (Ironwood) in the upper understorey. Finally we have Ironwood Forest, with *Cynometra* occurring at all sizes and a general paucity of species as those found in the earlier stage of Mixed Forest die out. A fourth forest type was also distinguished by Eggeling, namely Swamp Forest, occurring along rivers that are seasonally waterlogged. It contains the spiky *Calamus* or rattan climber (studied in Budongo Forest by Okia 2001) as well as wild palms (*Raphia farinifera*) and riverine tree species such as *Pseudospondias microcarpa* and *Mitragyna stipulosa*. The species distribution of Budongo Forest and the development of its climax vegetation have been further studied by Walaga (1993).

Eggeling's views on the succession of Budongo Forest have not gone unchallenged. It has been pointed out that they do not take account of the possibility of human disturbance in the recent history of the forest, and they do not make allowance for the possibility that particular soil conditions or rainfall gradients may be responsible to a greater or lesser extent for the occurrence of particular species in particular locations. Also, perhaps most importantly in the case of Budongo, they do not account for a possible role of elephants in bringing about the dominance of *Cynometra* because elephants do more damage to young trees of other species than to *Cynometra* (Shiel 1996, 1999*a*, *b*, 2001, 2003; Shiel and Burslem 2003).

Timber

Eggeling's study provided a useful model for understanding the ecology of the Budongo Forest. It also provided a scientific basis for the management of this forest during the 40-odd years (1930–1970) when British foresters were in charge. Much silvicultural work was undertaken during this period, mainly centred around the idea that by removing the dominant species, *Cynometra alexandri*, from the forest, Mixed Forest would be able to flourish without being overtaken and as a result the mahoganies, *Khaya* and *Entandrophragma*, which thrive in Mixed Forest and are much less common in *Cynometra* forest, would grow in abundance.

Four sawmills were established in Budongo Forest during the time of the British Protectorate, and a huge amount of revenue was generated by this one forest. Synnott (1972) reported that at that time 'over 10 million board feet, about 24,000 cubic metres, mainly mahoganies' were being produced annually by the Budongo mills. The value of this timber was over 10 million Ugandan shillings per annum on the local and export markets (in 1972 there were 14 Ugandan shillings to the pound sterling, so this was the equivalent of £714 285 then, or perhaps about 10 times that amount today). In addition to the mahoganies, *Milicia excelsa* was a popular timber tree of the highest quality and indeed the licence fee for cutting *Milicia* was higher than for the mahoganies, which occupied second place in the fee structure (R. Plumptre, pers. comm.). *Cynometra alexandri* was also cut once heavy sawmilling machinery using saws with hardened teeth had been introduced (this species is durable, being hard and close-grained with a dark, red-brown appearance. It is thus excellent for flooring, and indeed the floor of the Royal Festival Hall in London is made of Budongo *Cynometra*. In the late 1960s,

Table 1.1: Timber production in Budongo Forest, 1930–1989 (from Plumptre 1996).

Decade	Total volume (m³)	Volume/area (m³/ha)	Percentage mahogany
1930–39	66016	32.2	71.3
1940–49	170080	42.9	68.3
1950–59	151334	42.1	74.9
1960–69	247110	25.1	65.8
1970–79	171836	38.0	66.0
1980–89	66251	24.9	61.9

equipment was introduced to Budongo to make mosaic parquetry flooring panels from *Cynometra*. Nevertheless, mahoganies were the main species cut, and Plumptre (1996) has shown that during the period from 1930 to 1989 mahogany production accounted for more than half the output from Budongo's mills, as is shown in Table 1.1.

Although this level of production sounds excessive, it was not so. The Budongo Forest Working Plans according to which the felling was carried out were carefully designed to prevent over-exploitation of timber species, which were felled according to strict rules with felling cycles of 40, 60 and even 80 years designed to enable resumption of felling after the recovery of each species. Today, certification has become an important method which is designed to enable timber species in certificated, well-managed, forests to survive. The process is becoming more widely accepted by the timber producers, retailers and indeed by the public which has in the past boycotted firms selling uncertificated hardwoods.

Sawmilling started in Budongo in 1926. Planting of mahoganies had actually started four years earlier, in 1922, but without success (Willan 1989). Great efforts were made to increase the stock of mahoganies in this forest, as well as outside it in plantations. Seeds were grown in nurseries and seedlings planted inside and outside the forest. Details of planting methods used inside the forest include planting in lines 150 ft apart, trees being planted in groups of three 80 ft apart. Game scouts were used to shoot animals that ventured near planted areas. A photograph showing lines of *Khaya anthotheca* (mahogany) striplings growing in Budongo Forest is to be found in Eggeling (1940*a*).

This method seemed at first successful and in 1946 some 1350 acres were stocked in lines 75 ft apart with trees 25 ft apart. Trees were planted mainly as striplings or stocks, the roots being inserted into a pit 2 ft $\times$ 1 ft. By the end of 1946 Webb (1948) was able to report that 'the whole area is a pleasing sight and every species planted has made good growth. For the first time *Entandrophragma utile* [one of the mahogany species] shows promise but final results will not be known until the coming dry season has passed.' The early optimism proved to be unfounded. A combination of factors, including the dry season, depredations by elephants (which, according to R. Plumptre, pers. comm., enjoyed walking down the lines of trees and eating the tops of saplings) and other browsing animals, diseases or other factors, mahogany regeneration by any other means than natural has never really succeeded in Budongo. The striplings eventually died or failed to thrive. By the time the British were ready to leave Uganda they had abandoned all the different methods of mahogany planting in favour of 'natural regeneration', i.e. leaving the forest to produce its own crop of young mahoganies, and the main silvicultural activity consisted of canopy opening and weeding by removal of understorey climbers and other impediments to the growth of young mahoganies. Man in the service of nature rather than the other way round.

By 1960 Budongo Sawmills, the largest sawmill in Uganda, was producing 600 tons of sawn timber each month. A series of 10-yearly Working Plans had led to a highly sophisticated cyclical system of selective logging, with minimum girths strictly adhered to and, depending on the species, 40-year, 60-year and 80-year rotations being started (the number of years being the time to elapse between felling and returning to the same part of the forest to fell again). Over a large area, trees of no commercial value such as figs, especially the so-called 'strangling' (epiphytic) varieties, together with *Cynometra*, *Celtis* spp. and *Lasiodiscus* were treated with a potent arboricide containing 2,4,5-T and 2,4-D dissolved in diesel oil. This was done so they would not compete with the mahoganies and in order to open the canopy and give the mahoganies more light, which was believed to stimulate growth. By mid-1966 a total of 80 000 acres had been treated, with a then current rate of treatment of 12 000 acres a year, which was to increase to 17 000 acres a year (Rukuba *Forestry in Uganda*, p. 223, quoted in Paterson 1991 p. 186; for details of the arboricide treatment see Dawkins 1954, 1955). The effects of arboricide treatment on the regeneration of species in Budongo Forest has been studied by Bahati (1995).

Elephants were shot because they ate the tender young mahogany saplings. When my wife Frankie and I were in the Budongo Forest doing our first chimpanzee study in 1962, we saw tree poisoning and elephant shooting. On our return to Britain we met with Colyear Dawkins, a British forester based at the Oxford Forestry Institute, who had

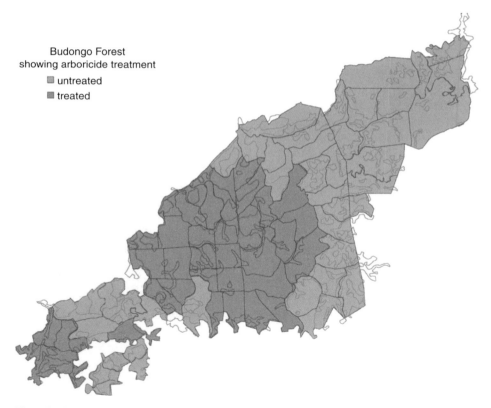

Budongo Forest
showing arboricide treatment
▨ untreated
▩ treated

Fig. 1.3:   Use of arboricide in Budongo Forest.

been very involved in the arboricide programme in Budongo Forest and whom we had met in Budongo (he stayed with us for a week in the Forest Department rest house on Busingiro Hill where we lived while we did our chimpanzee studies). He was as concerned as we were that the use of arboricides on figs and other tree species that provided food for the chimpanzees (which the Ugandan Forest Department had a mandate to protect) would reduce their food supply and cause a reduction in their population. He supported us in a letter we wrote to the Ugandan Forest Department complaining about the poisoning of fig trees but to little avail: we received a reply stating that chimpanzees were adaptable and could eat other species.

The net result of the selective logging for mahoganies and other valuable timber trees, combined with the use of arboricide on *Cynometra* and other species of 'weed', was apparent many years later and was dramatic (as indeed it was planned to be). Plumptre (1996) compared the distribution of forest types (an expanded version of Eggeling's basic forest types described above) in 1951 and 1990 using maps produced by the Forest Department's mapping section.

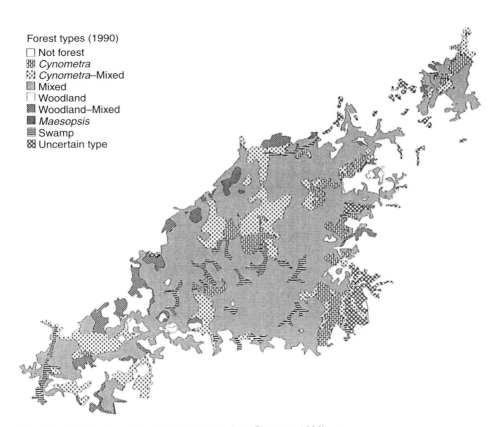

Forest types (1990)
☐ Not forest
▦ *Cynometra*
▨ *Cynometra*–Mixed
▨ Mixed
☐ Woodland
▨ Woodland–Mixed
▨ *Maesopsis*
▤ Swamp
▧ Uncertain type

Fig. 1.4:   Distribution of forest types in 1990 (from Plumptre 1996).

The 1951 figure showed a predominance of *Cynometra* forest; 40 years on this was much reduced and Mixed Forest predominated, as can be seen in Fig. 1.4. This was exactly what the British intended; they had wanted to reverse the succession described by Eggeling and they had succeeded. The object of the whole exercise was to increase the forest's crop of mahogany trees in the long run by providing them with the conditions (Mixed Forest) in which they would thrive. In this the British were less successful. What they in fact succeeded in doing was to increase a variety of species of fruiting trees such as figs (Tweheyo 2000) with very good results for the forest's fruit-eating animals.

Fortunately for them — the chimpanzees and monkeys, the hornbills, bats, civets and all the other fruit-eating species in Budongo — the efforts to kill off the fig trees and other 'weed' species met with as little success as the efforts to increase the stock of mahoganies. This was actually foreseen by Synnott (1972) who wrote 'most of the fig species in Budongo start life as epiphytes and can maintain themselves outside the range of arboricides. On the present 80-year rotation of harvesting these trees will easily be able to reach maturity . . .' As we shall see in Chapter 4, the net effect of the use of arboricides, even on figs, was actually to increase their number. Today Budongo Forest has higher densities of trees producing fleshy fruits and of fruit-eating primates in its logged and arboricide-treated forest areas than it does in its unlogged and untreated Nature Reserves.

The forest today

The forest today is the result of two processes: natural changes and changes resulting from human interference. Natural processes have created a forest with a large number of species competing with each other at the individual level. Each seed that falls and germinates finds itself in immediate competition with all the trees and seedlings and saplings that have gone before it. Very few seeds actually make it to maturity: Synnott (1975) studied the fate of 3000 *Entrandrophragma* seeds in plots of various densities in various parts of the forest without treatments or protection, and a further 2000 in a treatment area where they were given extra water, fertilizer and reduced root competition. More seeds were planted in nursery conditions outside the forest, with three different light/shade routines, and some further seeds were planted in laboratory conditions. In all the natural forest plots, a majority of seeds were eliminated by small rodents before or shortly after germination. Most of the rest of the seedlings were eaten within a year or two by bushbuck, but there were other causes, such as one under a fallen branch, and two buried under buffalo dung. At the end of the study none survived. *Cynometra* seeds do better because they are rarely eaten by rodents and the seedlings grow more successfully in the shady conditions on the forest floor (Synnott, pers. comm.).

The causes of mortality are many: seeds and all the later stages of development of woody plants are attacked in the soil by viruses, bacteria, fungi and larger organisms. If seeds manage to germinate they become food first for mice and shrews, and later if they throw out leaves these are browsed by duikers and most other mammals. At all stages airborne viruses, fungi and bacteria attack leaves so that if you look around you at the leaves in the lower canopy you see everywhere the signs of disease. Arthropods abound in the leaf litter on the ground, up every tree stem and into the canopy. In the case of larger trees, woodpeckers and other birds feed on insects under bark and make holes in trees, thereby damaging them by allowing bacteria and other disease-bearing organisms to enter. Trees themselves compete for light and soil nutrients, so that many fail to thrive and remain static at an early stage of development before dying. And when a large tree finally dies and falls, it brings several healthy neighbours down with it.

Many trees have uses as timber, notably of course the mahoganies *Khaya* and *Entandrophragma* but also *Milicia* and *Maesopsis*, which are used for furniture and house construction, *Cynometra*, the ironwood tree which was used for flooring, *Cordia*, which has a long history of being used for dugout canoe construction, and so on. Kityo and Plumptre (1997) have described the uses and qualities of many forest tree species. All these species have at various times been removed from the forest by man, with consequent damage at the felling sites and the skid routes into and out of the forest from and to the sawmills. Today, as more than a decade ago (Reynolds 1993), the removal of Budongo's mahoganies continues sporadically,[1] illegally, as teams of pitsawyers, often working by night, continue to cream off the most valuable trees in the forest, even in the Nature Reserves where such activity is truly criminal and damages the forest scientifically as well as commercially and physically. But the biggest single change to the composition and dynamics of the forest was the widespread use of arboricides, already discussed.

Despite (and to some extent because of) the depredations of man, Budongo continues to be a healthy forest. Here and there we can find a hole in the canopy that has not filled in after logging. Babweteera (1997, 1998; Babweteera *et al.* 2000) made studies of gaps in the forest resulting from felling operations (pitsawing) and found that the forest can regenerate well if a single tree is removed and the gap formed is less than 400 m^2 and the canopy is not opened by more than 25%. However, if two or more adjacent trees are removed and the gap is greater than 400 m^2 or the canopy is more than 25% opened, the amount of disturbance is too great for the forest to recover naturally with saplings pushing up to fill the gap in the canopy. What happens is that we get a tangle of climbers such as *Momordica foetida* and other ground vegetation leading to a bushy undergrowth that is too thick and too shady for seedlings to penetrate. Such places can easily be found in Budongo, where several trees have been removed close by one another. But the forest has been fortunate: the mahoganies that were sought for timber (and indeed other timber species) mostly grow well spaced out one from another, so timber extractions have been, and continue to be, selective, i.e. for single trees at any one logging location, with consequent recovery of the forest. Rukondo (1997) studied changes in tree diversity in relation to canopy opening at Budongo, and Mwima *et al.* (2001) found that regeneration of mahoganies (*Khaya anthotheca*) was poorer in gaps left after removal of parent trees than in the surrounding forest and suggest that this is due to absence of seed rain

---

[1] For example, see the *New Vision* article published on Thursday 24 April 2003 entitled 'Budongo Plunder' which concerns illegal logging of mahoganies in Budongo Forest's Strict Nature Reserves.

from the parent tree. Plumptre (1995) has shown that parent trees of *Khaya* need to have attained a dbh (diameter at breast height) of 80 cm before maximum seed production takes place. It would thus introduce a long delay in mahogany regeneration if an excessive number of large trees was removed and this is exactly what happened in the 1990s. We do not yet know how long it will be before Budongo recovers a population of mature mahogany trees.

2. *The Sonso community*[2]

> It remains the most exciting thing to be walking fast in the forest, moving with a travelling party of chimps, we stop, they call, drumming, answers come from west, they move, we move... (22 March 1999, 8.30 a.m.).

The word 'community' is widely used in chimpanzee studies to describe the collection of individuals that share a single range or 'territory'. Is it appropriate? What is a community?[3] We normally think of it as a collection of people who live in one place, such as a village community, though it can mean a dispersed group such as 'the community of equals' or 'the community of pigeon fanciers'; in this latter sense the emphasis is on something shared by all members of the community. In the chimpanzee case the former meaning predominates: the community lives together in one place, but there is also a sense of sharing: for example, information about food sources is shared, some kinds of food such as meat and large fruits are shared, there are shared elements of culture, and there is sharing in the social and sexual sides of life.

Counting chimpanzees in a community

Because all the chimpanzees in a community are not found together, it is not immediately, or even quickly, possible to say how many of them there are. This is an extreme form of the problem faced when working out the size of any group of animals: in the case of birds, we have 'bird counts' and the problem of obtaining accurate numbers is considerable. With primates, the problem is less extreme in terrestrial species such as baboons or macaques: in such species the group lives together and moves together at some time during the day, as in the case of Hamadryas baboons leaving their sleeping quarters (Kummer 1968), and so can be readily counted. In the case of group-living forest monkeys the problems are greater: even though the group lives together you often cannot see all of them and so the number of individuals in a group increases with the

---

[2] A complete list of the members of the Sonso community to date can be found in Appendix A.

[3] The *Oxford English Dictionary* gives a number of definitions of 'community' of which the most relevant to the chimpanzee case are the following: 'The quality of appertaining to or being held by all in common; life in association with others; the social state.'

length of observation until asymptote is reached and, however long observation continues, no more individuals are encountered. This may take hours or days or even in some cases longer, depending on the species, its dispersion pattern and the thickness of the vegetation. In the case of fission–fusion species such as the South American red spider monkeys in which groups constantly split up and re-form (Chapman *et al*. 1993) a new level of difficulty is encountered: where does the group begin and end? This is only solved after a prolonged period of study, when the borders of the range of the animals are established.

In 1962 my wife Frankie and I spent 8 months watching chimpanzees in the Busingiro part of the Budongo Forest. Jane Goodall was already working at Gombe in Tanzania when we went to Uganda. Apart from an early study by Nissen (1931) in an area of open country in Guinea, West Africa, during which sightings of chimpanzees were few and far between, and a more recent one by Kortlandt (1962) in which he watched, filmed and did interesting field experiments on chimpanzees in a populated area near Beni in E. Congo, there had been no prolonged field studies of chimpanzees, and none in the primary habitat of these apes, the rainforest. In our 1962 study, written up as a book (Reynolds 1965) and in the form of a scientific chapter (Reynolds and Reynolds 1965) we did not cut trails through the forest, which made finding and observing chimpanzees arduous. Despite this we obtained a lot of information and were able to piece together the fission–fusion system of chimpanzees that had never before been described, as well as their dietary preferences, their social and sexual behaviour.

When I returned to Budongo in 1990 the first priority was habituation. It took five years of daily following and observing them by our field assistants and researchers before we knew the number of chimpanzees in the Sonso community; or, to put it another way, before the shyest individuals allowed us to observe them (Reynolds 1997/8). Our field assistants were in the forest for 7 or 8 hours a day.[4] The first individuals they identified were the big, confident males; later they made progress in identifying the females and younger animals which were shy and disappeared at the first sign of human presence.

The process of discovering how many chimpanzees we had in our Sonso community is graphically illustrated by Fig. 2.1. From this figure it looks as if the Sonso community was growing over the first five years, but it was not: our knowledge of it was growing. Let us consider this process in more detail.

As mentioned, one of our first actions on establishing the site at Sonso was to employ four field assistants. These were Zephyr Kiwede, Geresomu Muhumuza, Dissan Kugonza and John Tinka. I met them first in September 1990 when I returned to Uganda after my initial visit in March of the same year. I was immediately struck by their lively enthusiasm for the work.

By the end of 1992 we had identified 20 individuals. At that time we had no idea that we had still named less than one half of the community. It was not until September 1995, three years later and five and a half years after the beginning of the project, that we had identified, without knowing it then, all 50 of the chimpanzees living in the Sonso range at that time. Some of the individuals in the Sonso community are illustrated in Figs. 2.3 and 2.4.

---

[4] Since January 2003 this has increased to 10.5–13.5 hours a day.

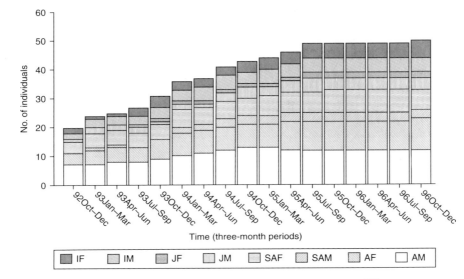

Fig. 2.1: Increase in numbers of Sonso chimpanzees over the years 1992–1996.
Key: F = female, M = male, I = infant, J = Juvenile, SA = subadult, A = adult

Fig. 2.2: The author with field assistants Geresomu Muhumuza and Zephyr Kiwede in the forest (photo: V.R.).

Fig. 2.3:   Some individuals in the Sonso community. (a) Black. (b) Maani. (c) Bwoba. (d) Duane. Fig. 2.3(a) by N. Newton-Fisher, other photos by the author.

Fig. 2.4:   Some individuals in the Sonso community. (a) Gashom. (b) Musa. (c) Bob. (d) Tinka. Figs. 2.4(b) and 2.4(c) by N. Newton-Fisher; Fig. 2.4(d) by S. O'Hara. Fig. 2.4(a) by the author.

After that, increases and decreases in the population of the Sonso community were due to births, deaths and migration rather than recognition of new members. The members of the Sonso community as it is now, including residents, infants born into the community, immigrants, and those who have died or disappeared, from the beginning of our studies to the time of writing, together with some details of each individual, are listed in Appendix A. In all, over the time since our studies began in the early 1990s, including all individuals, many of whom have died or disappeared, we have named 85 chimpanzees. To date the community size has fluctuated between 48 and its present size at the time of writing, which is 62.

Habituation

The process of overcoming the natural fear animals have of people, and reaching the point where they no longer flee when we human beings arrive, is called *habituation*. At some chimpanzee study sites, notably the Japanese-run project in the Mahale Mountains of Tanzania, to the south of Gombe, and to some extent at Gombe itself, habituation was achieved with the use of food put out for the animals, a technique known as provisioning. This technique of gaining the confidence of animals was developed by Japanese primatologists in Japan in the 1950s during their pioneering studies of Japanese macaques.

Tutin and Fernandez (1991) investigated the process of habituation to researchers by gorillas and chimpanzees at the Lopé Reserve, Gabon. Despite the extra difficulty and time taken to find the animals, they decided not to use provisioning, which affects the apes' natural feeding behaviour and, in the case of gorillas, can lead to charging displays by adult males. Instead their method was to follow the apes and gradually break down their fear through frequent contacts. Grieser-Johns (1996) highlighted a particular difficulty with habituating chimpanzees (at Kibale Forest), namely that because of their fission–fusion social system one does not see the same individuals each day and so the process of gaining familiarity takes longer than it would otherwise. She also noted that adult females were shyer than adult males, as we found at Budongo. The researchers at Kibale, like those at Lopé, have not used provisioning.

At Sonso we took the decision early on not to provision the chimpanzees. While this makes habituation harder, it has become apparent that provisioning chimpanzees has effects on their feeding ecology because it steers them in their daily movements towards the provisioning site and thus their natural foraging methods are disrupted. Also, in some cases it has caused outbreaks of aggression because the concentration of food in one place has increased the level of feeding competition between the animals. Power (1991) has argued at book length that provisioning can have widespread and quite devastating effects on the behaviour of chimpanzees, leading to violence between adults and against infants. What really seems to happen if food is artificially provided is that the increasing competition leads to outbreaks of violence, as Power suggested. However, there have now been many documented cases of violence between adult chimpanzees, and violence directed against infants, in communities that have never been provisioned (Wilson and Wrangham 2003). In consequence of our decision not to attract our chimpanzees with food, we had to habituate them by the slow process of finding and following them on a daily basis, remaining with them constantly until they became accustomed to our presence and were convinced that we were harmless.

Before leaving the subject of habituation we should ask whether there is a downside to the process, and indeed there is. Wallis and Lee (1999) and Woodford *et al.* (2002) have studied the risks of habituation, focusing on disease transmission. The great apes are susceptible to many human diseases which can cause severe morbidity or death. The commonest routes for transmission are respiratory and faecal–oral. The danger increases the more habituation has been achieved. The respiratory route can be very dangerous. Five chimpanzees died of suspected pneumonia in the Kasakela community at Gombe National Park, Tanzania in 1968 (Goodall 1986) and a further nine died from a similar disease in 1987 (Wallis and Lee 1999). In the neighbouring Mitumba community a respiratory disease killed at least 11 chimpanzees in 1996. Similar events have occurred at Mahale Mountains National Park in Tanzania (Hosaka 1995). It used to be the case that visitors to gorilla tourism sites were encouraged to hold a baby gorilla in their arms. Although it would be a peak experience for any tourist, it was a highly dangerous moment in the life of the infant gorilla. In 1988 no less than 81% of the gorillas in seven tourist and research groups in the Virunga Volcanoes region showed signs of influenza-like symptoms, six females died and 27 cases were treated with penicillin. As a result, a vaccination programme was implemented and 65 gorillas were vaccinated using a dart-gun (Woodford *et al.* 2002). A study by Kalema (1995) compared two groups of gorillas at Bwindi Impenetrable Forest, one habituated tourist group and one unhabituated group. She examined faecal samples and found higher levels of parasites in the habituated group, probably due to contact with human excrement in the forest.

As awareness of this problem has grown, measures have now been taken to reduce the risks, and most sites where habituated apes can be seen have rules limiting the proximity of visitors to the apes. At Budongo, sick visitors, especially those with respiratory diseases such as the common cold or intestinal diseases such as diarrhoea, are not allowed into the forest. Inside the forest, we adhere to a 7-metre rule: you must not go closer than 7 metres to any chimpanzee. This may actually be too close. A study of the Sonso community (Phillips 2003) found that some of our chimpanzees responded by glancing or moving away when observers arrived on the scene. Lone individuals were particularly wary of oncoming humans. Females were more affected than males. Phillips recommends that observers should normally remain at a distance of 16 m or more from the chimpanzees, unless closer proximity is demanded for particular reasons.

In most cases we do not get very close to our chimpanzees, but in the case of the adult males, most notably the senior males Duane and Maani, closer proximity could be achieved and we have to make conscious efforts not to get too close. We also have a rule that all human faeces must be safely disposed of. The health and diseases of the Sonso chimpanzees are discussed in Chapter 3.

There is no doubt, as my colleague Duane Quiatt has often pointed out to me, that the presence of human observers has some kind of effect on wild chimpanzees, but this is not often studied. Bethell (1998) did so, and found that females and males had different strategies for coping with observer presence. Males tolerated closer human proximity, and had higher levels of vigilance towards humans than did females, who nearly always kept a greater distance from observers but directed less vigilance towards them. Duane, the alpha male, though easy to approach on the ground and very tolerant of human presence, paid more attention to observers than did other males. Juveniles paid more attention to observers than did subadults, who in turn paid more attention than adults, indicating that familiarity with humans led to a loss of interest in them.

There is a different, and much longer term, danger associated with habituation. Once a group of wild animals has lost its fear of human beings it is at risk from unscrupulous people who want to collect specimens, sell babies as pets or to circuses, and these days even eat the apes. The whole basis of the BFP has been, from the outset, to protect these chimpanzees from such people. When we started in 1990, mothers were being shot and their babies taken to Entebbe for sale. Since that time they have been largely safe, at least in the area where we have been working. In neighbouring Congo, however, they are not at all safe. Not only do the tribes of the Congo eat ape meat (as do many of the peoples who live in and around the great Congo forest from the border of Uganda right across Central and West Africa) but they continue to shoot mothers and sell the offspring to traders, or indeed anyone who is willing to buy a young chimpanzee. We know this because of the constant trickle of infant chimpanzees from the Congo, confiscated in Uganda, which find their way these days to the wonderful sanctuary set up by the JGI at Ngamba Island in L. Victoria. Even as I write, a new baby has arrived at Ngamba. This is a trade that is repugnant to us in the west, but is commonplace in many African countries where attitudes to animals, including chimpanzees, are very different from ours. So, returning to our theme, habituation is dangerous. Often when I stand among a group of chimpanzees of our Sonso community and delight in their lively independence of us humans in their midst, I fear for their lives if we should ever leave them to the forces of the marketplace.

In 2002 we introduced a rule that no more than four researchers may be involved in observing our chimpanzees in the forest at any one time. We made this rule to avoid harassing them and causing them to become either aggressive towards us (this has happened at some other sites, especially in relation to the dominant male), or unduly bothered by our presence as can happen in the case of some of the more naturally timid members of the community.

Data

BFP has a number of large chimpanzee data files on its computer. One pertains to party composition, one to social behaviour. Researchers have their own files. The amount of stored data is therefore very great. We have data for less than 15 years but there is still too much for any single person to analyse. Even finding all the data, dispersed over many universities, would be a considerable task. In the preparation of this book I have used my own data and, with permission, data collected by other researchers and senior scientists. I have also drawn from data files compiled by our field assistants, such as our meteorological data files or files that contain data on the species of trees that provide chimpanzees with food, or the distribution of those trees in the forest. I have also used maps and photographs, and last but not least the copious notes in my own field notebooks.

Range and density of the Sonso community

The BFP camp, located in a large clearing in the heart of the Budongo Forest made originally by Budongo Sawmills Ltd, is in the middle of the range of the Sonso chimpanzee community. We could not have found a better location if we'd tried. Our trail system radiates out in all directions from camp, and we find Sonso chimpanzees to the north, south, east and west of it (see Fig. 1.1).

Reynolds and Reynolds (1965) estimated the range size of the Busingiro community to be 6–8 square miles (15.5–20.7 km^2), and we estimated that within that range there were some 60–80 chimpanzees. This gives a density of 3.9 chimpanzees/km^2. A later study by Sugiyama (1968: 243) put the density of the chimpanzees in the Busingiro region higher, at 6.7/km^2.

The home range[5] of the adult male Sonso chimpanzees was investigated by Newton-Fisher (2002*b*, 2003) over a period of 15 months in 1994–95. This study was based on the ranging behaviour of adult males, who move over the whole home range of the community. Wrangham (1979) first suggested that male chimpanzees at Gombe Stream Reserve had a home range that was 1.5 to 2 times greater than that of females, because males jointly patrol the borders of the community's territory whereas females tend to avoid the border areas. In addition, at Gombe, Wrangham found females to be less sociable than males, so that females were dispersed relative to males within the defended territory. The same difference between male and female range use was found by Chapman and Wrangham (1993) in the Kanyawara community of Kibale Forest Reserve, Uganda. This pattern of chimpanzee range use has been much debated and may not apply at all chimpanzee sites, e.g. at Taï in West Africa described by Boesch (1991) and Boesch and Boesch-Achermann (2000), where female and male ranges are the same. Females have never been found to have larger home ranges than males and so Newton-Fisher's focus on males is wholly justified. All males need to be included, because use of the home range is not equal from individual to individual (Newton-Fisher 2003).

Home range is defined by Newton-Fisher as that part of the range that is habitually used, i.e. it does not include the less frequently used parts of the range where it borders on the ranges of other communities. He calculated the home range at the time of his

---

[5] We should remember that the home range of a community of chimpanzees shifts from year to year: at the time of writing the Sonso chimpanzees seem to be extending their range in a southeasterly direction and may be reducing it to the southwest (N. Newton-Fisher, pers. comm.).

study using three methods of analysis (minimum polygon, fixed kernel and adaptive kernel), as between 6.78 and 14.51 km², giving a density for the 46 individuals in the community at that time of between 6.8 and 3.2 individuals/km². This is somewhat higher than the density estimate of 3.9/km² we found for Busingiro by our much cruder method, and is likely to be more accurate.[6]

It is also higher than the density found for the whole of the Budongo Forest. Andrew Plumptre conceived the idea early in our research of making a study of the whole of the Budongo Forest and of including in this a survey of the chimpanzees of the forest, based on nest[7] counts. The survey was carried out in 1992 and was the first accurate survey of chimpanzees in a large area (previously, single communities had been counted accurately but not larger areas), and all the more remarkable for being in a forest habitat (Plumptre and Reynolds 1996). With expertise based on his own prior survey work on gorillas, Plumptre organized our field assistants to conduct transect-based counts of nests in eight areas of the forest. These areas were chosen to represent the different forest types and the logging history of the forest. Analysis of the results using adjusted 'standing crop' counts gave a figure of 650–890 chimpanzees for the whole forest, giving a density of 2.12–2.22 individuals/km².

We can also draw a comparison with Kibale Forest, to the south of Budongo in Uganda. Range use by the Kanyawara community was studied there by Chapman and Wrangham (1993). Using three methods (minimum polygon and two sum-of-cells methods) they calculated the home range as 7.8–14.9 km², giving a density for the 41 individuals in the community of 5.2–2.75 individuals/km². This again is somewhat lower than Newton-Fisher's density for Sonso.

The higher density found by Newton-Fisher (3.2–6.8 individuals/km²) than by Reynolds and Reynolds (1965) and Plumptre and Reynolds (1996) suggested to him that the area of the Sonso community might be particularly rich in chimpanzee foods, and there is support for this as we shall see in more detail in Chapter 4. Another reason given by Newton-Fisher for the higher density of the Sonso chimpanzees is the relatively lower level of timber exploitation at Sonso compared with Busingiro, and here again he may well be right. At Busingiro both at the time of our study in 1962, and at the time of the later studies in the 1960s, there was a considerable amount of legal and illegal pitsawing for mahogany going on (as it still does today). This has not been the case in the Sonso area during our studies from 1990 onwards, mainly because the presence of the BFP has tended to keep illegal pitsawyers away from our study site.

---

[6] All measures of home range (or total range) are necessarily estimates (N. Newton-Fisher, pers. comm.).

[7] The word 'nest' to describe the structure made by chimpanzees each night and sometimes in the daytime has been criticized by Prof. Toshisada Nishida (pers. comm.), who prefers the word 'bed'. Indeed there are problems with the word 'nest', which in many species signifies a structure made to lay eggs or rear young. This is not applicable to chimpanzees. However, the word can also mean a retreat or a place of safety. The *Oxford English Dictionary*, after the first definition which is a place for rearing young, gives a second definition as follows: 'A place in which a person (or personified thing) finds rest or has residence; a lodging, shelter, home, bed etc., especially of a secluded or comfortable nature; a snug retreat.' This seems to cover the chimpanzee case. For that reason, and because the word 'nest' is widely used in the literature on the subject, I have maintained this usage here.

It thus appears likely that we have a high population density of chimpanzees at Sonso. However, there is an important proviso to any estimate of density from range size. Often, and this is the case at Sonso, the full extent of the range is not known. It is very hard in the forest to determine where the patrolled borders of the Sonso range are, because the chimpanzees move silently on patrols. A. Plumptre (pers. comm.) stresses the need to map several contiguous communities before calculating density, in order to determine the full extent of their ranges. Taking this into consideration, it may be that lower density figures are more accurate than higher ones, though at present this remains uncertain.

Community size

The size of any community changes over time according to the number of births and deaths, emigration and immigration. What is the size of the Sonso community? Table 2.1 shows its size, by age–sex groups, over an eight-year period, beginning at 1996 when all individuals were known. The table shows a growing community with, until recent years when some of our younger males have matured, a declining number of adult males.

A detailed analysis of birth, death and migration, and how these affect the size of the whole community over time, shows two things. First, the different age–sex classes did not gain (through birth) and lose (through death) evenly. For seven years, until 1997, our adult males seemed immortal; we had a good number of sturdy adult males, the core of the community. Females came and went but the males remained. From 1997 we started losing adult males: Chris last seen August 1997, Kikunku last seen July 1998, Zesta killed (see Chapter 8) November 1998, Vernon last seen June 1999, Magosi found dead July 1999, Muga last seen March 2000, Andy found dead July 2000. Seven adult males lost over a three-year period. This was a heavy blow because adult males provide the defence of the community. In four cases, Chris, Kikunku, Vernon and Muga, we do not know what happened to them.[8] They may have died in snares or been killed by adult males from other communities. None of them was sick at the time they disappeared. For a while, during these three years, the Waibira community to the north was making raids

Table 2.1: Size of the Sonso community, broken down into sex–age classes, 1996–2003.

	1996	1997	1998	1999	2000	2001	2002	2003
AM	15	15	14	13	11	10	13	16
AF	17	18	18	17	16	17	21	22
Juv	6	12	12	6	7	11	12	11
Inf	13	8	9	13	14	11	9	10
Total	51	53	53	49	48	49	55	59

Key: AM = adult males, AF = adult females, Juv = all juveniles, Inf = all infants.

[8] It is possible that Vernon was speared while crop-raiding. This emerged in 2003, when our fine adult male Jambo was speared while raiding a sugar cane field at Nyakafunjo. See Chapter 9 for details.

into Sonso territory and our chimpanzees appeared frightened of them. We feared at that time that all our males might be killed, and the females taken over by the Waibira males from the north, as happened at Gombe when the Kasakela community attacked the Kahama community further south (Goodall 1986). But this did not happen; there was a considerable amount of illegal logging in the area between Sonso and Waibira and it may be that the northern community was forced to move further north, away from the loggers and away from our study community.

Second, the migration data confirm a feature of the behaviour of a few of our chimpanzees, namely disappearance followed by reappearance. This was first noticed in the behaviour of an adolescent female, Mukwano. Female emigration is the norm in the philopatric society of chimpanzees. But Mukwano did not simply emigrate at adolescence; she disappeared and returned. She was present from April 1992 when she was first identified and named until March 1993, then she disappeared and reappeared 20 months later in November 1994. She has remained in the community since that time, disappearing for short periods. In November 2002 she gave birth to a son, Monday, and in April 2003 she disappeared with him, returning without him in July 2003.

A second chimpanzee to go missing for a while was Jogo. Jogo was a juvenile male named in November 1992; he disappeared in October 1993 but reappeared in April 1994. He was around until October of that year, then disappeared again and was not seen until March 1995, after which he resumed association with members of the community. In April 1997 he disappeared and has not been seen since.

In September 1997, Emma immigrated into the Sonso community. At that time she was a juvenile aged about 6 years. She remained in the Sonso community until January 2003 since when she has not been seen. Thus Emma emigrated at the age of around 12 years.

Another female, a subadult or adult whom we named Juliet in January 2002, had immigrated into the Sonso community not long before. She remained at Sonso and had her first maximal sexual swelling in January 2003. Soon afterwards she left the community and has not been seen since. She was therefore in the Sonso community for around a year.

These individuals who come and go may be following a strategy of belonging to two communities, or it may be that they come to Sonso at a time when their normal home community is undergoing some kind of trauma or disintegration and they move to and fro before settling at Sonso.[9] In theory, females who move between two communities increase the number of mating partners available to them. Alternatively, in theory, females who move between two communities and become known to males in both can avoid the danger of infanticide to their infants. We know from all other sites that adolescent females frequently move from their natal community to another where they settle and have their young. The only exceptions are certain high-ranking females who remain in their natal communities, for example Flo at Gombe (Goodall 1986). At Sonso there are some non-adolescent, parous females whom we know to have moved between communities; we consider these cases in more detail in Chapter 5.

---

[9] I am grateful to Melissa Emery Thompson for this suggestion.

The migration process is also found in females of other communities studied. For example, Goodall (1986: 87) gives details of 12 immigrant females, Nishida *et al.* (1990: 73–76) give details of 18 immigrant and 13 emigrant females to and from Mahale M Group, and a further nine females who transferred back and forth (multiple transfers). Boesch and Boesch-Achermann (2000: 16–17) confirm that the same female immigration–emigration process happens in the Taï community; indeed, at Taï nearly every subadult female in the community transfers, which is not the case for the Sonso community or for Gombe, where some high status females remain in their natal community for life. In the case of immigrant females, some described by Nishida *et al.* (1990) failed to integrate, and this is true of Sonso too; indeed, it may be a failure to integrate that leads to multiple transfers and maybe also to the decision by some females to emigrate after they have given birth to a new infant, which can lead to infanticide (see Chapter 7).

Births

Since our study began in 1990 there have been 28 births. Of these, one was premature and the baby died (in 1997, mother was Kigere), a second was stillborn (in 2002, mother was Melissa), while three others died in infancy (in 1996, Zip, offspring of Zimba; in 2001, Ben, son of Banura; and in 2002, Monday, son of Mukwano). The annual distribution of the 27 full-term births is shown in Fig. 2.5.

The sample size is not large enough to determine whether there is a birth season at Sonso. We can compare the Sonso distribution with the distribution of births by month at other sites. According to Goodall (1986: 108), 'there is no evidence of a birth season as such', but she notes that her sample size ($N = 53$) is small; the one for Sonso is even

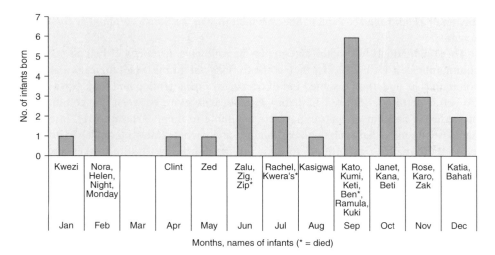

Fig. 2.5: Distribution of 27 full-term births by month, 1991–2003.

Table 2.2: Sonso birth rate 1994–2002 (data compiled by C. O'Hara and L. Bates).

Year	No. of births	No. of females	Birth rate
1994	1	12	0.08
1995	2	15	0.13
1996	3	17	0.18
1997	6	18	0.33
1998	3	18	0.17
1999	1	17	0.06
2000	1	17	0.06
2001	4	18	0.22
2002	4	17	0.24
Average	2.78	16.56	0.16
St. dev.	1.72	1.94	0.09

smaller ($N = 27$). Boesch and Boesch-Achermann (2000: 21–22) also found no evidence for a birth season based on a sample of 75 births, and no birth season was found for the chimpanzees of Mahale (Nishida *et al.* 1990).

However, Goodall did find evidence of a non-random distribution of oestrous swellings (see her Fig. 5.6) and this is the case at Sonso too (Wallis 2002*a*). This interesting phenomenon is dealt with more fully in Chapter 6.

The birth rate at Sonso can be calculated by dividing the number of infants born each year by the number of females in the community. This is done in Table 2.2, which shows the Sonso birth rate from 1994 to 2002, together with the mean birth rate (0.16) and the standard deviation (0.09) for the whole period.

The inter-birth interval (IBI) is a measure, in months, of the time between one infant and the next. This is longer when the first infant survives than when it dies. For Sonso the mean IBI for 10 intervals was 67 months (range 58–84 months).[10] For mothers of more than one infant where the preceding infant died we have three cases from Sonso:

1. Kigere gave birth to a premature stillborn infant in early September 1997. She gave birth to Keti, an infant female, in September 1998 (IBI one year).

2. Zimba's infant Zip disappeared soon after birth in August 1996. She gave birth to Zig, a male infant, on 7 July 1997 (IBI 11 months).

3. Melissa's son Mark disappeared in August 2002. She gave birth to her next infant, Monika, a female in July 2003 (IBI 11 months).

A rule of thumb for Sonso therefore is: IBI around 5 years when the baby survives, but around 1 year when it dies.

At Mahale, the IBI when the first infant survived was 69.1 months for 33 births, and when it died was 14.4 months (Nishida *et al.* 2003); the IBI at Sonso was shorter in both

---

[10] Thanks to Melissa Emery Thompson for these calculations.

cases. At Taï the IBI when the first infant survived was 69.5 months, when it died the IBI was 12.9 months (Boesch and Boesch-Achermann 2000). Again the IBI was shorter at Sonso. Data for other sites on the IBI when the first offspring survives are: Bossou 61.2 months, Gombe 66 months and Kibale 86.4 months. All these intervals are longer than the equivalent ones at Budongo.

Emery Thompson *et al.* (in press) in a study of reproductive parameters (including endocrinological ones) of wild chimpanzee females, compared the IBIs of Sonso females with those of Kibale's Kanyawara community. They found that whereas there was considerable variation in IBI at Kanyawara, at Sonso the IBIs were more clustered around the mean. This indicates more regular birth spacing at Sonso and may perhaps reflect the more reliable or more productive food supply. Comparing peripheral females (those seen infrequently, defined as seen in less than 15% of all scans) with central ones, they made the surprising discovery that, in a reversal of the Kanyawara pattern, peripheral females at Sonso have slightly shorter IBIs than central ones. The reasons for this are not yet clear but may relate to the fact that food is not in short supply for peripheral females, or that they are less stressed socially, or both. Unlike Gombe and Kanyawara (Kibale) chimpanzees, 'central' and 'peripheral' Sonso females do not show any obvious distinctions in core area. Thus they are socially peripheral, but not necessarily spatially so. Distinct core areas would be surprising given what appears to be a relatively small community range and the high density of *Broussonetia* and large fig trees in the central, logged area.

The birth of Katia

It is highly unusual to observe a chimpanzee giving birth, perhaps because females tend to hide themselves away at this vulnerable time. However, our field assistant Zephyr Kiwede did observe a complete birth which, even more unusually, took place high in a tree. The events are described in full in Kiwede (2000) from which these extracts are taken:

At 07:30 on 30 December 1998, chimpanzees of the Sonso community were heard calling. Many chimpanzees were found feeding on ripe figs *(Ficus mucuso)*. A nulliparous female named Kewaya (KY) was sitting on a branch of the tree at a height of 30 metres above the ground just below the crown. At 08:07 KY was not feeding but appeared to be in pain. At 08:38 she pushed her left hand into her vagina for 40 seconds. The right hand of KY was deformed and lacking functionality due to a snare injury. When she removed the hand it was wet. She licked her fingers. At 08:42 she pushed the hand again into her vagina. When she removed it, a quantity of fluid came out from the vagina. It was assumed that the amniotic sac had broken. She moved half a metre on the same branch and sat, then stood up quadrupedally, then moved back, sat again and again moved back. She pushed her hand into her vagina several times. At 08:50 while removing the hand from the vagina, fluid emerged again. She remained 'unsettled' and pushed her left hand into her vagina. At 08:55 when she removed the hand, some more fluid (a smaller amount than before) came out of the vagina. At this time she stopped pushing her hand into her vagina, but she placed/pressed her hand on her vagina as if she was waiting for something. KY repeatedly touched her vagina and moved about in the tree, changing from one branch to another. At 09:12 she started showing abdominal contractions with the right (functionless) hand placed on the vagina.

At 09:22 the vagina opened widely: the infant's head was seen to come out first. The infant emerged fully at 09:23:20 and KY caught it with her left hand. She immediately pulled the infant on to her belly and hugged it . . . At 09:40 the placenta emerged, KY moved her left hand to her vagina, collected the placenta and started to feed on it immediately. At 09:46 she finished feeding on the placenta and started licking blood from her hands. She did not eat the umbilical cord and she left it hanging. At 09:48 she grasped the umbilical cord from the distal end and started pulling it. At 09:48:30 she was seen biting the umbilical cord. At 09:49 she stopped and after 30 seconds she started feeding.

The social aspects of this birth are interesting and we shall return to them in Chapter 5. For the moment the following aspects are of note:

- the mother was nulliparous yet fully competent before, during and after the birth;

- this was true despite that she had only one functional hand (the left);

- despite this she gave birth 30 m up in a tree;

- the placenta was eaten within 25 min of the birth, earlier than appears to be the case at Gombe (Goodall 1968: 223).

Age and age groups

How long do chimpanzees live? From our Sonso data we do not know. Goodall (1986), Nishida (1990) and Boesch and Boesch-Achermann (2000) give figures of 42–44 years of active life and the possibility of survival up to 50 years, and there seems no reason not to accept these figures for Budongo.

We know accurately the ages of the individuals born into the community. The ages we have allocated to our older animals, those already in the community when we began our studies, are approximate and are the result of two processes: comparison with chimpanzees at other study sites where ages are known; and intra-community comparisons, i.e. deciding on the basis of darkness of the face (chimpanzees' faces become darker with age), graying hairs on the back (chimpanzees go grey around the rump as they get older), vigour and muscle mass (chimpanzees tend to suffer from muscular wasting as they get older).

Applying these principles we can age the individuals in our Sonso community with reasonable accuracy. Dates of birth, known and estimated, together with margins of error, are given in Appendix A.

For most purposes of analysis, age groups suffice. At Sonso we distinguish the following age groups: Adult, Subadult, Juvenile and Infant. We categorize and define these age groups as shown in Table 2.3.[11] These definitions have been arrived at by

---

[11] This table was revised in December 2003. I am grateful for help provided by S. O'Hara, C. O'Hara, M. Emery Thompson and L. Bates.

Table 2.3: Age categories.

Age category	Sex	Age where known	Alternative criteria where age not known
Infant	Both sexes	Birth to end of 4th year	Dependent on the mother
Juvenile	Both sexes	5th to end of 9th year	Increasing independence from mother
Subadult	Male	10th to end of 15th year	Beginning to increase in dominance Testes becoming larger
	Female	10th to end of 14th year	Sexual swellings begin, may be irregular No offspring yet
Adult	Male	16 years +	Testicular development complete Face black Dominates females and challenges males
	Female	15 years +	Normally has offspring

reference to the categories used by other field workers, e.g. Goodall (1986), Boesch and Boesch-Achermann (2000) and Kawanaka (1990). The object is not just to reflect distinct changes in the life cycle but also to make the Sonso categories comparable with those of others to enable cross-site comparisons to be made.

Infertility

There is for all female chimpanzees a period of adolescent infertility[12] when they nevertheless show sexual swellings. Evidently the senior males are aware of the disjunction between these early swellings and infertility because they take little interest in such females. This period was most marked in the case of Kalema, a late juvenile female at the start of our study. Kalema was continually swollen for 873 days (2 years 143 days), from 8 July 1992 to 30 November 1994. Her swellings waxed and waned from stage 1 (small) to 4 (maximal) during this period but never disappeared (stage 0). Although Kalema and her age-mate Kwera (whose swellings began at the same time as Kalema's and were irregular but not constant like Kalema's) stayed away from parties containing adult males and kept very much to themselves, the males took little sexual interest in either of them. However, the adult male Maani was observed copulating with Kalema on 19 April 1994. We know from subsequent genetic studies (see Appendix C) that Maani was not the father of Kalema's subsequent infant but it was around that time that she conceived (by the adult male Black). Kalema continued to show swellings during her pregnancy, she was maximally swollen on 30 November 1994, showed a reduced swelling (stage 2) on 2 December 1994, and within two weeks had given birth to a daughter, Bahati, first seen on 16 December 1994. After the birth her swellings ceased. Her age-mate Kwera kept in parallel giving birth to a son, Kwezi, on or around 7 January 1995, the father being Muga, the son of the alpha female Nambi.

---

[12] Nishida *et al.* (2003) found the median length of this period in four females at Mahale to be 2.8 years.

Demographic structure

What conclusions can we draw from the demographic structure of the Sonso community? Perhaps the most important thing in the long run for any community whether of chimpanzees or humans is its productivity, the number of surviving offspring. The structure of the Sonso community at the time of writing is as follows: 11 adult males, 19 adult females, 5 subadult males, 5 subadult females, 4 juvenile males, 7 juvenile females, 3 infant males, 7 infant females, plus one newborn not yet sexed. Total: 62 individuals.

At the time of writing we have more female infants (7) than male infants (3) which is not good. If all our young females emigrate when they reach adolescence this will seriously deplete the community, and the lack of males will weaken the community's strength. Infant males provide the future core of the community, its defence against any ambitious neighbours who may in the future mount raids on Sonso. This has been the pattern at Gombe (Goodall 1986) and Mahale (Nishida 1990; Nishida *et al.* 2003) where studies have gone on for three or four decades, though it has not happened at Taï in West Africa (Boesch and Boesch-Achermann 2000). Since the deaths and disappearances of seven of our adult males referred to above, the Sonso community is already rather weak numerically. There are nearly twice as many adult females (19) as adult males (11) largely due to the high male mortality referred to earlier in this chapter. An excess of females is a feature of the Taï Forest community and also of Mahale, so our Sonso community is coming to resemble these. Inter-community strife is not a feature of the Taï chimpanzees, but it is of Mahale, and we have seen evidence of it in Budongo.

Males who go missing temporarily

On a number of occasions, adult males have gone missing for days or even weeks at a time, and then returned. We have never discovered where they have been. These cases are not thought to be consortships with females in the Sonso community since evidence of missing adult females is lacking. However, we have shreds of evidence that females from other communities may be involved. Bwoya (BY), a fully adult central Sonso male, disappeared from 22 September to 26 October 1999. He disappeared again on 10 November 1999, last being seen with an unidentified female in full oestrus, heading southeast. He was seen in Sonso again on 5 January 2000. Nkojo, another central male, had a series of disappearances in 2001–2002. He was frequently seen until 31 March 2000, after which he was seen on the following days only (he would normally be seen most days): 26 April, 30 May, 8 June, 20 June, 17–22 July, 29–30 September, 1–24 November, and then in 2001 11–18 January, after which he was seen regularly again. In such cases we have little idea what these lone male wanderers get up to. On 15 August 2003 Nkojo, who had become a full-time resident again, was seen for the last time; he has not been seen since and is now presumed dead, though there is a slender chance that, with his history of short-term disappearances, he may yet be found alive one day.

3. *Morbidity and mortality*

Kigere + Kadogo + new dead baby — miscarriage — c. 4 months? — hands and feet not fully formed — she draped it over the crook of a branch y'day evg & it's still there this morning, with flies buzzing around it. Some hair on head & arms but legs are white skin & bare of hair. Kigere feeding nearby. Ruda also here with her new baby Rachel, + Bob. F. sur (11 Sept 1997, 7.58 a.m.).

Intestinal parasites

Kalema (1992, 1997) undertook the first, pioneering field study of the intestinal parasites of the Sonso chimpanzees (see Fig. 3.1), over a two-week period in September 1992. Kalema, using the McMaster method to count eggs and a light microscope to identify them at the Sonso field site, identified two types of helminth in 28 chimpanzee faecal samples from an area of forest within the range of the Sonso population: *Oesophagostomum stephanostomum* and *Strongyloides fulleborni*. Five samples from another, recently logged area contained no evidence of helminths. Eggs and larvae of these helminth species occur on the ground in the forest and are ingested incidentally by chimpanzees, whereupon they develop into larvae in the large and small intestine, some of which are subsequently excreted and can be found in faeces. An unidentified trematode, possibly *Dicrocoelium dendriticum*, was also found in one sample. Rates of infection are shown in Table 3.1. Heavy infections with *Oesophagostomum* sp. can cause severe diarrhoea. One of the samples had 230 *Oesophagostomum* eggs per gram faeces; this was the most heavily infested sample. However, the consistency of the faeces was firm.

A six-week field study was conducted by Barrows (1996). She sampled all five diurnal primate species living at Budongo, with a focus on chimpanzees. The five primate species studied were: chimpanzees, baboons, black and white colobus monkeys, blue monkeys and redtail monkeys. In addition samples were studied from employees of the BFP who entered the forest as part of their work. Samples were collected immediately after defaecation, from identified individuals, and initially examined at camp using a variety of methods to detect a wide range of endoparasites. They were subsequently analysed in detail in the UK.

The results of this study are shown in Tables 3.2 and 3.3. Table 3.3 shows the prevalence of infection for the chimpanzee population. The results were analysed with respect

Fig. 3.1:   Gladys Kalema examining chimpanzee faeces at Sonso.

Table 3.1: Rates of infection found by Kalema in Sonso chimpanzees.

Parasite	Per cent infected ($N = 28$)
Oesophagostomum stephanostomum	82.1
Strongyloides fulleborni	64.3
Both the above	57.1

to age, sex and dominance rank. Female chimpanzees had higher levels of infection than male chimpanzees; levels of helminth infection decreased with age; and subordinate chimpanzees had higher levels of infection than dominant chimpanzees.

Like Kalema, Barrows points out that heavy infestation of helminths can produce severe diarrhoea, but this was not seen in the Sonso population during this study (although occasionally it does occur). She also notes that a severe reaction to migrating larvae in the lungs can produce pneumonia and suggests that this may be a possible cause for the coughing that occasionally occurs. Tapeworms and hookworms can likewise cause damage but levels of infection appear to be under control at Sonso. More helminthic parasites were found in chimpanzees and baboons than in the other primates studied, which may be because they spend more time on the ground. Frequencies of parasites in the human sample were all extremely low. Strongyloides, found by

Table 3.2: Intestinal parasites found in primate species in the range of the Sonso chimpanzees, and in humans working at the BFP (from Barrows 1996).

Parasite	Chimpanzee (N = 126)	Baboon (N = 4)	Redtail (N = 3)	Blue (N = 5)	Colobus (N = 6)	Human (N = 13)
Trichuris sp.	Yes	Yes	Yes	Yes	Yes	Yes
Strongyloides sp.	Yes	Yes	Yes	Yes	Yes	—
Oesophagostomum stephanostomum	Yes	—	—	—	—	—
Unidentified strongyles/hookworm	Yes	Yes	Yes	Yes	Yes	Hookworm
Bertiella sp.	Yes	—	Yes	—	Yes	—
Trichostrongylus	—	Yes	—	—	—	Yes
Ternidens deminutus	Yes	Yes	—	—	—	—
Capillaria sp.	—	—	Yes	—	—	—
Giardia lamblia	Yes	—	—	—	—	Yes
Entamoeba coli	Yes	Yes	Yes	Yes	Yes	Yes
Entamoeba hystolitica	Yes	Yes	Yes	Yes	Yes	?
Unidentified *Entamoeba* spp.	Yes	Yes	Yes	Yes	Yes	?
Troglodytella	Yes	—	—	—	—	—
Chilomastix mesnili	Yes	Yes	Yes	Yes	Yes	?
Iodamoeba buetschlii	Yes	—	Yes	Yes	—	?
Blastocystis hominis	Yes	—	—	—	—	Yes
Balantidium coli	Yes	—	—	—	—	—
Unidentified protozoon	Yes	Yes	—	—	—	—
Unidentified protozoon	Yes	Yes	—	—	—	—
Endolimax nana	—	Yes	—	—	—	—

Table 3.3: Prevalence of infection with each parasite for the chimpanzee population (N = 126) (from Barrows 1996).

Parasite: helminths	No. samples infected	Samples infected (%)	Parasite: Protozoa	No. samples infected	Samples infected (%)
Trichuris	6	5	*Giardia lamblia*	9	7
Strongyloides	43	34	*Entamoeba coli*	72	57
Oesophagostomum stephanostomum	95	75	*Entamoeba hystolitica*	78	62
Unidentified strongyles/hookworm	62	49	*Entamoeba* sp.	54	43
Bertiella	47	37	*Troglodytella abrassarti*	96	76
Ternidens deminutus	39	31	*Chilomastix mesnili*	24	19
			Iodamoeba buetschlii	48	38
			Blastocystis hominis	72	57
			Balantidium coli	1	0.8
			Unidentified sp.	36	29
			Unidentified sp.	18	14

Ashford *et al.* (1990) in Batwa pygmies and gorillas in Uganda, was not found in the human samples at Sonso.

Barrows concluded that 'Although the results show that Budongo's primates are infected with a reasonably wide variety of helminths, over the time period of this study parasite burdens, which in some cases might be considered to be reasonably heavy, appeared to be well tolerated by chimpanzees.' This conclusion is borne out by visual inspection: the majority of the Sonso chimpanzees (there are exceptions such as the adult male Tinka) have shiny coats and appear to be in very good condition.

Medicinal plant use

The Sonso chimpanzees, like chimpanzees in many parts of Africa,[13] use particular plants for self-medication, and this may be one way in which they maintain good health in the face of parasitic infections (Huffman *et al.* 1996; Huffman 1997, 2001). At Sonso, a species of *Commelinaceae* is used for this purpose: *Aneilema aequinoctiale* (Fig. 3.2). This is a herb that grows up to around 1.5 m high on river banks and at roadsides in the forest. Bakuneeta (pers. comm.) first observed an adult female chimpanzee self-medicating using the leaves of this species by the side of a logging road that runs northwards through Budongo Forest from the sawmill. The observation was made early in the morning. The chimpanzee was taking leaves into its mouth and pressing them in the mouth rather than chewing them, before swallowing them. He also found whole or part whole undigested leaves of *Aneilema* in chimpanzee faeces. Infants learn this behaviour by close observation of their mothers (Huffman 2001). Since then, other individuals have been seen self-medicating in the same way.

This herb is not only used in Budongo. Huffman *et al.* (1996) record it among a number of plants used by the chimpanzees of Mahale and Kibale. The three samples described were obtained from faecal samples. The presence of nematode larvae was noted in the samples with *Aneilema* leaves.

Huffman (1997, 2001) notes that all the various species used in self-medication by chimpanzees, bonobos and lowland gorillas (totalling 35 species as of 2001, found at sites across the continent) share the characteristic of having leaves covered in trichomes. We collected leaves of *Aneilema aequinoctiale* from Sonso and photographed them under an electron microscope at Oxford University. Trichomes were indeed present (see Fig. 3.3).

Huffman suggests that these leaves made rough by trichomes, consisting of material that does not easily digest, are rapidly passed through the chimpanzee's intestine and this physical purge acts to remove the nematodes by excretion with the faeces. Huffman (2001) includes a picture of a leaf of *Aneilema aequinoctiale* together with a worm of *Oesophagostomum stephanostomum*; this was one of 20 worms expelled in one dung sample together with 50 folded *Aneilema* leaves.

---

[13] Use of such leaves has been reported at 13 sites from Bossou in West Africa right across to Budongo in East Africa. The majority of these sites are physically isolated from each other.

Fig. 3.2:   *Aneilema aequinoctiale* (photo: V.R.).

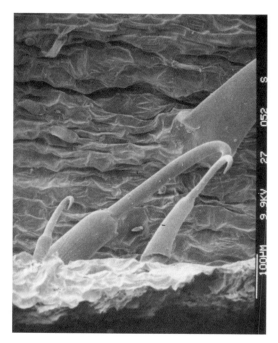

Fig. 3.3:   Electron micrograph of surface of leaf of *Aneilema aequinoctiale* showing trichomes (photo: Oxford University Electron Microscopy Laboratory).

Parasite control is achieved not only by the physical action of leaves being swallowed whole, however. As Huffman points out, some of the plants used by chimpanzees in this way, in particular species of the genus *Vernonia*, are also used by human beings and for their livestock as medicinal treatment for such illnesses as malaria, dysentery, intestinal worms and general stomach upsets. Some people also cook parts of the plant as a tonic food which enhances appetite and gives strength (Huffman 2001). Likewise, a female chimpanzee at Mahale observed by Huffman with symptoms of diarrhoea and a heavy parasite load, ate the *pith* of *Vernonia amygdalina* after which the output of *Oesophagostomum* eggs in her faeces declined over a 20-h period (Huffman *et al.* 1993). Steroid glucosides and sesquiterpene lactones have been found in the leaf, stem, pith and roots of *Vernonia amygdalina* (Huffman 2001) and have been demonstrated to be among the compounds that are medicinally active in this species.

One other plant deserves a mention in the present context: *Khaya anthotheca*, the Budongo mahogany. From time to time, at irregular intervals, chimpanzees avidly remove flakes of the bark of this tree and scrape the inner side of the flakes with their incisor teeth. Reynolds *et al.* (1998) found the bark to be very high in condensed tannins, with a bitter or astringent taste. However, a clear medicinal role for this bark, which is also used by human populations to control disease, has not been demonstrated in the way it has for *Aneilema*. Possibly its high antioxidant activity serves to scavenge for free radicals in the foods eaten (Greenham, pers. comm.). Our studies of *Khaya*-eating have also led us to study the gum that is present under and around flakes of bark, which we have found to have a high level of polysaccharide. We are currently examining whether this may be attractive to chimpanzees. Such gum may be the tree's response to attack by insects laying eggs under its bark. This has been found to lead to the production of gum in *Acacia* trees, and this gum provides an important part of the diet of galagos (Bearder and Martin 1980). According to T. Wagner (pers. comm.) many species of beetles use damaged areas of forest trees for nutritional purposes, feeding on phloem sap, while some, e.g. Nitidulidae, lay eggs on the 'wound' and their larvae feed on the rotting cambium between bark and wood. This may be of interest to chimpanzees.

Soil-eating

The Sonso chimpanzees eat termites and at other times they eat the soil of *Cubitermes* termite mounds. Fifteen cases of geophagy by 12 individuals (4 females, 8 males) have been observed to date. Chimpanzees used their hands and teeth to break open the mounds of *Cubitermes speciosus*, including both active and inactive mounds (see Fig. 3.4). In 60% of cases, termites were eaten along with the soil. This activity is not a common one, frequency of observation being 0.79 instances per 100 h of observation. The peak time for this behaviour to occur was between 0900 and 1000 h.

It is likely, though not as yet proven, that this is done for medicinal reasons (as a preventive or as a cure) but we do not currently know. Parasite studies by Huffman and Pebsworth (pers. comm.) showed that all individuals involved in termite-soil-eating were infected by at least two nematodes, *Oesophagostomum* sp. and *Strongyloides fullebornii*, and

Fig. 3.4:   Part of a Budongo Forest termite mound with termites (*Cubitermes speciosus*). Mnason Tweheyo observed soil and termites being eaten from this mound in September 2001 (photo: V.R.).

a protozoan species, *Troglodytella abrassarti*. Details of the above studies and of the physical, chemical and mineralogical properties of termite soil are described in Tweheyo *et al.* (in press). Termite soil is rich in iron, and consists of very fine particles of pharmaceutical grade clay mineral belonging to the kaolinite, halloysite and meta-halloysite group. It may thus be that in eating termite mound soil chimpanzees are mirroring the behaviour of human beings who take kaolin as a cure for diarrhoea and other digestive upsets. An additional possibility that remains to be investigated is that there may be microbial materials in the ingested soil, located on the inside of the chambers of the termite mounds, that are important pharmaceutically.

SIVcpz status of the Sonso chimpanzees

During 2003, we were invited to join a project analysing urine and faecal samples to determine the SIVcpz status of East African chimpanzees of the subspecies *Pan troglodytes schweinfurthii*. SIVcpz is the immediate precursor of human immunodeficiency virus type 1 (HIV-1) and may indeed be the source of this virus. The study was part of an ongoing project to determine the SIVcpz status of chimpanzees in E. Africa. Prior studies had included Kibale National Park where 31 members of the Kanyawara

Table 3.4: Deaths and disappearances.

Age-sex group	Number
Adult/old males	10
Adult/old females	6
Subadult males	0
Subadult females	2
Juvenile males	5
Juvenile females	3
Infants	9
Total	35

community and 39 members of the Ngogo community had been studied; none was found to be positive. At Gombe National Park, 15 members of the Mitumba community, 51 members of the Kasakela community and 10 members of the Kalende community were studied; 7 individuals were found to be positive.

At Sonso all 40 non-infant members of the community were studied from urine samples and all were found to be negative. Our chimpanzees have a clean bill of health as far as SIVcpz is concerned.

Death rate

Nishida *et al.* (1990: 86) list a number of causes of mortality: senility, death of the mother, aggression and disease. Losses of Sonso individuals to date have been as shown in Table 3.4. At the time of writing this covers a period of 10 years (1993–2003) so this community has lost an average of 3.5 animals a year.

Death of a young adult male

In many cases we do not know the cause of death for sure, though there are indications. Such is the case of the young adult male Andy (AY) (see Fig. 3.5). Andy was a well integrated male in the community and was popular with researchers as he sometimes made day nests in front of our camp, putting his head up every now and then to see what we were doing. Because he spent much time near houses close to our camp site, he may have died as a result of catching a respiratory disease from a local resident living near our camp. The discovery of his cadaver was written up by three students then at Sonso: Melissa Emery Thompson, Lori Oliver and Kim Duffy. I copy it with minor changes here.

Background

On Tuesday 18 July 2000, Andy had been observed by L. Oliver to be coughing. Other individuals, including Janie and Maani, were observed with a similar sustained cough, which the student researchers found abnormal and worrisome. One of these chimpanzees, Maani, was also exhibiting diarrhoea on this date.

Fig. 3.5: Andy (photo: V.R.).

Discovery

The carcass was found approximately 100 m behind Budongo School by a man living there and reported to the BFP field assistants on the morning of 23 July 2000. He was alerted to the carcass by the strong smell and the frequent visits of a local dog to the area. This young adult male chimpanzee (estimated 15–20 years) had recently frequented a mango tree very near to his place of death and was very likely going to or returning from feeding there. He was last seen by the Budongo staff on Tuesday 18 July 2000 in apparently good condition. Judging by the appearance and odour of the body, it was estimated by the field assistants to be approximately two days dead. The cadaver was positioned face down with the right arm behind the back and the left arm missing from the elbow with the humerus exposed. A small clearing was cut and the cadaver was dragged a short distance onto wire mesh; more mesh was placed over the cadaver and fixed into place with wooden stakes. This was intended to keep scavengers, including the aforementioned dog, away from the carcass while disturbing it minimally so that a necropsy could be performed.

Condition

At the time of discovery, there was no apparent cause of death. The left arm was likely removed after death as large amounts of blood were not seen. The arm was found by Karamagi Joseph on the morning of Monday 24 July approximately 4 ft up in a tree; this site was approximately 200 m from the corpse. That afternoon, large amounts of chimpanzee hair were identified in two deposits of dog faeces, one less than 50 m from the arm and one on the Budongo Forest Project campsite. As only the humerus remained on the carcass, the arm was apparently severed at the joint, though the proximal ends of the radius and ulna were missing from the recovered arm. Approximately 6 in. of flesh were missing from the site of the severance. No other injuries or abnormalities (other than moderate oedema thought to be related to decomposition) were noted at the time of discovery or when the cadaver was recovered on 25 July after veterinary assistance failed to arrive. By the time of recovery, considerable autolysis had occurred.

Unfortunately, despite many efforts to obtain the services of a veterinarian to perform a necropsy, none was available and the carcass was eventually cleaned by our students and staff and preserved at Sonso.

Death of an adult female

A second case of death was actually observed by a number of our field assistants as well as many of the members of the Sonso community. This was the case of Ruda (RD), an adult female.

On 29 October 2001, RD, the mother of two offspring, Bob (BO, a juvenile male) and Recho (RE, an infant female), was found dying, surrounded by a large number of members of the community, 26 in all. She was very emaciated and died in the following night, with members of our staff in attendance. The circumstances of her dying are of interest and I report them here as written up by field assistant Zephyr Kiwede:

At 10.15 hrs in block 30 Zephyr and Melissa came across RD, she was on the ground calling. Black (BK, adult male) was hitting her. Her body was covered with faeces, she had serious diarrhoea. Kalema (KL), Kewaya (KY) and Bob (BO) were feeding on *Ficus* fruits nearby and when they heard RD calling (a strange call) they immediately climbed down, we followed them to where RD was lying. Bahati (BH) and Katia (KA) were also present.

At 10.38 BK went away, leaving the other chimpanzees, some giving alarm calls. At 10.40 BK returned with Duane (DN, alpha male) who looked very scared, he could not move close to RD, just stayed away while calling. At 10.45 Maani (high ranking adult male) arrived and tried to hit Ruda to see if she could get up, in vain.

10.46 BK moved close to RD and smelled her, then moved back, all other chimpanzees were giving alarm calls.

10.50 DN looked carefully at RD, then ran away scared, returning at 10.54 to sit nearby looking scared and calling.

10.57 BK and MA displayed near RD and hit her, BK turned her round, she could just scream loudly.

11.01 BK pant hooted while turning RD aggressively, smelled her body and moved away.

11.07 All the chimpanzees left except BO and RE who stayed behind looking at their dying mother. BO moved close to her and smelled her, then moved away.

At this point Zephyr had to leave Ruda; other field assistants stayed; Zephyr returned at 17.20 and continues his report as follows:

17.20 BO and BK made strange calls, another group replied from north.

17.23 BO and BK made strange loud calls again, another group replied again from north. RE was sitting 5 m from her mother looking at her, RD was still calling but not loud. Further strange calls from BK and BO. Tinka (TK, low ranking adult male) arrived but did not go close to RD.

17.36 BO screamed loudly, which I termed 'crying'. We really felt so sad when BO started screaming joined by RE at 17.39. RE was screaming in such a way you would know there was a problem, she continued up to 18.00.

18.38 chimpanzees started nesting around the spot. We could not see whether RE nested with her brother BO. Because RD was still breathing we had to stay out in the forest to make sure that strange animals at night did not celebrate on it.

On 30 October 2001, the following day, the chimpanzees that had nested around RD got up and left the site. Inspection showed that RD was dead. Later in the morning her carcass was carried to camp. Veterinarians at UWEC and Makerere University had been informed and would arrive later that day to conduct a post-mortem.

A necropsy was conducted by four Ugandan vets and one Australian vet who came up to Budongo for the purpose. It revealed that the cause of death was likely to have been heart failure secondary to severe chronic peritonitis. Many sacs of pus were found in the abdomen, RD had large liver abscesses, emphysema, and at least three types of intestinal parasites including one tapeworm. A full report of the necropsy by Wayne Boardman can be found in Appendix D and a picture of the post-mortem in progress at Sonso camp is shown in Fig. D.1.

Death of an old male

Bwoya (BY), an old male, died after a long illness during which he became very emaciated. He was seen by Zephyr Kiwede and Melissa Emery Thompson on 3 September 2001 looking very sick and feeble, and he was seen again on 28 September 2001 with a runny nose, still looking emaciated and weak (see Fig. 3.6).

Fig. 3.6: Bwoya shortly before he died (photo: M. Emery Thompson).

Notes on his death and skeletal remains were made by Melissa Emery Thompson, Zephyr Kiwede and Juliet Craig as follows:

> On 24 December 2001 three men from the Forest Department found a dead chimpanzee in block I19, just off the Royal Mile and near to the border with Nyakafunjo. The corpse was very smelly but not greatly decomposed so death probably occurred 20–22 December. We were informed on 7 January 2002 and the skeleton was retrieved on 8 January 2002 and brought to camp. At this time the corpse was reduced to bones and hair. All long bones and the skull were recovered along with more than half of the smaller bones — some animal activity had taken place as the skull was found approx. 4 m away from the rest of the body. The right ankle was found, but the right foot appears to have been carried off by scavengers.
>
> We saw BY once in November 2001 looking very near death — completely emaciated and moving very slowly. Geresomu saw him briefly in early December. Since that time BY has not been seen. All other chimpanzees in the community have been seen since 8 January except BY and Zana.
>
> Skeletal characteristics also indicate that this was BY. The cranium and canine teeth were quite large in comparison with two existing adult male skeletons at Sonso, MG and ZT. However, the long bones were slightly shorter than either of these two skeletons. Thus the individual was an adult male of short stature, consistent with BY. Pathologies of the skeleton were noted: both femoral heads show extreme bone loss, only small nubs remain; the acetabula of the pelvis show compensatory bony growth — the sockets are filled with lumpy bone; the epiphyses of all long bones, as well as the proximal epiphyses of the metacarpals and metatarsals, have loss of bone fusion — so much that the distal end of the right tibia was completely separate.

Takahata (1990*b*) mentions that at Mahale old chimpanzees become increasingly solitary; this certainly fits with our observations at Sonso, of Matoke, Magosi and Bwoya. Huffman (1990) discusses the question of old age in Mahale in detail; he points out that there is a decrease in activities such as travelling, and in social activities such as grooming. The reasons are partly related to a loss of fitness and partly to sickness. I felt sorry for each of our old individuals at Sonso as they got less active. I recall seeing Magosi (who had once been the alpha male) in his latter years, sitting on the trail and looking up into the trees where there was an oestrous female with a number of males (Fig. 3.7): the spirit was willing but the flesh was weak.

Jambo (JM), an adult male, was speared by a sugar cane guard on 4 May 2003. This happened during the writing of this book. The case is described in Chapter 9.

Comparison with other sites (Goodall 1986; Boesch and Boesch-Achermann 2000; Nishida 1990) shows that Sonso has so far been spared the fatal epidemics that have ravaged some other sites such as Gombe (polio), Mitumba, near Gombe (influenza) and Taï (first ebola, more latterly anthrax[14]).

---

[14] For details, see Leendertz *et al.* (2004).

Fig. 3.7: Magosi sitting on trail looking up into trees (photo: V.R.).

Human–ape disease transmission

In recent years the threat to ape communities posed by contact with human beings (and vice versa) has come to be increasingly recognized, and during the writing of this book a 3-day meeting was organized by Christophe Boesch at the Max Planck Institute of Evolutionary Anthropology in Leipzig to discuss the issue. As a result of this meeting, a new network of wildlife veterinarians and laboratories is currently being set up to investigate the extent of the problems and to find ways of preventing this kind of disease transmission in the future.

Homsy (1999) has listed the large number of pathogens of humans to which chimpanzees are known to be susceptible, and the equally large number of pathogens of chimpanzees to which humans are known to be susceptible. A very large number of viruses, bacteria and parasites are involved. Since the Leipzig meeting referred to above, strenuous efforts are now being made to identify pathogens before epidemics take their toll of wild (and, of course, captive) populations. In some cases, such as ebola, extreme care has to be taken in collecting samples, let alone performing necropsies, in the field. For example, it is necessary to take all the equipment required to the site where the animal has died, set up a mobile laboratory on the forest floor, and wear fully protective clothing; these conditions are extremely demanding of time, energy and resources yet are essential if severe and highly contagious diseases are to be effectively diagnosed and preventive measures taken.

In the case of the polio and influenza epidemics at Gombe and Mitumba, transmission by humans may have been responsible (Wallis and Lee 1999). What the future holds for our Sonso chimpanzees we do not know. As mentioned before, we operate a strict 7-m rule (human visitors, staff and researchers must not approach closer than 7 m to any chimpanzee) at Sonso. In 1991 Gladys Kalema visited our project and warned us of the dangers we might one day face.[15] Recently much has been written about human transmission of diseases to apes (see Wallis and Lee 1999; Woodford *et al.* 2002), especially with regard to gorillas where the dangers arising from tourism are greatest. We have not allowed tourists to visit our chimpanzees at Sonso and this has probably saved them from a number of disease outbreaks, but we are not complacent because, however careful we are, the forest is constantly being visited by local hunters setting snares and others collecting firewood or forest products, and at any time a disease outbreak with the possibility of large-scale mortality could happen.

Non-fatal epidemic

There has been one serious non-fatal epidemic during the period of our studies (i.e. since 1990). This was a respiratory disease that gradually spread over the community. We called in veterinary support and as a result a full report was written by Gladys Kalema, Veterinary Officer to the Uganda Wildlife Authority, and Wayne Boardman, Wildlife Veterinarian at the Uganda Wildlife Education Centre, Entebbe. We had been very concerned the epidemic might pose a threat to the lives of our community members; mercifully it did not. Their report is presented in Appendix D.

Other communities have been less fortunate. At Mahale where such respiratory infections occur periodically and can spread widely in the community, they have caused, directly or indirectly, many deaths. Nishida *et al.* (2003) report that over a 19-year period of study, disease was the most common cause of death, accounting for 48% of all deaths. From 1993 to 1994, 11 chimpanzees were suspected to have fallen victim to a flu-like epidemic.

Injuries

A number of the Sonso chimpanzees have sustained non-snare-related injuries over the course of our studies:

1. A leg or foot injury to Tinka (TK), a low ranking, fairly old, adult male (Events Book entry by Emma Stokes, observations by her, James Kakura and Zephyr Kiwede):

   At 07.43 hrs on the morning of 22/10/97, TK was observed feeding on *Broussonetia* leaves, seated in the lowermost fork of a tree. Whilst moving within the tree he was unable to move his left foot voluntarily, and locomotion was enabled only by lifting the foot with the left hand and placing it in front of the right foot. This made arboreal locomotion a very slow and unsteady process, and both

---

[15] It has since been recommended (Homsy 1999) that the minimum distance be increased to 7.5 m.

hands were used to steady himself between each successive foot movement. There appeared to be no outward signs of injury on the left foot, and the cause of the injury was unknown. He remained in the lower part of the tree for the entire feeding bout, and then made a day nest in the same tree fork until climbing down (again requiring the use of the left hand to position the left foot) at 12.08, and moving out of view on the ground.

On 24/10/97 TK was again seen feeding on the leaves of *Broussonetia*, this time on the ground. There was still no voluntary movement of the left foot, and in addition the foot was held off the ground whenever possible. It was possible that he was unable to climb to feed arboreally.

By 10/11/97 TK was still limping, but facultative as opposed to obligatory use of the left hand in assisting movement of the left foot was observed. He was seen feeding on the flowers of *Broussonetia*, located in both the middle and lower sections of the tree.

By 5/12/97 TK was capable of complete voluntary movement of the left foot and was not seen to use his left hand to aid locomotion, but he still had a detectable limp. By 21/1/98 his limp had gone and he was moving with apparent ease.

2.   A wound to Zefa (ZF), a young adult male (Events Book entry by Katie Fawcett, observations by her and Geresomu Muhumuza):

On 23/4/98 at 08.31 it was observed that ZF had what appeared to be a puncture wound on his outer left calf. When we arrived Kwera (KW, adult female) was grooming ZF and ZF was grooming and licking his wound. At 08.40 Zefa left to the north, 08.44 KW followed.

3.   A wound to Vernon (VN), an adult male (Events Book entry by Katie Fawcett, observations by her, Geresomu Muhumuza, Emma Stokes, Kakura James, Paula Pebsworth and Karamaji Joseph):

6/6/98 at 08.20 an apparently superficial wound was observed on the lower left leg of Vernon as he was climbing into a *Cynometra* tree. The wound was estimated to be approximately 5 in. × 2 in. on the outside of the lower leg just above the ankle. The hair and skin appeared to have been scraped off giving the appearance of a graze. The wound seemed to be causing some discomfort as he did not appear to place his whole body weight on the left foot while climbing. Later at 10.15 the same morning he was again observed, this time limping badly while walking on the ground, however he was keeping up with the rest of the group.

9/6/98 VN was observed feeding in a *Broussonetia* tree, his wound was still visible but he did not seem to be in any discomfort.

8/7/98 VN was observed walking on the ground limping badly. The surface wound appeared deeper and now 'raw'. There were no signs of scab formation. There was yellow colouration in the centre parts of the wound, the rest of the wound being red. This indicates the possibility of infection. It was expected by all observers that by now the wound should have healed.

20/10/98 VN's wound still not healed. No longer yellow but still red and raw. No weight put on left foot when walking.

This wound subsequently healed up fully and without visible trace.

4. A wound to Duane (DN), the alpha male. (Events Book entry by Zephyr Kiwede, observations by him, Juliet Craig, James Kakura and Raymond Ogen):

15/2/02 DN had the intention of copulating with Zimba (ZM) who was in full oestrus. DN started chasing ZM, ZM was screaming a lot, DN ended up not succeeding in the attempt. He waited for another opportunity. The chimpanzees moved to another tree where DN tried again, chasing ZM, this turned into real fighting. Immediately Maani (MA, a senior male) joined the fight in favour of ZM. DN was chased by MA and ZM up in the tree. After this fight we saw DN limping and blood oozing from his right foot. He had sustained a big cut between his first and second toes. Jambo (JM, a senior male) groomed DN very much after this incident, then JM stopped grooming DN and moved to MA and they both groomed each other. After a while DN wanted to join them but as DN approached, MA walked off.

5. A wound to Maani (MA) a senior male (Events Book entry by Zephyr Kiwede, observations by him, Joseph Karamaji, Raymond Ogen and James Kakura):

22/2/02 MA was seen with an open wound which looked like a bite, he was limping and not using his left hand, he was hanging it down and looked to be in great pain because he was restless when seen. He lay on the ground and raised his leg up, and then the hand, we suspected that it could have been revenge from DN following the events (above) with ZM. It may be that JM had become over-friendly with DN and angered DN by siding with ZM against DN on 15/2/02.

27/2/02 MA was feeding in a *Ficus barteri* tree, he climbed down. On reaching the ground he started licking his wound and eventually picked some leaves of *Acalypha*, chewed them a bit, and started using them to clean his wound. He passed the leaves across the wound three times then dropped them and moved away.

6. Finally, while this book was being written, the alpha male Duane (DN) was injured as a result of a fall. This was observed by Nick Newton-Fisher and Geresomu Muhumuza who reported as follows:

18/2/04 DN and ZF, in the middle of some unseen altercation, fell approx. 60 feet from a branch and thudded on to the ground. There was nothing to break their fall. ZF appears uninjured — he may have landed on DN — while our alpha male has injured his left leg. Geresomu and I stayed with DN for a while afterwards; putting any weight on his leg appeared to cause a lot of pain, but he did eventually manage to hobble away, with several rests. He also has a laceration to the knuckle of his third finger, left hand, which predated the fall.

As it turned out, DN's injuries were minor and within a week he was fully recovered; ZF was unhurt.

4. *Diet and culture at Sonso*

> Bob + Rachel — the small and tiny orphans — wandering alone. But fine. They join up with Nkojo & Bwoba. We follow all 4 for ?1 km along trails and through blocks. Somewhat punishing pace but we keep up with them & arrive at destination — a *Morus lactea* tree to the west of camp. Now a med. sized gp (10–15 indivs) all feeding on *Morus* fruits, small but abundant. 8.33 Calls to SW, ours reply. 8.40 Nkojo, Black descend and move towards SW. Others stay. Kalema + Kumi. 8.44. Jambo down, Nambi down, then others, now all are descending, quite a long and difficult descent but each makes its own route down. In some cases takes 2 mins to descend (14 March 2002).

The chimpanzee's day is 12–13 hours long, and of those hours about half, some 6–8 hours each day, are spent foraging and feeding. They start the day feeding in the company of the chimpanzees that have nested nearby, then either move to a new feeding site with one or more of them or head off on their own to join a new group. Sometimes they are guided in their movements by calls coming from groups in other areas of the forest. If they are on a tree with an abundance of ripe fruits they pant-hoot in chorus. If they hear a chorus from over the canopy they stop feeding and listen attentively, sometimes replying. We consider the vocalizations of the Sonso chimpanzees in more detail in Chapter 6.

If food is scattered widely then individuals can forage most successfully on their own or in small groups, thus reducing competition. In these circumstances, chimpanzees tend to be rather quiet; on the other hand, if food is concentrated in one location, as when a single *Ficus mucuso* tree is fruiting heavily, there is enough food for many chimpanzees to gather there to feed. Indeed, their vocalizations are loudest and their pant-hoot chorusing is most frequent when many of them are concentrated in a large party on a single tree. We deal with the topic of party size in relation to food availability in Chapter 5.

Forest types and chimpanzee foods

As described in Chapter 1, there are four forest types: Colonizing Forest, Mixed Forest, *Cynometra* Forest and Swamp Forest. These forest types have different tree species compositions and therefore the foods of chimpanzees differ from one forest type to the other. Thus the composition of the forest determines the range of feeding possibilities available to the chimpanzees. It would be tedious to describe the foods found in each forest type but here are some examples of differences:

- Colonizing Forest is characterized by *Broussonetia papyrifera* (a recently introduced species) and *Maesopsis eminii*, both of them much liked foods.

- Mixed Forest contains the majority of food species: *Celtis gomphophylla*[16] and *Celtis mildbraedii*, as well as the many *Ficus* species.

- *Cynometra* Forest is characterized by the climax species *Cynometra alexandri* which provides food at one time of the year only, normally in the dry season (January–February).

- Swamp Forest is where the chimpanzees find the fruits of *Pseudospondias microcarpa*, the pith of *Calamus deerata* and the soft woody inside of the stems of dead Raphia palms, *Raphia farinifera*.[17]

Of these forest types, Mixed Forest was shown by Plumptre (1996) to be the most common at the present time, constituting about one-half of the forest in the 1990s. This contrasts with its lesser extent in the 1950s before the bulk of commercial logging for mahoganies and certain other species, and before the arboricide programme (see Chapter 1). The increase in the extent of Mixed Forest was a blessing for all the forest fruit-eating species, including the primates, and has meant that Budongo Forest is unusually rich in species of fruiting trees such as figs that provide chimpanzees and monkeys with their preferred foods. Budongo as a whole can today be seen as a food-rich forest for primates, and this is confirmed by the higher densities of monkeys found in logged forest and arboricide-treated forest than in unlogged forest such as the Nature Reserves (Fairgrieve 1995a; Plumptre and Reynolds 1994). Tweheyo (2003) and Tweheyo *et al.* (2004) have shown in a 14-month study that most chimpanzee feeding instances occurred in logged forest, more than in any other forest type.

Whether chimpanzees, like monkeys, have benefited *numerically* from the logging and arboricide is less clear. They ought to have done, because of the increase in their food supply. However, chimpanzees may be less able than monkeys to tolerate the impact of strange humans working in the forest; a survey of chimpanzees in the forests of western Uganda (Plumptre *et al.* 2001) has shown them to be less numerous in Budongo Forest than in Kibale Forest where the food supply may be less rich, but where human impact over much of the forest has been less profound. The effects of logging on the primates of different forests can differ in a number of complex ways (Plumptre and Grieser-Johns 2001).

Food types and food species

The main source of our data on the foods eaten by the Sonso chimpanzees is the ongoing record of foods seen eaten, recorded by our field assistants and researchers in the course

---

[16] Formerly *Celtis durandii*.
[17] Budongo Forest does not contain the oil palm, *Elaeis guineensis*, common in swamp forests of West Africa. The only area in which it is found in Uganda is Bwamba in the southwest.

of their daily observations.[18] From the start of the BFP in 1990, we have collected samples of food species and many of these were photographed fresh for our Plant Identification Album, and subsequently dried and mounted by Andy Plumptre. In cases where the identification was uncertain they were identified at Makerere University's Herbarium, by Perpetua Ipulet, Tony Katende and others, after which they were placed in the BFP's Herbarium at Sonso.[19]

In 1997, Rhiannon Meredith and Susie Whiten made a list of 76 species of plants that provided food for the chimpanzees, based on records collected by our field assistants and researchers. An updated version of this list is included in Appendix B and shows the months they had been seen eating these species as well as the types of plants and parts eaten. It can be seen that whereas some species are rarely eaten, others are eaten in every month of the year. These records are based on observational data and do not imply anything about seasonality of the plant species or that chimpanzees do not eat certain species in months with blanks on the table; they were simply not observed eating them in those months.

Chimpanzees were seen to eat parts of the following kinds of plants: trees, shrubs and herbs, climbers and epiphytes. They fed on the following plant parts: fruits, leaves, flowers, bark, seeds, stem/pith, gum and wood. Of these, ripe fruits and young leaves make up the bulk of the diet. For very many species, including figs (e.g. Tweheyo and Obua 2001; Tweheyo and Lye 2003) chimpanzees eat the young leaves which are sugar- and protein-rich. Bark and gum are eaten from *Khaya anthotheca*, the so-called 'Budongo mahogany'; bark is also taken from other species including *Ficus exasperata and Cynometra alexandri*. In both the latter cases the inner, sap-covered surface of the newly stripped bark is licked, while in the case of *Khaya*, flakes of bark are removed and the inner surface is scraped off with the incisor teeth. This happens at irregular intervals, and when it happens many chimpanzees do it. The bark contains high levels of condensed tannins (Reynolds *et al.* 1998); why the chimpanzees occasionally binge on it remains unknown. There can be long periods when chimpanzees do not eat it at all; from the late 1990s until the time of writing this book for example, a period of 5 years, no *Khaya* bark-eating has been seen.

Insects and honey

Infrequently, in addition to the plant foods mentioned above, the diet is supplemented by animal foods. The Sonso chimpanzees, like the chimpanzees of Kibale, Mahale, Taï and Gombe, occasionally kill monkeys and eat them. Meat-eating will be dealt with later in this chapter. They also eat fig wasps in the course of eating figs, and occasionally, caterpillars. This happened for example in April 1997 when there was an explosion of caterpillars in the forest and the Sonso chimpanzees fed on them avidly. While I was

---

[18] For a list of food species of the Sonso chimpanzees see Appendix B.

[19] In 2001 the Herbarium, which had been in constant use by students, was checked and where necessary reorganized by Wilma van Riel. It is currently in good condition but requires the attentions of a botanist to progress further. Details of the Herbarium are available on the BFP's website, www.budongo.org.

writing this book, Kennedy Andama (field assistant) and Catherine O'Hara (research student) observed Black (adult male) feeding on wasp larvae:

> On Tuesday 29 April 2003 at 11.50 a.m., Black (BK) was followed to block F0 where he met Duane (DN). After 30 min of mutual grooming and resting, DN displayed and moved off to block ED. BK started to follow but delayed in block F0 where he was heard to break some branches of a small tree. When he emerged from the block he climbed up on to a low level horizontal branch close to the trail and was observed to eat something he had carried with him and that resembled honeycomb. He appeared to interject the eating of the comb with eating of leaves of an adjacent climber (unidentified). The leaves were being wadged together with the comb. BK remained eating for 15 min until all the substance he was holding was finished. DN remained close by with Wilma (WL) (with maximal swelling) but neither showed any interest in what BK was eating.
>
> After BK finished eating Kennedy (field assistant) returned to the place where BK had been heard breaking branches and found a damaged wasp nest in the tree with some thin honeycomb-like structures on the ground underneath. Within the nest matrix were white/transparent eggs containing wasp larvae. It appears that this was what BK had been feeding on.

Very occasionally we have come across the Sonso chimpanzees eating honey. An incidence reported by Sean O'Hara (pers. comm.) was as follows:

> Last Friday afternoon [in April 2003] we came across Jambo who was seated eating a large honeycomb. Nick, Janie and Janet were sitting on the ground beside him trying to put their faces close but Jambo ate it alone. He was there for several minutes. When they moved off we found a small piece of the honeycomb on the ground and there were a couple of live bees. About 4 m away was a *Celtis zenkeri* tree with a sloping trunk. About 5–10 m up was a hole in the upper surface and we could see bees flying around the entrance. Their defence seemingly wasn't fierce. (This was recorded in the Events Book by two Field assistants, Monday and Ogen.)

Food preferences

Among the main factors that determine chimpanzees' food preferences are the properties of the foods themselves and the nature of chimpanzees' taste preferences. We know rather little about the latter. Some research on chimpanzees' sense of taste has been done (see Hladik and Simmen 1996 for a review). The question of taste in another species is complicated by the fact that we have to deduce what tastes they like and dislike from their reactions (eating rapidly and wanting more versus spitting out and avoidance); this is normally done in experimental situations to avoid confounding factors. Thresholds of ability to taste can also be documented in experiments in captivity. Nishida *et al.* (2000) describe the taste of a large number of chimpanzee food species to humans, noting that chimpanzees are able to tolerate species that to humans taste unpleasant, bitter or astringent as well as others that taste neutral or sweet. They refer to studies that show that the foods of *Cercopithecine* monkeys are less palatable to humans than those of chimpanzees and that these species have developed more effective internal mechanisms for the detoxification of secondary compounds such as alkaloids than have chimpanzees or humans. However, chimpanzees can tolerate foods that are more astringent and contain more alkaloids than humans can.

When working with wild chimpanzees the study of food preferences has to be operationalized. We can gain a handle on food preferences in the wild by determining the *frequency* with which the animals are observed feeding on particular kinds of foods. For the Sonso chimpanzees, Plumptre and Reynolds (1994) obtained the order of food preferences by recording what species and plant parts were eaten during a series of half-hour scans over a two-year period of observations. The results were as shown in Table 4.1.

Food preferences were also determined by Newton-Fisher (1999*a*) independently of Plumptre's study. His conclusions, also based on scan sampling, were very similar (see Table 4.2).

A third independent study (Fawcett 2000) recorded the Sonso chimpanzees' food preferences (see Table 4.3).

Finally, food preferences of the Sonso chimpanzees were determined by Tweheyo *et al.* (2004) and showed that over a 14-month period (June 2000 to August 2001) chimpanzees spent most of their feeding time on *Broussonetia papyrifera*, *Ficus sur*, *F. mucuso*, *F. exasperata* and *F. varifolia* confirming the findings of other studies and establishing the importance of fig species in the diet.

These studies show that the preferred ('favourite') foods of the chimpanzees of the Sonso community are *Broussonetia papyrifera* (Fig. 4.1), *Ficus* spp. and *Celtis mildbraedii*. These findings are very specific to the Sonso community. It is unlikely that the same food preferences would be found in other communities, even in Budongo Forest, because *Broussonetia papyrifera* (the paper mulberry tree) is rarely found in other parts of the forest, and is lacking in most of the other forests of Uganda. It is an exotic species introduced by the British in the 1950s around the Sonso Sawmill, to see if it would grow well there, in the hopes that it might provide wood pulp for paper production at the mill. It failed to grow to the size needed for commercial purposes (it remains a small tree in Budongo). But it certainly did take hold where it was planted,

Table 4.1: Food preferences of Sonso chimpanzees (based on Plumptre and Reynolds 1994).

Species	Score	Rank	Food types
Broussonetia papyrifera	1436	1	RF, FL, YL
Ficus sur	1408	2	RF, UF
Celtis durandii	662	3	RF, UF
Cynometra alexandri	628	4	Seeds
Maesopsis eminii	541	5	RF, FL
Celtis mildbraedii	385	6	YL, RF
Ficus mucuso	379	7	RF, UF
Khaya anthotheca	310	8	Bark
Ficus exasperata	215	9	UF, RF
Cordia millennii	174	10	RF, FL

Table 4.2: Food preferences of Sonso chimpanzees (based on Newton-Fisher 1999a).

Species	All items % time feeding	Fruit only % time feeding	% time eating fruit
Ficus sur	23.0	23.0	35.6
Broussonetia papyrifera	22.7	4.0	6.2
Ficus mucuso	9.8	9.8	15.1
Maesopsis eminii	9.4	9.2	14.3
Celtis durandii	8.4	7.4	11.5
Celtis mildbraedii	4.6	0.1	0.2
Khaya anthotheca	2.9	0.0	0.0
Croton macrostachys	2.8	2.8	4.3
Ficus exasperata	2.2	1.5	2.4
Cordia millenii	1.7	1.4	2.1
Desplatsia dewevrei	1.3	1.3	2.0
Cynometra alexandrii	0.9	0.7	1.0
Ficus sansibarica (= brachylepis)	0.9	0.9	1.4
Cleistopholis patens	0.8	0.1	0.1
Raphia farinifera	0.6	0.0	0.0
Ficus natalensis	0.5	0.5	0.7
Ficus varifolia	0.5	0.0	0.0
Terrestrial Herbaceous Vegetation	3.2		
Climbers	1.5		

Table 4.3: Food preferences of Sonso chimpanzees (based on Fawcett 2000).

Food Species	Rank
Ficus mucuso	1
Celtis mildbraedii	2
Ficus sur	3
Mildbraediodendron *excelsum*	4
Broussonetia papyrifera	5
Cynometra alexandrii	6
Ficus exasperata	7
Ficus varifolia	8
Celtis wightii	9
Ficus sansibarica	10
Celtis zenkeri	11
Morus lactea	12

around the edges of the large clearing in the forest where the sawmill was located, and today has grown outwards from the forest edge into the sawmill clearing for up to 100 m in places, so that around the mill there are large stands of *Broussonetia*. It is not invasive and does not seem to replace other species inside the forest (Mbogga 2000), it grows outwards from the forest edge into the grassland. Chimpanzees eat the young leaves, flowers and fruits of this rather spindly tree, stripping and breaking many of the branches.

Fig. 4.1: (a) *Broussonetia papyrifera* fruit (arrowed) and leaves on stem (photo: V.R.). (b) Chimpanzee feeding on *B. papyrifera* flowers (photo: V.R.).

From the above it will be clear that *Broussonetia* is a new food for the Sonso chimpanzees. If we didn't know it was an exotic, and the time and place of its introduction, we might conclude that it was an established item of diet of chimpanzees, something they had evolved with and were by now genetically adapted to. The opposite is the case. We approach the question of *why* they like this new food in the next section.

The most frequently eaten fig species is *Ficus sur*. This is a smallish fig, rarely reaching the size of a golf ball. It hangs in pedicles from the trunk and branches of the tree. A fruiting *F. sur* tree is a sight to behold. Pedicles, each holding from 20 to several hundred figs, stick sideways out of the tree trunk and then droop down under the weight of the fruit (Fig. 4.2). There are many thousands of figs on a single tree. They start life green, then turn yellow and finally gain a reddish tinge before they drop. While the figs

Fig. 4.2:   Chimpanzees feeding on *F. sur* (photo: T. Mnason).

are still green, chimpanzees find them and cursorily eat a few fruits before passing on. When they are yellow the feeding begins in earnest. On a tree they like, the figs rarely get a chance to turn red.

They don't always like these figs, however. I recall a *F. sur* tree the chimpanzees passed by regularly. At first the fruits were unripe and green and they ignored it. Later the fruits turned a brownish colour and fell. I tested a fruit and found it to be dry. This was a poor fruiting of the tree. Fig trees fruit sporadically, not seasonally, some fruitings being heavy and luscious, others light and dry.

F. sur fruits non-seasonally whereas *B. papyrifera* has two distinct fruiting seasons coinciding with the two drier periods of the year (Fawcett 2000). However, chimpanzees often feed on young leaves of *B. papyrifera* and these are available all the year round. *F. sur* is plentiful in the range of the Sonso chimpanzees and in most months there is a fruiting tree to be found. So both species are reliable sources of food, and can properly be described as keystone foods for this community.

There is no doubt that the Sonso chimpanzees like eating *F. sur*, but if a *F. mucuso* tree is fruiting heavily in their range they forget all about *F. sur* and focus all their attention on *F. mucuso*. The figs of *F. mucuso* are twice the size of *F. sur*, and can be extremely juicy and succulent (even if full of sticky latex). Unfortunately for the chimpanzees, there are far fewer *F. mucuso* trees than *F. sur*.

Factors underlying food preferences: sugars and tannins

Chimpanzees eat some species of fruits and leaves frequently, some occasionally and some never. In the early 1990s we decided to investigate what factors were common to the most eaten species and not to the less eaten or avoided ones. We also wanted to investigate this with respect to fruits on a single tree: normally the ripe ones were preferred but sometimes unripe fruits were eaten. What were the factors involved?

We had one clue to go on from the start: many fruit-eating species (ourselves included) prefer ripe fruits because we find them sweeter. So one thing to measure in the eaten and not-eaten fruits was sugar content (fructose, sucrose, glucose and galactose). But there was an interesting problem when it came to figs. Ripe figs as well as unripe ones of the species eaten by chimpanzees often taste horrible to humans — full of latex, they are astringent to the point where we have to spit them out. This astringency is found in many fruits when they are unripe, and declines as the fruits ripen, while the sugar content, at first low, rises so that the ripe fruits lose their astringent taste and become sweet tasting. The reason is that many species (including figs) depend on animals eating them to disperse their seeds far from the parent tree where they have a better chance of survival. As a result the fruits that contain seeds become fleshy and attractive to eat, but not immediately: the edibility quotient of the fleshy outer part is closely geared to the development of the seeds inside. Natural selection has favoured fruits that evolved mechanisms to ensure they did not get eaten when unripe with underdeveloped seeds that would fail to germinate when excreted, and it has favoured mechanisms that attract animals to eat them when the seeds are ripe and ready to germinate. Seeds of some species are crushed when eaten but rapid eaters like chimpanzees swallow many seeds intact. In some tree species (e.g. *Maesopsis*) the seeds are too hard to be crushed and come through in the faeces intact. Such seeds have the advantage of being deposited away from the parent tree and away from all but a few competitors of their own species.

Do seeds that have passed through the gut of chimpanzees germinate faster or survive better than those that have not? This and other aspects of seed dispersal by primates was studied by Chapman (1995) who concluded 'evidence suggests that the seeds primates disperse that are not found by rodents or secondarily dispersed by dung beetles probably are capable of germination if the conditions are right. When researchers take seeds from primate dung and attempt to germinate them in controlled settings, evidence typically suggests that the passage through the frugivore gut has improved the rate of germination and reduced latency to germination' (p. 77). A study at Sonso of the germination of tree seeds taken from chimpanzee dung vs. seeds picked up on the forest floor found better germination in the former (Plumptre *et al.* 1997, pp. 76–7). A study of *Uvariopsis* seeds at Kaniyo Pabidi in the northeast of Budongo Forest had the same result: 160 plots were studied, each plot being planted with 10 seeds. Half of the seeds in each plot were taken from chimpanzee dung in the forest, the other half from seeds found under the parent trees. There was higher germination of seeds that had passed through the chimpanzee gut (Muhumuza, pers. comm., based on Plumptre, unpublished data).

While trees (and other fruit-producing plants) have been evolving mechanisms to ensure that animals would eat them at the appropriate time and not before, animals have evolved taste preferences for sweet-tasting foods because the sugars they contain are a rapidly available source of energy. Thus natural selection has favoured animals evolving a liking for sugars. A second evolutionary development has been an aversion to tannins. Tannins are sometimes called antifeedants and have anti-digestive properties which they achieve by binding to proteins in the diet of the animal and making them inedible. Too much tannin in the gut leaves an animal unable to digest its food properly and in the case of some species can have serious consequences up to and including death. Thus natural selection has favoured animals that avoid tannin-rich foods.

However, the ripe fruits of *F. sur* and other fig species remain astringent even when ripe and this poses a problem. Why should they remain astringent when ripe? Do they contain high levels of tannins? There were clues in the literature that they did: a study by Wrangham *et al.* (1993) of the chimpanzees of Kibale Forest had found high levels of tannins in figs, particularly in the seeds.

I approached Jeffrey Harborne[20] of Reading University with this problem, and over the next several years, together with Jenny Greenham, we investigated it. It was during this investigation that I realized just how complicated the relationship between plants and animals is. I make no claim that we eventually solved the question we set ourselves, but at least we made some progress.

We decided the critical test would be to take 'fruits seen eaten' and compare them phytochemically with 'fruits not eaten', 'fruits rejected' and 'fruits left on the tree'. Here, immediately, we hit the first set of problems: at the point of collection. If a fruit is eaten, how can it then be collected for analysis? And if a fruit is not eaten but left on the tree, how can you then collect it? Fortunately for us, chimpanzees are wasteful feeders, especially when it comes to figs. When they reach out and break off or pull in a branch of figs they often drop one or more pedicles or single fruits to the ground. There are enough such accidents to make collection of 'fruits seen eaten' (i.e. fruits that would have been eaten if they had not been accidentally dropped) a fairly simple process. On the second question, pedicles of *F. sur* fruits contain both riper and less ripe fruits. If, having broken off a pedicle and eaten the fruits that were wanted, the chimpanzee then dropped the pedicle apparently on purpose and the 'rejected' fruits fell to the ground, we collected them.

We repeated similar procedures, adjusted to the particular tree and species, for all the other fig species, and for as many other species as we could, over a three-year period, always distinguishing between fruits eaten and fruits not eaten. Every fruit was taken to our drying area at camp and carefully dried, marked with the species and date, and when enough samples were dry they were packaged up and sent to the Reading laboratory for analysis.

We undertook two main analyses: for sugar content (measuring fructose, sucrose, glucose and galactose) and for proanthocyaninins (condensed tannins). The results were as shown in Table 4.4 and Fig. 4.3.

---

[20] Sadly, Prof. Jeffrey Harborne died on 21 July 2002, during the writing of this book.

Table 4.4: Sugar and condensed tannin levels found in foods eaten and not eaten (from Reynolds et al. 1998).

Item	Mean eaten (n = 209)	Mean not eaten (n = 19)	SD	t	P
Tannins	15.54	12.37	11.28	0.33	n.s.
Sucrose	13.64	14.11	22.33	−0.28	n.s.
Fructose	23.63	8.84	25.50	3.41	0.002
Glucose	28.73	8.47	26.67	3.44	0.002
Total sugars	68.01	31.42	54.79	2.81	0.009

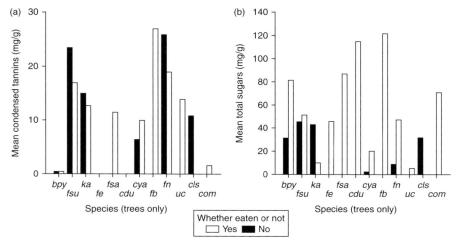

Fig. 4.3: (a) Condensed tannin content found in foods eaten and not eaten. (b) Total sugar content found in foods eaten and not eaten (from Reynolds et al. 1998).
Bpy = Broussonetia papyrifera, fsu = Ficus sur, ka = Khaya anthotheca, fe = Ficus exasperata, fsa = Ficus sansibarica, cdu = Celtis durandii, cya = Cynometra alexandri, fb = Ficus barteri, fn = Ficus natalensis, uc = Uvariopsis congensis, cls = Caloncoba schweinfurthii, com = Cordia millennii. Species are shown in order of preference from left to right.

These results were statistically significant and showed that, for fruits, flowers and leaves on the same tree, fruits of the same species, and fruits of different species, there were two rules:

1. Chimpanzees chose to eat foods with a higher sugar content in preference to fruits with a lower sugar content.

2. Chimpanzees chose foods irrespective of the level of condensed tannins.

The sugar finding was entirely as expected; but we had no explanation for the tannin finding except to conclude that chimpanzees have a much greater tolerance of condensed tannins that humans do. Human children are highly sensitive to tannins at first, but if they encounter them in the diet regularly their sensitivity wears off and they develop an improved ability to secrete protein-rich mucins from the mucosal surfaces of the mouth, which go into their saliva and serve to neutralize tannins. For example, children who regularly drink red wine (even diluted) develop salivary mucins and are more readily able to tolerate red wine in later life. In the UK we don't normally give wine to children but we drink it in adult life, developing salivary mucins as a consequence. We may find wine a bit astringent at first, especially dry wine with low sugar content, but in the end we come to neutralize the tannins, and to tolerate and like such wine. Another example is tea and coffee, both are rich in tannins. If we don't like the tannins we neutralize them with milk, which is rich in proteins, and we can go one step further and add sugar to make them sweet.[21]

So to some extent chimpanzees may develop a tolerance to the tannin-rich figs they eat almost daily in Budongo. It seems likely that they produce salivary protein-rich mucins, as we do. But the high levels of tannins they eat may also reflect a genetic factor. Assuming they have been eating figs and other tannin-rich foods over many thousands, even millions of years, this is to be expected. However, while they can tolerate the high level of tannins found in figs, they do reduce the number of tannin-rich seeds they swallow. To do this they make 'wadges' — a wadge is a bolus of compacted fig seeds that is kept in the mouth while the flesh of the figs is masticated and the juice is swallowed, after which the wadge is spat out. Thus we see that chimpanzees can tolerate tannins in the mouth but make an effort to avoid swallowing this antifeedant secondary compound.

Returning to *Broussonetia*, we found that the much favoured young leaves, flowers and fruits of this species were characterized by high sugar levels and a complete absence of condensed tannins. This absence helps to explain why this species is so readily eaten.

Food availability

Availability of food is the combination of the amount — abundance — and the spatial arrangement — dispersion — of food within a given location at a given time. Over time food availability varies according to what the various food species are doing — whether shedding their leaves or growing new young ones, or flowering or fruiting. Some tree species are highly seasonal, others not at all. Examples of important seasonally fruiting food trees in Budongo are *Cynometra alexandri*, *Celtis* spp., *Pseudospondias micro-carpa* and *Maesopsis eminii*. Non-seasonal species include the figs. The list of foods of chimpanzees in Appendix B shows the months in which these species have been seen providing food.

Plumptre *et al.* (1997) recorded the diet of the Sonso chimpanzees between December 1994 and December 1996. As shown in Fig. 4.4, this study showed that ripe fruit formed over 50% of the diet for much of the time except during the December–February dry season. In most dry seasons the chimpanzees feed heavily on the seeds of *Cynometra alexandri*, as they did in 1995–96. However, in the corresponding period for 1994–95 *C. alexandri* failed to fruit and more unripe fruit was consumed in this period

---

[21] I am indebted to the late Jeffrey Harborne for this information. Little is known about chimpanzee mucins and in particular no work has been done on their possible role in neutralizing tannins so this must remain speculative. I am grateful to Tony Corfield (pers. comm.) for help with this topic.

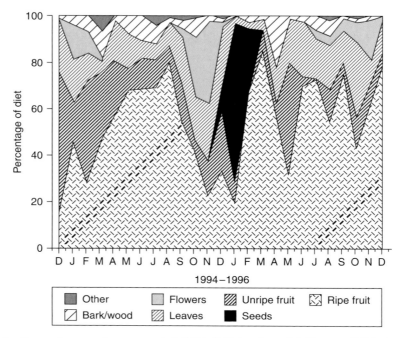

Fig. 4.4: Sonso diet over a two-year period, 1994–1996 (courtesy of A. Plumptre, WCS, from Plumptre *et al.* 1997).

than is normally the case. Thus fluctuations occur from year to year in what is available. However, there is no month in which food is scarce, and even in the drier times of year some ripe fruit is available.

With regard to food abundance, because you cannot find all the trees of a given species in the forest (this can be done in a very small patch of forest but not for a region of 15 km², the range of the Sonso chimpanzees) some kind of sampling technique is necessary. Having estimated the number of fruit trees you then need to have a method for estimating the quantity or biomass of the fruit on each tree. This can be done by placing plastic squares or buckets under the tree and calculating the weight or mass of fruit falling on them, then extrapolating up to the estimated area or volume of the whole crown of the tree. Or it can be done by eyeballing the crown, counting the number of fruits in a small section and then extrapolating up. Alternatively, a more indirect method can be used. Chapman *et al.* (1994) showed that the diameter of a tree at breast height (DBH) is a good predictor of its fruit abundance for trees in Kibale Forest and we have also used DBH to estimate fruit abundance at Budongo (Newton-Fisher *et al.* 2000).

Our work has led us to conclude that the Sonso chimpanzees live in an area of forest with high food, and especially fruit, abundance (Plumptre *et al.* 1997; Newton-Fisher 1999*a*). This contrasts with the situation at Gombe where there is a period each year of food scarcity (Goodall 1968) and at Kibale (Wrangham *et al.* 1993) where chimpanzees at times have to resort to feeding on terrestrial herbaceous vegetation (THV) as a fall-back food in the absence of ripe fruits. Additionally (and to us at Sonso strangely) chimpanzees at Kibale feed on figs as a fall-back food, not as a primary food (Wrangham *et al*. 1993; Conklin and Wrangham 1994). This is perhaps because fig trees at Kibale are less common than at Budongo, possibly because they have not had the benefit of so much selective logging (Reynolds 1992). As we have seen, in the area of Budongo where we have been working, *F. sur* is one of the two favourite food species. If all figs are considered together, then figs become the most preferred food type of all. Newton-Fisher (1999*a*) writes, 'It may be better to regard figs as a staple, rather than fallback, food.'

Food availability and potential food availability for the Sonso chimpanzees have actually been measured in three separate studies using three different methods. The work of Plumptre *et al*. (1997) has already been referred to. It was based on transect walks and phenology records over a two-year period. Newton-Fisher's measures of availability were based on a 15-month study, during which he recorded the time spent feeding and foods eaten by the chimpanzees in his study sample (Newton-Fisher 1999*a*). Reynolds (unpublished data) studied the number of chimpanzee food trees in a large number of blocks in the trail system to the north, east and west of Sonso camp, regardless of their phenological state, and obtained a measure of potential food availability by multiplying the number of stems of each species by their DBH. All three methods make certain simplifying assumptions, but all three indicated that this area is rich in food for chimpanzees (see Chapter 1 for why this might be so).

An interesting area of study on which a start has now been made is the question of how, in practical terms, chimpanzees find their way from one food source to another. Subjectively, it has always seemed to observers that they do in fact know where to go when they finish at one feeding site and head off to another. Sometimes the distance they travel is great, as much as several kilometers across the forest. How do they know where to go, and how do they find their way? This topic forms the subject of a current study at Budongo (Bates, 2005).

Seed dispersal by chimpanzees

From 1991 to 1994 Chris Bakuneeta, studying for his Ph.D. at Makerere University, washed dung samples collected from the Sonso chimpanzees to determine what seeds were excreted. These were collected, and where possible identified, and are housed in the collection in the Herbarium at Sonso camp. Many of these seeds have never been identified. Chimpanzees tend to swallow a lot of seeds; they are good dispersal agents because they move quite large distances (often several kilometers a day) thus taking seeds far from their parent trees. And they deposit a large lump of manure with the seeds when they defecate, so that the seeds are able to benefit and have an advantage over those that simply fall to the forest floor beneath the parent tree.

A large number of the seeds dispersed by chimpanzees are from trees that have uses for the timber industry. Kityo and Plumptre (1997) list the tree species in Uganda that have uses of one kind or another as timbers; the list reads like a list of chimpanzee foods. There is one tree species that is primarily, maybe only, dispersed by chimpanzees. This is *Cordia millennii*, a fine forest tree, present in Budongo and some of the other forests in Uganda where it is also dispersed by chimpanzees, e.g. Kibale Forest at Kanyawara (Chapman, pers. comm.) and Ngogo (Mitani, pers. comm.). The wood of this tree has been traditionally (and still is) used for the manufacture of boats on Lake Albert. Kityo and Plumptre (1997) in their handbook on timber species of Uganda write of this species:

> This light brown, open grained timber is greatly sought after for boat-building, both for planked canoes and for dugouts. It is soft and easy to work, is highly durable in water, very impermeable to water movement and, therefore, slow to dry and to absorb water. It also shows extremely low shrinkage on drying and consequently does not crack easily in the sun and with changes in humidity. Its properties are ideally suited to boat-building and since it is available only in small quantities it is best used for this purpose.

Katende *et al.* (1995) add that boats made of *Cordia* float if overturned.

The reason why chimpanzees are the main or even sole dispersal agents of this tree is found in the nature of its seeds. The fruits look from the outside like large oak acorns. Eggeling (1940*b*) describes the fruits as follows: 'Fruit ovoid, $1\frac{1}{2}$–$1\frac{3}{4}$ in. long, about $\frac{3}{4}$ in. diam., cupped by the enlarged calyx.' These fruits begin life green and later turn yellow. The outer husk is shiny and hard, but inside the flesh is sticky and slippery, the consistency of sticky jelly. This sticky jelly is very firmly attached to the seed, which is large and very hard and is located in the middle of the fruit; the seed is about an inch long and thus a large item to swallow.

Chimpanzees are very fond of this species, eating it at both the green and yellow stages. They spit out the husk and most of the large seeds after eating the flesh, but evidently the slippery jelly causes a number of seeds to be swallowed, with the result that when they have been feeding on *Cordia* we find one or more seeds (as many as six in a large sample) in their dung. Monkeys also eat *Cordia* but do not swallow the seeds; in a one-year study of blue monkeys (*Cercopithecus mitis*) at Budongo, Fairgrieve (1995*a*) only once found a *Cordia* seed in the dung. At times it has been suggested that hornbills and fruit bats disperse *Cordia* seeds but this has not been borne out by research findings. For hornbills the fruits are not eaten (Kalina 1988), and in the case of fruit bats (a) they prefer soft fruits to eat, and *Cordia* fruits have a hard outer husk when on the tree, and (b) fruit bats find their food by smell and Cordia fruits do not smell when on the tree (Robert Kityo, pers. comm.).

In 1995 Chris Bakuneeta, Kirstin Johnson, Bob Plumptre and I wrote a paper about human uses of tree species whose seeds are dispersed by chimpanzees in the Budongo Forest (Bakuneeta *et al.* 1995). Of the many tree species whose seeds are dispersed by chimpanzees, not all are used by people, but a number of them are, and we were interested in showing the contribution made by chimpanzees to the local human population

Table 4.5: Seeds of trees used by local people that are dispersed in chimpanzee faeces in the Budongo Forest (from Bakuneeta *et al.* 1995).

Species	Use as timber	Local uses
Chrysophyllum albidum	*Furniture, boats*	Fruits chewed to reduce body pains
Cleistopholis patens		Seeds kept to expel bad spirits
Cordia millennii	Joinery, boats	Roots boiled, mixed with beer, result drunk to cure urinary infections and worms
Cynometra alexandri	Flooring	
Ficus spp.		Bark used for coverings, mats; roots pounded for soap; juice of fruits used to cure boils
Maesopsis eminii	Joinery	Inner bark powdered and inhaled for influenza; bark used as purgative
Mildbraediodendron excelsum	Heavy construction	
Morus lactea	Heavy construction	
Pseudospondias microcarpa		Branch wards off evil spirits
*Psidium guayava**		Fruit eaten; leaves pounded for diarrhoea; leaves used with other species for cough; roots pounded for indigestion
Ricinodendron heudelotti	Balsa substitute	

* Introduced species.

by assisting with the survival of these trees. We obtained the information on which species were used by interviewing 224 local residents of nine villages at the southern edge of the forest over a period of three months (Johnson 1993). The tree species involved, and their uses, are shown in Table 4.5.

Fruit-sharing

Normally, feeding is a group activity for chimpanzees (though occasionally they eat alone). Each animal feeds itself, except when a mother shares food with her infant (see Assersohn and Whiten 1998 for Budongo; Silk 1978 for Gombe; Hiraiwa-Hasegawa 1990 and Nishida and Turner 1996 for Mahale) or her sick mother (Goodall 1986). It is very unusual for healthy adults to share vegetable food with one another. This however has been seen at Budongo and also at a few other sites, e.g. Gombe when *Strychnos* fruits were shared (reported in Goodall 1986) and at Taï Forest in Ivory Coast where *Treculia africana* fruits were shared.

In July 1997 a student at Budongo, Emily Bethell, together with two BFP field assistants, observed the adult male Muga carrying with him a large fruit of *Treculia africana*, approximately 20 cm in diameter. These are among the largest fruits in Budongo Forest and can measure up to 45 cm in diameter and weigh up to 35 lb (Eggeling 1940*b*). Chimpanzees eat the seeds and spongy pulp of the fruit.

Muga was seen sitting down with two other adult males, Magosi on one side and Maani on the other. A subadult male Andy sat behind him. Magosi was eating the fruit and Muga was staring at his face from a distance of 30 cm. A further 4 minutes later Magosi pulled off a hand sized piece of fruit and handed it to Muga. Muga took the piece and Magosi grunted. Muga finished his piece and stared at Magosi again, at which Magosi gave him another piece, grunting again when it was accepted. After this, Muga began to help himself, breaking off pieces of fruit while Magosi continued to hold it. This continued with Muga taking more and more pieces until he took the remaining piece from Magosi and started feeding on it. This piece was about one third of the size of the original fruit. Then Magosi took it back and Muga continued to take pieces from it as before. 26 minutes after the start Magosi put the remaining fruit down and walked away. Muga continued to feed on it and was then approached by Andy but turned his back on him. Andy began to groom Muga's back but stopped after a short time, and 5 minutes later Muga laid the remaining part of the fruit down in front of Andy who fed on it. 20 minutes later all three males left the area, abandoning the fruit core (Bethell *et al.* 2000).

As Bethell *et al.* point out, the sharing described above resembles the way chimpanzees share meat (Boesch and Boesch 1989). It is better described as active sharing than as passive sharing. Active non-familial sharing has been reported by McGrew (1975) in a situation when bunches of bananas were artificially supplied to chimpanzees, with 'adult males giving bananas to unrelated females'. Since the observations described above, Nakamura and Itoh (2001) have described two further cases of fruit-sharing at Mahale, Tanzania. The first involved sharing two lemons between two young adult males, Bonobo (odd name for a chimp!) and Alofu, both aged 15 years, after which a female in oestrus, Ako, joined them and slowly took one lemon that Alofu had put on the ground in front of him. In the second case, the fruit concerned was *Voacanga africana*, one of the largest fruits in Mahale, about 1 kg in weight and 15 cm in diameter. A 19-year-old male, Dogura, approached a lower ranking male Masudi (22 years old) who was chewing a mouthful of *Voacanga* pulp and begged by extending his hand to Masudi's mouth and continued to do this 26 times, obtaining some of the pulp 15 times with his index finger. As the authors note, these large and sometimes aggressive adult males were wholly tolerant of each other on this occasion.

On 9 January 1998 two adult females from the Sonso community, Zimba and her daughter Kewaya, were observed by Zephyr Kiwede, Emma Stokes, James Kakura and Clea Assersohn sharing an unripe fruit of *Desplatsia dewevrei*. These fruits are about 20 cm in diameter with a very hard outer husk which is difficult to break open. Zimba (with her infant Zig clinging to her belly) found a *Desplatsia* tree with four fruits only on one branch, she detached the branch and started feeding on a fruit. Kewaya approached her and extended an open hand towards Zimba's mouth in a begging gesture. Three times Kewaya was successful in taking fruit out of Zimba's mouth, after which Zimba turned away and rejected her solicitations. A short while later Kewaya resumed and succeeded a further seven times; the observers classified this as 'tolerated scrounging'. Zimba dropped a fruit and Kewaya descended to the ground and then started to feed on it. Five minutes later, Zimba transferred some of the food to her infant daughter,

Fig. 4.5: Rachel sitting beside a hole in a *Raphia* palm tree, before entering to eat the dead wood. The hole can be seen on the left (photo: S. O'Hara).

Zig. The main feature of interest in this story is the transfer of food between the two adult females, in this case a mother–daughter pair.

Also at Budongo we have a case of a transfer from a low ranking adult male, Tinka, to an unknown (immigrant) juvenile female. This took place on the morning of 21 April 1998, being observed by Emma Stokes, James Kakura and Joseph Karamaji. Tinka was feeding on ripe fruits of *Ficus mucuso* when an unknown juvenile female approached him screaming in fear, having been chased down from above by the adult male Magosi. Tinka held a fruit in his hand and had just put one in his mouth. He appeared startled by the rumpus at first but then sat back down again, pulled the fig out of his mouth and gave it to the juvenile, who sat next to him and ate it. Tinka then ate the fig he was holding in his hand. No further interaction between Tinka and the juvenile female was observed.

The final plant-sharing we have seen in Budongo is the sharing of the soft pithy wood from the inside of dead *Raphia farinifera* palms. This is an extraordinary kind of food as it is unclear what benefit the chimpanzees derive from it. What we find, in the Swamp Forest, are dead tree stumps, up to 15 ft high or even higher, of *Raphia*. Occasionally one of these stumps has a large hole in the bottom of it, just above ground level (Fig. 4.5). This hole is visited by chimpanzees who put one or both arms or their head or even their whole body into the tree and emerge with some of the inner pith or bark, which they chew to extract the juice, then spit out as a wadge. There is some competition between the chimpanzees for who is allowed to do it and who gets the benefits, but there is also sharing.

On 3 October 1998 Joseph Karamaji and a student, Paula Pebsworth, observed Maani and Bwoya, both adult males, feeding on *Raphia* pith. They were later joined by the adult male Muga and subadult male Zefa. Zefa extended his hand and begged from

Muga who gave him some of the pith. Then Clea and a juvenile joined the group, Clea screamed and was permitted by the males to scrape for pith. An hour after the start of observations, Vernon joined the group and began displaying, whereat all the others left the tree. Soon Muga and Zefa returned, Zefa begging from Vernon who three times gave him a part of a chewed wadge. Then Zefa groomed Vernon, before starting to feed on some inner pith loosened and dropped by Vernon.

A most fascinating possibility here is that the chimpanzees are obtaining a mild alcohol with the pith. Local people use the tree for this purpose. They cut down a living tree, remove a 4–6 ft length of the bark exposing the sap inside, and leave it to ferment on the forest floor in the sunlight, returning a week or two later to take the fermented sap which they bottle and either drink or sell as 'palm wine'. I have tasted it and it is fairly pleasant, rather fizzy and a bit sour but definitely drinkable. Whether any alcohol remains in the pith after the tree has died and rotted away we have not yet discovered.

The eating of pith is not confined to Budongo. At Mahale, chimpanzees also eat the woody pith of dry dead trees such as *Pycnanthus*, *Ficus* and *Garcinia*. They make a hole at the bottom of the tree just as observed at Budongo. *Raphia* is not found at Mahale (for these facts I am grateful to T. Nishida, pers. comm.).

Termite-eating and sharing

Eating of termites is well known in chimpanzees at some sites, notably Gombe, where they eat termites of the genus *Macrotermes* using twigs and branches as tools (Goodall 1986). At Budongo we don't have the familiar large termite mounds found in more open country, and termite-eating, while it occurs, does not form a large part of the chimpanzees' diet. However, here and there in the forest one comes across a termite mound, sometimes between the buttresses of a tree or next to a tree buttress. At such sites the Sonso chimpanzees occasionally feed on *Cubitermes* termites, breaking open the mound to do so with hands and teeth. No use of sticks or other tools for ant or termite-eating has so far been seen in Budongo, whereas this is common in other sites such as Gombe, Bossou, Taï and Assirik (McGrew 1992).

Newton-Fisher (1999b) has described termite-eating and sharing by Sonso chimpanzees:

> The sharing of termites was observed on 28 August 1995. From 08.15 hours Field assistant Geresomu Muhumuza and I had been watching a relatively large mixed-sex foraging party of eleven individuals. After 10.15 hours the party began to fragment. Four of the community's twelve adult males (MG, MA, JM, KK) and a juvenile male (GS) travelled north, away from the other party members, along one of the trails ... (past) a small termite mound. This mound housed *Cubitermes speciosus*.
>
> At 11.15 hours, first MG, and then JM wrenched lumps of earth from the mound and fed on the termites. The chimpanzees used their lips to pick the termites from what had been the inside surface ... MG moved 2 m to the north of the tree carrying his lump of termite-ridden earth. KK pant-grunted softly as he approached MG, who in response divided the fragment of termite mound in two, handing one part to KK. Both males sat and fed on the termites. No tools were used to acquire the termites (Newton-Fisher 1999*b*: 369–70).

Meat-eating

When Frankie and I made our first study of the Budongo chimpanzees, in 1962, we did not once see meat-eating, nor did we imagine that chimpanzees would eat meat. We concluded that they were strict vegetarians, except for invertebrates. I saw and photographed an adult male eating ants; he let them crawl up the trunk of a tree, on to his hand and half way up his arm where he promptly ate them. I photographed this Budongo method of ant-eating and this picture was used on the cover of the English edition of *Budongo: A Forest and its Chimpanzees* (1965). Besides ants, we knew that chimpanzees included fig wasps in their diet — the tiny wasps found inside figs that have a symbiotic relationship with the fig, pollinating it and gaining food at the same time. This certainly did not amount to meat-eating. Our study at Busingiro was followed by those of Sugiyama (1968) and Suzuki (1971). Suzuki saw meat-eating at Busingiro (he was also the first person to see and describe infanticide, something very different, which we consider in Chapter 7). And meat-eating was described by both Goodall (1968) and, together with meat-sharing, by Teleki (1973) at Gombe.

Suzuki (1971) observed the Busingiro community of chimpanzees feeding on a young blue monkey on 13 May 1968, and on a young black and white colobus monkey on 30 May 1968. In both cases several chimpanzees were involved, and meat-sharing occurred in the second case.

My first reaction to these descriptions of meat-eating by chimpanzees was that this was abnormal behaviour, because we had never seen it in our 1962 study. Goodall described instances of the killing of a variety of species by chimpanzees — colobus monkeys, baboons, duikers, bushpigs and other mammals. Time has proved me wrong: meat-eating is natural for chimpanzees, not in the least abnormal. Examples of meat-eating appeared in print from other sites where long-term studies were taking place, and in particular from the West African site at Taï Forest in Ivory Coast. It now seems likely that all chimpanzees eat meat, and this fact is widely known thanks to television.

Having said that, it does appear that the Sonso chimpanzees do not eat meat often. Why did we not see it at all in our study in 1962? Those who have studied wild chimpanzees over a decade or more concur that it is only after several years of study that this behaviour becomes apparent, as if the chimpanzees were shy of doing it when human observers are around, or need to be well habituated before they kill and eat animals in the presence of human observers. Our study in 1962 was simply too short for us to see meat-eating. The Sonso chimpanzees have been fully habituated for many years and so we do now see it. But even when we follow the adult males all day we rarely see it. This appears to be a real difference between our chimpanzees and those of Kibale or Gombe or Taï, which eat meat regularly. It goes with other differences: our Sonso males do not appear to be very good at hunting and have perhaps not developed their hunting skills as much as have chimpanzees at some other sites.

Indeed, sometimes the Sonso chimpanzees seem to pass up opportunities to catch monkeys and eat them. For example, in September 2003, our alpha male Duane encountered a live blue monkey in a snare. At first he threatened it but then 'he just prodded it and sat a couple of metres away feeding on leaves. The monkey vocalized but obviously could not run away; this appeared to confuse Duane a little. After a while he went [towards it] again and had a few stabs at it but then backed off and eventually left' (S. O'Hara, pers. comm.). Possibly Duane considered the adult monkey to be too dangerous even though it was caught in a trap. It is noticeable that most of the monkeys caught by chimpanzees are juveniles which are less dangerous to catch and easier to kill.

We have seen chimpanzees engaging in stand-offs with black and white colobus monkeys in which the chimpanzees have come off worst. The two species meet in the canopy and the monkeys threaten the chimpanzees. On one occasion two adult male colobus monkeys threatened and charged four adult male chimpanzees in a tree, with the astonishing result that all four chimpanzees retreated and left the tree. This occurred the day after a faecal sample we collected in the same place had yielded a number of bones from a young monkey's tail. We concluded that the chimpanzees might have eaten a young colobus monkey and now the furious colobus adult males were retaliating and keeping the chimpanzees away from their females and young.

The first record of meat-eating at Sonso was made on 13 August 1994.

At 7.30 a.m. field assistant Zephyr Kiwede found a group of males feeding on the bark of *Ficus exasperata*, after which they moved towards another party of chimpanzees feeding on *Cordia millennii* nearby. On the way, they made a lot of noise. Two adult males, Maani and Vernon, climbed up into a tree, then a third adult male Duane approached on the ground, beating a tree buttress with his feet as he arrived. He climbed up into the tree. Zephyr noticed that a blue monkey had been trapped up in the tree by Maani and Vernon, and now Duane was climbing up towards it. As Duane got close to it, it jumped into a neighbouring tree. Maani, Vernon and Duane gave chase but lost it. Maani and Vernon gave up, but Duane continued the chase, moving on the ground. The monkey was moving up in the trees and Duane climbed up towards it but it moved away and he climbed down again.

Soon afterwards, Zephyr saw a blue monkey — it was an adult male eating a young one (a case of blue monkey infant killing, see Fairgrieve 1995*b*). Duane was watching carefully from the ground. The blue monkey dropped some meat, a hind leg and part of the back and ribs, and then left. The meat got stuck in some branches. Duane climbed up slowly and got the meat. He climbed up further to where the blue monkey had been and then started eating the meat. Vernon and Maani called from nearby and Duane replied. Then Vernon arrived and climbed up to join Duane. He begged for some meat and was given some by Duane, he swallowed it immediately and begged again, ate again and begged a third time but this time Duane hurriedly finished the meat and avoided giving Vernon any more. Maani arrived too late to get any, Vernon and Duane were licking blood from their hands. Then the three males climbed down, to find three more adult males, Bwoya, Nkojo and Jambo, under the tree. They had arrived too late and missed the meat.

This incident was unusual in that the chimpanzees did not catch the monkey, they ate the meat of an infant already killed by an adult blue monkey. Whether the male blue monkey dropped the meat by accident or as a sop to Duane we don't know.

Just 8 days later we saw meat-eating again, on 21 August 1994. This time the observers were two visiting students from the UK, Bob Smith and Jo Lee, later joined by Zephyr Kiwede and Nick Newton-Fisher. They were attracted by the sound of a chimpanzee screaming intermittently.

After a minute there was suddenly a great deal of screaming and buttress drumming. The students found the adult male chimpanzee Muga sitting low in a tree eating a piece of meat 8 cm long. They then saw Duane holding the back part of a black and white colobus monkey including the tail and back legs. The monkey was a subadult, they estimated, not fully grown. Duane was eating the meat and they heard his teeth crunching on the bones. Vernon approached Duane and put his face near the meat but Duane did not give him any.

A search on the ground revealed a monkey limb and the head of the monkey which had not been eaten. Several chimpanzees were around on the ground, and in the trees nearby was the group of colobus monkeys from which the prey had come; they had kept quiet and hidden during the entire episode.

The third species of forest monkey found in Budongo is the redtail, and as it happened the next observation of meat-eating was of that species, in May 1995. The observers this time were Zephyr Kiwede, Geresomu Muhumuza and Nick Newton-Fisher. Maani was first seen carrying part of a redtail in his mouth as he approached a party of chimpanzees feeding on Maesopsis fruits. Vernon displayed at Maani who dropped the carcass, at which Vernon climbed down to get it, and ate it.

The next animal we observed to fall prey to our chimpanzees was a blue duiker. This happened on 27 January 1999, observed by a student, Kate Arnold, and field assistants Kakura James, Geresomu Muhumuza and Joseph Karamaji. They were watching a group of males when they noticed a blue duiker moving along the ground straight towards the chimps. At 8.45 a.m. it was caught by Bwoya who killed it by biting it in the middle of its head. From then until 9.52 a.m. it was eaten, with the usual begging and sharing, by Bwoya, Maani, Andy, Duane (who was given a piece without begging), Jambo, Nkojo (all males) and Kutu, an adult female, who was given meat by Duane. Finally Muga begged for the skin from Duane who gave it to him, then Vernon displayed at Muga and took the skin but Muga screamed and begged for it back and Vernon gave it to him.

This was an excellent display of meat sharing by our males (see Teleki 1973 for a full analysis of this very interesting topic at Gombe). Of particular note was the fact that on two occasions Bwoya, a subordinate male but nevertheless the killer of the duiker and thus with first rights to it in the conventions of chimpanzees, gave meat to Duane, the dominant male, without Duane having to beg for it. Also, Duane passed on some of his meat to the adult female Kutu (who was not in oestrus).

A second blue duiker killing, by either Bwoya or Duane or both, occurred on 19 July 1999. The observers on this occasion were field assistant James Kakura and two students, Kate Arnold and Siddhartha Singh. At 11.11 a.m. a duiker in distress was heard screaming and Bwoya was found with part of a duiker carcass, a larger part being in the possession of Duane. Almost immediately other chimpanzees joined them and sat in a huddle. Meat was obtained by Maani and also by Nora, the juvenile daughter of Nambi, who was present but without meat. Nora obtained her piece by taking it out of the mouth of the dominant male Duane! She later obtained a second piece by taking it from Bwoya. This indicated a privileged status for Nora, in keeping with the fact that her mother Nambi was the top ranking female in the community. Maani also took meat from Duane's mouth on one occasion and from his hand on another, and later still some wadged (i.e. mixed and chewed) meat and leaves; on this occasion he held Duane under the chin as he removed the food from his mouth. Duane, in keeping with his role as a good alpha male, shared willingly. All the meat and skin of this duiker had been eaten by 13.11 p.m.

Eating and sharing a blue duiker was seen again on 19 October 2003, when Geresomu Muhumuza, Nick Newton-Fisher and James Kakura came upon a small party of chimpanzees feeding on the remains of a (possibly young) blue duiker. The meat had already been divided between the alpha male Duane, who had the majority, and another high ranking male, Maani. Mukwano, an adult female in full oestrus, successfully tore a strip of flesh from Duane's share of the kill. Bwoba, a junior adult male, ended up with the head after Duane had discarded it. The meat was eaten with leaves.

Most interestingly from a comparative perspective, the Taï chimpanzees studied by Boesch and Boesch-Achermann, which have the most highly co-ordinated hunting techniques of any community so far studied, do not eat duiker meat or appear to consider duikers as prey. These authors write

> chimpanzees encounter them [blue duikers] at least three or four times daily. Adult Taï chimpanzees were never seen to make any intentional movement to capture one, even when a duiker happened to be running towards them and they had to step aside to avoid it ($N = 20$). At most they made a soft bark and it fled (Boesch and Boesch-Achermann 2000: 169–70).

They go on to describe how on a three occasions young chimpanzees engaged in rough play with young duikers, in all cases the duikers were released 'Taï chimpanzees never seemed to consider duikers as prey' (p. 170).

Co-ordinated hunting

The redtail and colobus monkey meat-eating described above may have been the result of co-ordinated hunting, but if so we had not seen it. The killing of a duiker by Bwoya was opportunistic. The killing of a duiker by Bwoya and Duane (assuming they were both involved) was a case of minimal co-ordinated hunting. The first truly co-ordinated hunt was seen by Zephyr Kiwede on 18 August 1996. Zephyr was with a group of 8 adult males, all moving quietly and slowly, looking up in the trees. Duane was leading. A group of black and white colobus monkeys was sighted in a *Cynometra* tree and the chimpanzees stopped, all looking up at them.

> After a short time Duane moved towards the tree and immediately everybody followed in different directions quietly while looking up. Duane climbed up a different tree from which he could easily reach the tree where the colobus monkeys were. When the monkeys saw Duane climbing towards them they started jumping towards a different tree, taking the same direction as each other. Vernon had followed Duane up from a different direction, as had Maani. The rest stayed on the ground looking up, seeing where the colobus would move. When the monkeys started running the chimpanzees on the ground started making the 'Waa' noise and Duane almost got an adult but it jumped away. When Duane missed he climbed down followed by Vernon, Maani and Kikunku who had

joined them. He immediately moved to the tree where the monkeys now were. When they saw Duane they moved to another tree, so then Duane climbed down again, I imagine because he couldn't jump as fast as the monkeys did.

Now Duane did not climb the tree the monkeys had moved to but went back to the first tree. He had sighted one monkey which had hidden in this tree. He immediately climbed up followed by Vernon and Kikunku from different directions. When the monkey saw Duane it jumped towards where the rest were, Duane almost got it but missed. This chase continued for about 15 min. but always missing, so Duane decided to rest on the ground while looking up still, with the rest of the males all looking up. As they rested the monkeys were moving fast in the treetops to escape. The chimpanzees gave up and moved in another direction.

This was a co-ordinated hunt, albeit a failure. So was the next one Zephyr saw, not long afterwards, on 13 October 1996. Both hunts were led by Duane, both were directed at black and white colobus monkeys, and in both cases the target was an adult. In the first case 8 chimpanzee males were involved, in the second 7. Both failed. I remember Zephyr telling me during my visit to Uganda in March 1997 that our chimpanzees were not good hunters. I agreed. Comparing them with the chimpanzees of Gombe or Taï they seemed to be pretty hopeless at it. But I was surprised they hunted at all, I had never suspected it.

A successful co-ordinated hunt by the Sonso chimpanzees occurred on 2 March 2000. This hunt was seen from start to finish, and the meat-eating at its finish was recorded on video thanks to a visitor to camp, Sarah Marshall, a friend of our student Julie Munn. Sarah was kind enough to send us a copy of the video tape so that any remaining doubts that our Sonso chimpanzees ate meat (and there were many local people who doubted it, though I was convinced by now) could see for themselves. (We later showed it to the Nyabyeya village community and they were very surprised.) On this occasion not one but two colobus monkeys were caught, killed and eaten.

At 9.28 a.m. Joseph Karamaji, Julie Munn, Sarah Marshall and Miriam Schiller were following a group of around 10 chimpanzees. At 9.34 they found Tinka sitting on the trail screaming, other chimpanzees were around in the understorey, and a group of 3 black and white colobus monkeys (an adult, a juvenile and an infant) were in a tree close by. Janie and Black climbed into the tree, and Black chased the infant. The adult colobus threatened Black who retreated and tried another route. At 9.35 the adult ran away and at the same time Duane climbed into the tree and caught the infant. At 9.39 Duane came half way down the tree and sat with Black and Janie, half hidden by leaves. The cracking of bone was heard as the colobus was eaten.

At 9.40 Duane and Maani climbed, using different routes, and chased the juvenile colobus. After 2 min it fell to the ground, Duane and Maani came down, and it was caught by Bwoya on the ground, he held it down on the ground with a sapling. The observers were unable to determine how it was killed.

At 9.45 Sarah Marshall began video recording and the following is an account of what takes place on the video she made. The chimpanzees had formed into two parties, only one of which could be seen clearly and filmed owing to the extent of ground cover. In this subgroup Duane had a piece of meat and he took a leaf to eat with it. Next, Bwoya entered the group at which Duane got up and moved out of his way, and Bwoya sat in his place. Nkojo arrived and Jambo was there too, chewing a piece of meat. Janie entered the group with her daughter Janet, and took a piece of meat, after which she begged for some more. Bwoya and Nkojo were now sharing meat, so that the three males eating meat were Bwoya, Nkojo and Duane. Black now entered the group and presented himself to Duane, at which there were vocalizations and all the meat-eaters got up with their meat, walked past Black, and left the spot, conspicuously not sharing with Black who left in a different direction. A short while after the spot was deserted, the subadult Gershom entered and took a small piece of meat from the ground.

Duane was the main individual holding the colobus. In a huddle around the monkey were Nkojo, Bwoya, Jambo, Nambi, Nora (Nambi's daughter), Janie and Janet (Janie's daughter). These were feeding on the juvenile colobus. All these individuals were observed (and filmed) eating the meat. The second group was hard to see but it was thought to consist of Black, Maani, Clea, Musa and Tinka; they were feeding on the infant colobus that had been caught first.

Leaves of *Alchonea floribunda* were picked and eaten with the meat. Apart from food grunts, there were very few vocalizations during the feeding. At 9.47 Duane moved off dragging the colobus remains and all group members disappeared out of sight in the swampy undergrowth. Two minutes later the subordinate subadult male, Gashom, arrived at the scene, found a small part of the colobus and took it away with him.

Newton-Fisher *et al.* (2002) have compiled a list of all the individual, opportunistic and co-ordinated hunts seen since the Sonso chimpanzees were first observed in 1990. The total is 17 incidents, one in 1994, six in 1995, two in 1997, three in 1999, three in 2000 and two in 2002. Twelve concerned arboreal monkeys of which *Colobus guereza* (the black and white colobus) was the most frequent. Additional details to those in the above paper are given in Table 4.6.

Table 4.6: Meat-eating incidents observed, 1991–2002 (from Newton-Fisher *et al.* unpublished data).

Observation no.	Date	Type	Prey	No. of prey	Captor if known	Consumers
1	21/8/94	C	*C. guereza*	1		DN, MU
2	5/10/95	C	*C. ascanius*	1		MA
3	22/5/95	H	*C. ascanius*			
4	15/7/95	H	*C. guereza*	1		
5	18/8/95	H	*C. guereza*			
6	10/12/95	C	Elephant shrew	1		DN
7	15/12/95	C	*C. guereza*	1		DN
8	9/3/97	C	*C. guereza*	2?		VN, DN
9	12/9/97	OC	Blue duiker	1	NJ	NJ
10	27/1/99	OC	Blue duiker	1	BY	BY
11	29/4/99	H	*C. guereza*	1		
12	19/7/99	C	Blue duiker	1		BY, DN
13	16/2/00	OC,C	*C. guereza*	1	BK	BK, AY
14	2/3/00	C	*C. guereza*	2	DN, BK	Several
15	15/6/00	CA	Rat	1	JM	None
16	29/1/02		Blue duiker	1		DN, CL, KW
17	19/2/02		*C. mitis*			NK

Key: C = *Consumption*, H = Hunt, OC = *Opportunistic Capture*, CA = *Carrying*

Concluding this section we can say that the Budongo chimpanzees do not appear to hunt frequently despite a relatively high density of potential prey (forest monkeys of all three species occur at high densities in the range of the Sonso chimpanzees, Plumptre and Reynolds 1994). The number of hunting and meat-eating episodes observed may, however, be far short of the actual number taking place because some successful hunts are sure to have gone unnoticed by observers. This was particularly brought home to Hugh Notman when he observed two of the Sonso males catch and kill a colobus monkey in total silence when other chimpanzees were nearby, apparently keeping quiet about it in order to maximize their own benefits, because they moved away from nearby chimpanzees with their kill. This could be called 'covert hunting' (Newton-Fisher *et al.* 2002).

At other sites, hunting appears to be more frequent and the prey species may be different too. In Kibale, for example, the predominant species taken by chimpanzees as prey is the red colobus monkey, *Procolobus badius*, which is absent from Budongo. This species is preferred to black and white colobus monkeys which are also present at Kibale. In Tanzania at both Gombe and Mahale (Hosaka *et al.* 2001), and in West Africa at Taï Forest, red colobus monkeys are the preferred prey. It may be that their larger group size (up to 100 in a group) is a response to chimpanzee predation, they may have a less successful method of escaping from chimpanzees in the treetops, or they may just taste better, but they are certainly the preferred form of meat at most other sites where hunting has been described, 80% or more of all prey eaten being red colobus at these sites (Hosaka *et al.* 2001). These and other aspects of red colobus–chimpanzee interactions are discussed thoroughly by Stanford (1998).

Crop-raiding

The last kind of feeding behaviour I want to mention is crop-raiding. Chimpanzees are occasionally partial to mangoes, paw-paws, maize cobs and sugar cane, all of which are grown in the villages around the forest.

The Sonso chimpanzees do sometimes raid crops, particularly near the village of Nyakafunjo to the south, and sometimes with disastrous results.[22] Having examined their dung over the last 12 years, we know that they live almost entirely on forest foods. Nevertheless crop-raiding does occur on mangoes and these days into the sugar cane fields near the forest edge.

The Nyakafunjo chimpanzees do much more crop-raiding than their Sonso cousins. They live much closer to people (the village people of Nyakafunjo) and a survey (Watkins 2001, in press) found that the people of this village see chimpanzees often, sometimes in the context of crop-raiding. We return to the subject of crop-raiding in Chapter 10.

In this context, mention needs to be made of the Budongo Forest chimpanzees that live outside the main forest block. Around the southern margins of the forest, where

---

[22] In May 2003, while this book was being written, our fine adult male Jambo was speared to death while crop-raiding at Nyakafunjo. A similar fate may have befallen Vernon in 1999.

rivers flow, are found strips of riverine forest. These used to join the main forest block but now no longer do so because of tree-cutting by the local population. In recent years we have started to study the chimpanzees living along the Kasokwa River. At the present time, 13 chimpanzees live permanently in Kasokwa Forest.

They live a very different lifestyle from the Sonso chimpanzees and others in the main forest block. They are much closer to human habitations, more so in recent years since the population to the south of Budongo has increased greatly, and they practise crop-raiding, particularly at times when few forest fruits are available in their narrow forest strips. Their predicament is that they cannot find enough food in the riverine forest, they cannot move easily to the main forest block, and nor can they move easily into the other forest fragments. As a result they move out into the farmers' fields and gardens to feed on sugar, paw-paws and mangoes. This has led to a deterioration in human–chimpanzee relationships in that area, and the situation seems to be worsening at the present time (Reynolds *et al.* 2003). We shall discuss this problem more fully in Chapter 11, where we look in more detail at the complex interactions of chimpanzees and people in the Kasokwa area.

Sonso culture

Among the various chimpanzee cultures described, the culture of the Sonso chimpanzees is marked by a shortage of technological sophistication compared with chimpanzees from some other sites. The termite fishing of Gombe and the nut-cracking of Taï and Bossou are missing. Are our Sonso chimpanzees somehow deficient, lacking in skill? I believe they are not; they have a variety of cultural characteristics including the use of tools, but their technology has remained very simple (see below, and also see Chapter 11 on the Kasokwa chimpanzees). One of the reasons is not far to seek: their habitat, as we saw in Chapter 1 and Chapter 4, is rich in their preferred foods, ripe fruits, and there seems to be no time of year when they run seriously short of forest foods.[23] But I suspect that is only part of the explanation. There is also the fact that East African chimpanzees have not developed the sophisticated stone tools used by West African chimpanzees to crack nuts, and of the East African chimpanzees it seems to be those to the south that have developed the greatest skills in termite fishing, while as we move towards the more northerly forests (and Budongo is furthest north of the major forests along the Rift Valley escarpment, even though there are two small forest populations in Rabongo and Otzi Forests which are further north than Budongo) the use of tools seems to decline. This may be because, as in the case of nut-cracking, the idea of using sticks to fish for insects has never reached Budongo.

---

[23] This implies that functional considerations are important in determining the development of cultural activities in chimpanzees. This is a hotly debated point in respect of both human and primate societies. Function is no doubt related to the evolution of culture itself; cultural variations are perhaps less determined by function. However, as for example Humle and Matsuzawa (2002) show, the relative aggressiveness of certain ant species determines the tool-length and techniques employed by Bossou chimpanzees when engaging in ant-dipping. Probably we have to conclude that while some cultural variants can be explained in functional terms, others cannot.

The Sonso chimpanzees share a large number of cultural features with chimpanzees at other sites, among them the following (Whiten *et al.* 1999):

- Play start (invite play holding stem in mouth)
- Drag branch (drag large branch in display)
- Leaf-sponge (leaf mass used as sponge)
- Branch-clasp (clasp branch above, groom)
- Branch-shake (to attract attention, court)
- Buttress-beat (drum on buttress of tree)

They also have some which are shared with some other sites but not all. These more particular cultural features of the Sonso chimpanzees (Whiten *et al.* 1999) are as follows:

- Food pound on to wood (smash food)
- Fly-whisk (leafy stick used to fan flies)
- Leaf-napkin (leaves used to clean body)
- Leaf-groom (intense 'grooming' of leaves)
- Leaf-clip, mouth (rip parts off leaf, with mouth)
- Leaf-clip, fingers (rip leaf with fingers)
- Leaf-inspect (inspect ectoparasite on leaf placed on arm or hand) (Assersohn *et al.* 2004)
- Branch-slap (slap branch for attention)
- Shrub-bend (squash stems underfoot)
- Rain dance (display at start of rain)

Precise definitions of these items are provided in Whiten *et al.* (2001), in some cases based on the very detailed ethogram of the Mahale chimpanzees (Nishida *et al.* 1999). More items, e.g. pocket-tuck (placing an object in a body fold of groin or neck and moving around with it held there — see below) are being added to the Sonso repertoire as time goes by. Interesting questions arise about the distribution of culture traits. Nut-cracking is exclusive to West African chimpanzee culture, and leaf-grooming is exclusive to East Africa. The role played by diffusion is explored in a paper (Whiten *et al.* 2001), and so also is the evidence for multiple origins. One item of culture is particularly indicative of multiple origins: the grooming hand-clasp (McGrew and Tutin 1978). In this behaviour, two individuals clasp hands overhead, grooming each other with the other hand.

As McGrew (1998: 319) states:

> The hand clasp occurs daily at Mahale but has never been seen in more than 35 years of observation at Gombe. It also is customary at Kibale and Taï but not at Bossou (Boesch and Tomasello 1998). At Kibale it is common both in the Kanyawara community and in the Ngogo community (Watts, personal communication). Interestingly, the pattern has now emerged in a captive group at Yerkes, where it is spreading from its first performance by a captive-born adult female (de Waal and Seres 1997).

We can now add Budongo to the list of sites at which hand-clasping does not occur; we have never seen it at Sonso. And from the distribution of this item of chimpanzee culture we can see that it has not spread by diffusion; indeed, the observation of its origin at Yerkes shows it is capable of independent origination at multiple sites.

There are other anomalies. Leaf-sponging (drinking water using a crumpled leaf to soak it up and then squeezing it into the mouth) is in one sense not a cultural trait because it occurs at all sites and is thus a universal (Whiten *et al.* 2001); however, these authors state that in time it may be considered as a cultural variant because of differences in the ways or the contexts in which it is done. This seems correct, in the same way that whereas eating itself is not a cultural activity for humans, how we eat, when we eat, and what we eat are all very cultural indeed. We shall return to leaf-sponging below.

Leaf-clipping with the teeth (biting bits off a leaf held in one hand) is of interest because this behaviour has different functions in East and West Africa: in East Africa, at Sonso and at Mahale (Nishida 1980) it is found in the context of courtship behaviour, with males (including subadult and even juvenile ones) doing this to try to attract the attention of sexually attractive females. In Taï it occurs before buttress drumming and at Bossou it is done to attract a playmate; at neither site does it occur in the courtship context (Boesch 2003).

In one cultural feature the Sonso chimpanzees differ from those at other sites: penile-cleaning is particularly common. Adult Sonso males engage in this to a greater extent than do males at other sites. Sean O'Hara observed 116 copulations of which penile-wiping occurred in 34.5% of cases, and penile-cleaning with leaves in 9.5% of cases. The use of leaves is more common at Sonso than at Gombe where the frequency is 2.9%, and at Kanyawara in Kibale Forest where it is 2.2%. At Mahale, Bossou and Taï penile-cleaning with leaves has not been observed at all (O'Hara and Lee, in progress).

The question of how cultural items are transmitted raises a number of problems and I shall not deal with them here. There is little evidence for teaching in chimpanzees, and much for observational learning. We have seen how chimpanzee infants learn from their mothers what foods to eat; most people are familiar with the way young chimpanzees at Gombe learn termite fishing by watching their mothers, and the way chimpanzees at Bossou and Taï learn nut-cracking by watching their mothers, so very likely most cultural activities are learned by observation of mothers and other individuals in the community. The actual process of learning is hard to document in the wild but Andy Whiten has devised an ingenious experiment to demonstrate how traditional processes are handed on in captivity and semi-captive situations. These, and the topic of social learning generally, are dealt with at length by Whiten *et al.* (2003).

There are many cultural items at Sonso that need further attention. 'Groin-tuck' and 'neck-tuck' occur when a small branch with leaves is tucked into a body fold while the animal is resting, grooming or feeding, and then carried from place to place if the individual moves; it is carried manually while walking quadrupedally or in the groin if brachiating. If this behaviour has a function we have yet to discover what it is, though a connection with nest-making was suspected on one occasion, observed by Lucy Bates (pers. comm., 24 February 2003): two juveniles and an infant, Beti, Rose and Janet, were playing with the leaves of a *Cordia* tree, breaking off the tips of branches. All three individuals tucked the branchlets into their groins as they brachiated and held on to them as they moved around in the tree. Rose used hers to begin a day nest, at which Janet, who had dropped her leaves, began to pull at Rose's nest. Janet reached out and tried to hit her, no physical contact was made and Janet moved away and initiated play with Beti. After a short while both Janet and Beti picked leaves, groin-tucked them, and added them to the leaf-nest-pile started by Rose.

Sometimes the branchlet is draped around the back of the neck and a version of it ('pearl necklace') was seen when the branch was from a fig tree. Many individuals have engaged in this behaviour at Sonso.[24]

The Sonso chimpanzees also eat the woody pith of dead *Raphia* palm trees, but in a very strange fashion: they make a hole in the bottom of the dead tree trunk while it is still standing in the swamp forest, and gradually increase the size of this hole until it is big enough for small chimpanzees to crawl into it. All the woody pith is extracted with the hands and teeth, and chewed thoroughly in the mouth before being spat out in the form of a bolus or wadge. Is there any alcohol in it? The pith of this species of tree is used by local people to brew palm wine but only when the tree is freshly felled. Is there still a remnant of alcohol in the dead trees? If so it would be a first observation of alcohol consumption for chimpanzees. Certainly they have something in common with humans drinking alcohol in that pith-eating is popular and individuals compete with each other for access to the tree. As before, this behaviour has been filmed by Sean O'Hara.

In regard to vocalizations, we shall see in Chapter 6 a number of cultural variations between the structure of calls at Sonso and the same calls at other field sites.

In Chapter 3 we discussed the health of the Sonso chimpanzees and their use of *Aneilema aequinoctiale* for self-medication. Learning which plant species to use, which occurs when infants watch their mothers self-medicating, is a cultural process, and we know that the plants used by chimpanzees for self-medication differ from site to site (Huffman 2001).

Tool-use, as stated earlier, is minimal at Sonso. A clear demonstration of this is the way our chimpanzees eat termites. They break open the mound with their teeth and hands, then eat termite soil and termites with their mouths. They eat honey without using a stick to probe the nest, just breaking into it with hands and teeth. Use of a stick has been seen several times as an invitation to play.[25] Nest-making involves a complex

---

[24] This and several other items of Sonso culture have been photographed and filmed by Sean O'Hara.
[25] This does not occur at Mahale (T. Nishida, pers. comm.)

series of manipulations of branches and leaves but is not known to vary systematically from site to site and has all the appearances of a universal. The same is true of drumming which, while it varies systematically between individuals, is not known to vary systematically between sites.

Leaf-sponging (already referred to above) is perhaps the commonest form of tool-use we see at Sonso; it too appears to be a universal, occurring at all chimpanzee sites,[26] but it may yet prove to have cultural components. It was first observed by Duane Quiatt at Budongo (Quiatt 1994), who went on to make a three-year study of it, from 1996 to 1999 (Quiatt, in press). A total of 111 instances of leaf-sponging were observed either by Quiatt or by BFP field assistants. Leaf-sponging was often used to collect open water, e.g. from the River Sonso. Another source of water was from tree-boles mostly accessible from the ground. The leaves had been lightly crushed in the mouth prior to use, increasing their absorbency and maybe releasing flavour. Quiatt also studied unassisted drinking; only 18 occasions were seen in the same period. Thus Sonso chimpanzees preferentially drink using leaf sponges; 38 members of the community used leaf sponges. Several instances were observed of mothers and infants leaf-sponging together; learning could occur in this context. *Acalypha* spp., common climbers in Budongo, were the species most often used (78/111 cases); the leaves were mostly hairy and thus held moisture well.

I will end with an example of just how restricted in distribution some cultural features are. While we were making our 1962 study of the Budongo chimpanzees living at Busingiro, in the southwest of the forest, we observed one individual male eating tree ants as described earlier: he sat on or hung from a branch close to the trunk where the ants were moving up and down, put out an arm so the hand was in the path of the ants, and after allowing a number of them to crawl up his arm, moved it across his mouth to eat them. This behaviour has never been seen at Sonso, which is a mere 3 miles due east across the forest from Busingiro. This is most extraordinary, knowing that females move from community to community. While I did not record details of ant-eating in 1962, it may be that this behaviour was confined to adult males, in which case it is easier to understand its lack of transmission to other communities. The Sonso chimpanzees have not been seen eating ants, several species of which are common in the forest including the 'army' or 'safari' ants of the species *Dorylus (Anomma) wilverthi* (Schöning, pers. comm.). The Sonso chimpanzees do, however, eat caterpillars and termites.

---

[26] At Mahale this behaviour is listed as 'present' (Whiten *et al.* 1999, 2001) and is thus not common, customary or habitual at this well-studied site.

5. *Social organization*

7.30 Into forest with Zephyr. 7.45 Kewaya + Katia. Nick on trail ahead. Tinka. Mukwano in oestrus 4. Janie + Janet. Nkojo with erection as Mukwano approaches up the tree. Kwera in oestrus 4 + Kwezi. Maani present. Fine adult male. Zefa displays & mates with Kwera who dashes off at the end of it. Bwoba. We are in block 4C. Feeding on Cya seeds (Cya is late this yr as the dry season was a month delayed and lasted till early March). 8.26 Kwera cop with Nkojo. Gashom. Duane descends tree forcefully. Black. Order of march along trail: Duane, Black, Nkojo, Maani, Zephyr (human), me. They stop to feed on *Chrysophyllum albidum* fruits (large fleshy yellow plum-sized fruits). NB Gashom has snare (new) right hand. 2 fingers tightly held together by the wire (17 March 2001).

In this chapter we shall be concerned with the principles of social organization of the Sonso community, that is to say the way the community itself is organized. In the next chapter we shall move on to the social behaviour of the chimpanzees themselves.

The whole group that lives in a range or territory (within which it splits up into subgoups) is called a 'community' (or especially in Japanese primatology, a 'unit group', e.g. Nishida 1968, 1990) and the temporary subgroups into which the community divides are called 'parties' (for a review of terminological issues see Van Elsacker and Verheyen 1995). How and why parties form and split up, re-form and split up again, is a subject that we shall approach in this chapter.

The problem of how to define parties remains a difficult one because of their temporary nature. Boesch (1996) points out that for different observers a party may consist of all individuals in sight of one another, or all individuals in auditory contact, or all individuals within a certain distance of each other. No fixed cut-off distance was used in the present study; the individuals were not always in the same tree but could be in adjoining trees or some might be on the ground and others in the trees above. The decision to include individuals in a party was in the last analysis subjective, and efforts to find an objective method have so far proved problematic; perhaps a reasonable working definition might be 'all the individuals that appeared to be aware of each other's presence' (R. Wrangham, pers. comm.). Some observers, e.g. Newton-Fisher, have used a 30 m diameter cut-off point for party size and this is useful for some purposes. Chapman *et al.* (1993) point out that quantifying subgroup size is made difficult by limited visibility, extent of habituation and observation methods. These factors are important in making comparisons between one study and another. However, where studies are long term it seems possible to attempt a few comparisons, especially where conditions of observation are similar.

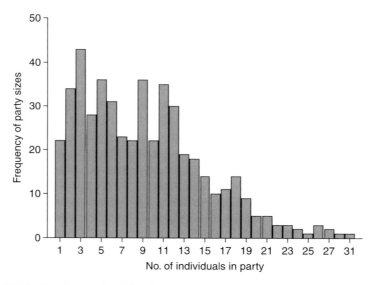

Fig. 5.1:   Distribution of party sizes (data for 2001).

Parties, food and social factors

The pattern of fission and fusion of parties is central to chimpanzee social organization. Explanations for this uncommon pattern, also found in spider monkeys in South America (Chapman 1990), have been sought in a number of factors. One is the patchy distribution of food, which occurs at many different places in the community range; formation of larger or smaller parties is a result of feeding competition. Individuals are not averse to forming large parties to exploit rich food patches (e.g. a large fig tree containing many thousands of ripe fruits), whereas when food is scarce or more dispersed they can find more food by splitting up into small parties.

A second set of factors affecting party formation can be collectively called 'social'. For example, sexually receptive adult females attract and are attracted by adult males. Non-sexual social factors are also important: alliance partners tend to move together, dominant males sometimes attract a retinue, friends tend to stay and move together, and kinship bonds are strong, with mothers and their infants always being together. Together, food abundance and social factors determine party size.

Party size[27]

A variety of methods has been used to collect data on chimpanzee party size (see Boesch 1996 for a review of some of these). We can contrast (a) continuous sampling over the period of the life of a party (i.e. from when it begins to when it ends), and (b) scan

---

[27] The parties referred to in this chapter are, except where stated, feeding parties, not parties of other kinds such as, for example, nesting or travelling parties.

sampling at a fixed interval, e.g. 2, 15, 30 or 60 min (i.e. counts of its members at regular intervals). In continuous sampling, data are collected by *ad lib* observation of individuals over a variable time period and the size of the party includes all members considered to belong to it from its beginning to its end; the result can be called *cumulative party size*.[28] In scan sampling all individuals seen at the sampling times are recorded and this results in a series of 'snapshot' party sizes. This method suffers from dependency of the results obtained from scan to scan because successive scans of a party include some or all of the same individuals. Therefore, when comparing parties from scan samples it is necessary to allow for data dependency. The size obtained by this method can be called *scan-based party size* and is invariably lower than cumulative party size based on continuous sampling.

For particular purposes, i.e. for measuring association frequencies between particular individuals, scan sampling must be used to be sure that the individuals are indeed together. At Sonso we currently collect data on parties using 10-min scans; cumulative data on party size are obtained from scan data by summing the individuals making up the scans over the life of substantive parties.

As always, data are initially recorded on checksheets in the field and entered into the computer later. All infants seen are recorded in the field and entered on computer, but for purposes of calculating party size dependent infants, i.e. infants whose presence in a party is determined by the presence of their mothers, are excluded.

To obtain mean feeding party size for the Sonso population an analysis was done on 2100 feeding parties observed on 854 observation days (0.58 of all possible days) over a 1467-day period. A mean of 2.46 parties (s.d. 1.95, range 1–15) was recorded per observation day. Mean cumulative party size was 7.01 individuals (s.d. 4.99, range 1–29). A subset of these parties was recorded by means of 30-min scans. A total of 5214 such scans were done, for which, after allowing for data dependency, mean party size was 5.01 (s.d. 3.57, range 1–26). Newton-Fisher (1999*d*) gives a comparable figure based on a separate set of scan data, again filtering the data to remove dependency effects: 5.70 ± 3.48 (median 5.0, range 1–19).[29]

Mean party size in other chimpanzee studies is in most cases lower than at Sonso, exceptions being Taï (Boesch 1996) and Ngogo in Kibale Forest, a very large community (over 140 individuals) that breaks up into large parties with a mean of 10.27 individuals (Mitani *et al.* 2002: 105). For all sites the frequency distribution is heavily

---

[28] This is the number of individuals forming a party over the period of time from the start of observations until the end of observations of that party. Observations end following a decision to observe another party, or the disappearance or disintegration of the party being observed. This method is similar, though perhaps not identical, to that of Matsumoto-Oda *et al.* (1998) who worked at Mahale. They wrote that 'parties are all individuals in a behaviorally cohesive unit that co-ordinated movement during the day'. There is no single method that can take account of all the complexities of subgroup size changes in fission–fusion societies, in particular the problem of when one party stops and a new one begins. A party begins either when chimpanzees arrive or when observation begins, and it ends either when all the chimpanzees leave or when observation is terminated.

[29] These are inevitably smaller means than means derived from cumulative sampling because cumulative sampling includes all the chimpanzees in the party, including those that join it during the life of the party, without reducing it for those that leave, whereas in scan sampling the party size is reduced when individuals leave.

skewed to the left, i.e. smaller parties are more frequent than larger ones. Maximum party size in Taï was 41 (53.9% of the community size of 76); in Sonso for 2001 it was 31 (65% of the community size of 48 at that time). Boesch (1996) argues that mean party size should be expressed as a percentage of the total community size because total community size limits the size parties can reach. Expressed in this way, he found that mean chimpanzee party size over the study period was between 9% and 21% of total community size. In the Sonso community in 2001 mean party size was 9.12 or 19% of the total community size which falls within the range found at Taï. The distribution of party sizes for the year 2001 is shown in Fig. 5.1.

For the same reasons that mean party size should be expressed in relation to the size of the community, when calculating the mean number of males and the mean number of females in parties, these should be expressed as a proportion of the total number of males and of females in the community. This is done, for example, by Emery Thompson and Wrangham (in press) and I have used this method in this book (see Table 5.1).

Party duration

We carried out an analysis of party duration: the length of time party size remained constant, i.e. how long, on average, it was before one or other of the party members left or new individuals joined the party. In a subset of 162 parties, the mean time parties remained constant was 19 min. Mean duration of parties at Taï was 24 min (Boesch and Boesch-Achermann 2000: 91).

Summarizing, parties at Budongo, over this four-year period, consisted on average of 5–7 individuals and ranged from 1 to 29 individuals, with a change in composition, on average, every 19 min.

Party composition

Parties can be made up of almost any age–sex composition. The smallest party consists of a lone individual. An analysis was done of the composition of parties observed during the period January–December 2001. Over this period, the composition of 482 feeding parties was recorded. During this period the number of individuals in each age class[30] who were present during the whole year was as follows:[31]

Males ($N = 18$)
9 adult, 2 subadult, 4 juvenile, 3 infant

Females ($N = 30$)
14 adult, 2 subadult, 3 juvenile, 11 infant

Total = 48

---

[30] Definitions of age classes can be found in Chapter 2, Table 2.3.
[31] Two adult females and two juvenile males were excluded from the analysis because they were not present for much of the year.

Parties were analysed to see to what extent they consisted of one sex only or of both sexes. Only parties consisting of adults and subadults were included in order to exclude the effects of kinship bonds on juveniles and infants. When parties consisted of one sex only, there were more all-male parties than all-female ones (42 (8.7%) all-male and 15 (3.1%) all-female). All-male parties were larger than all-female ones; all-male parties numbered from 1 to 8 individuals with a mean of 2.44, whereas all-female parties numbered from 1 to 2 with a mean of 1.06. By contrast, there were 355 (73.65%) parties consisting of both sexes: such mixed parties numbered from 2 to 28 (excluding infants) with a mean of 10.22. Mixed parties are thus the norm at Sonso and constitute the great majority of parties seen.

Much has been written about the tendency for female chimpanzees to be more individualistic and less gregarious than males (see, for example, Wrangham 1979, 2000*a*; Nishida 1990; Pepper *et al.* 1999; Emery Thompson and Wrangham, in press). In order to take account of the different proportions of males and females in the community, numbers of adults + subadults of each sex were converted into percentages of the total in each of these two age classes in the community (see Emery Thompson and Wrangham, in press). The results for both sexes are shown in Table 5.1 and in Fig. 5.2. As can be seen

Table 5.1: Statistics for 482 parties, January–December 2001, showing mean % of males and mean % of females in parties according to party size (range 1–28).

Party size	Males (%)	Females (%)
1.00	3.02	2.24
2.00	4.77	3.55
3.00	8.95	4.09
4.00	8.84	7.24
5.00	12.16	8.51
6.00	16.52	8.57
7.00	20.26	11.21
8.00	20.91	13.97
9.00	29.78	12.50
10.00	33.63	13.57
11.00	35.42	15.00
12.00	35.71	18.33
13.00	36.31	19.76
14.00	43.75	19.44
15.00	46.87	20.00
16.00	50.42	22.67
17.00	40.18	30.71
18.00	47.73	28.18
19.00	57.81	26.25
20.00	50.00	34.00
22.00	54.17	36.67
23.00	53.12	37.50
24.00	50.00	40.00
25.00	62.50	40.00
26.00	62.50	45.00
28.00	62.50	55.00
Mean	24.02	12.65

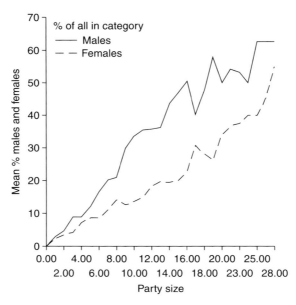

Fig. 5.2: Line graph for 482 parties, January–December 2001, showing mean % of all males and mean % of all females in parties according to party size (range 1–28).

from the table and, more clearly, from the line graph, the proportion of all Sonso males represented in parties regardless of size exceeded the proportion of all females.

Sonso females are also sociable, especially if they have infants (as is normally the case). Bates (2001) made a two-month study of association patterns between Sonso females, in which she compared lactating females who had young infants with non-lactating females who had no offspring or whose offspring were independent. There was little difference in the association patterns of the two groups. Both associated with other females and adult males and there was no difference in their average party sizes. Bates found no tendency of lactating females to associate preferentially with other lactating females and form *exclusive* nursery groups. Such groups, therefore, when they occur, probably reflect the numerical frequency of females with infants in the community.

Parties consisting of females and young only, without males, constituted 11.6% of all parties. Such parties were also described by Reynolds and Reynolds (1965) and Sugiyama (1968) who called them 'nursery' parties or 'mothers groups', which is something of a misnomer as it implies that females without young would be excluded from such groups. This, as we have just seen from the work of Bates (2001), is not the case; females without young are found in parties together with mothers and their infants. By contrast, parties consisting of adult males and infants only were never seen throughout the whole year.[32] By far the commonest party composition at Sonso, as we saw above, was the mixed party.

---

[32] Rarely, males are seen alone with infants, the mother being absent; the consequences may be disastrous for the infant.

Within mixed parties, mothers with young infants do not show any fear of, or distance themselves from, adult males. Things are different in some communities, e.g. Gombe (A. Pusey, pers. comm.) where females with new infants avoid adult males. A striking case of the lack of avoidance of adult males by new mothers was seen in the case of the birth of the infant female Katia to Kewaya, a resident nulliparous female, on 30 December 1998.[33] This birth was observed and recorded in detail by our senior field assistant, Zephyr Kiwede (Kiwede 2000), and was described in Chapter 2. For present purposes it is of interest to note the social situation at the time. The closest individuals to Kewaya during the birth were Kigere, an experienced mother, and her two offspring, Kadogo and Kato, who were approximately 1.5 m away from Kewaya. During the course of the next hour, three further adult females (Nambi, Kutu and Kalema), all with their offspring, joined Kewaya in the tree. At 10.30 two adult males, Muga and Andy, joined the group. None of these chimpanzees, female or male, showed any particular interest in the newborn infant. During the following week, while the *Ficus mucuso* was in fruit, Kewaya continued to feed in it and all of the adult males of the community joined in these feeding parties at one time or another; at no point did Kewaya avoid any of the males, nor did any of them show any interest in the new baby. In this particular case, therefore, the newborn daughter of a resident female who was with her mother was entirely safe in the presence of adult males. Indeed, we have not seen any avoidance of adult males by any of our resident females when they had new infants, but this relationship has yet to be studied in detail at Sonso.

A second, and also interesting fact concerning Kewaya's birth was that her own mother, Zimba (as we know from genetic data), was not present either during the birth or on the following days. Perhaps she disapproved?

Effects of oestrous females on party size and party type

Wallis and Reynolds (1999) and Newton-Fisher (1999*d*) found that in the Sonso community, more males were found in parties when oestrous females were present. The same has been reported for a number of other chimpanzee communities (Matsumoto-Oda 1999 for Mahale; Anderson *et al.* 2002 for Taï; Hashimoto *et al.* 2001 for Kalinzu; Wrangham 2000*a* for Kanyawara). It has also been suggested that females not in oestrus may seek the company of oestrous females, an action shown to stimulate the resumption of postpartum cycles (Wallis 1985, 1992*b*) and initiate the first full anogenital swelling in adolescent females (Wallis 1994). Food abundance is not the cause of these larger parties (Anderson *et al.* 2002: 96–7; Hashimoto *et al.* 2001: 953).

I analysed the 2001 sample to determine the effects of the presence of one or more females in oestrus on the percentage of all community males present in the party. The

---

[33] In fact Kewaya is thought not to have immigrated but to have remained in her natal group, though this is not certain. She was present from the start of our observations in 1991, at which time she was a juvenile; this was seven years before she gave birth. Her name means 'wire' in Swahili, from the wire that was attached to her right wrist.

Table 5.2: Number of oestrous females present in 483 parties (data for January–December 2001).

No. of oestrous females	No. of parties	Per cent of parties
0	239	49.5
1	139	28.8
2	77	15.9
3	24	5.0
4	2	0.4
5	2	0.4
Total	483	100.0

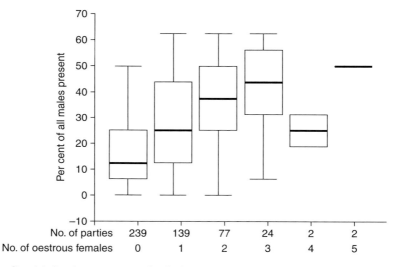

Fig. 5.3: Boxplot showing percentage of males in community in parties without and with oestrous females.

distribution of numbers of oestrous females in parties is shown in Table 5.2. One or more oestrous females was present in 244 of the 483 parties seen. The percentage of males present in these parties is shown in Fig. 5.3.

As can be seen, the mean and absolute percentage of males rose as more females entered the party. In the few cases where 4 or 5 oestrous females were present (two cases each) we can discount the results because of the small sample size. The overall impact of oestrous females on males is very marked, with an increase from 25% to 45% of all males present as the number of oestrous females increases from 1 to 3. A correlation of these two variables gives $r = 0.332$, with $p < 0.01$.[34]

[34] The effect of food abundance has not been analysed for the above data; it may be that part of the increase in the number of males in this data set was due not to the presence of oestrous females but to the presence of plentiful food. The influence of food supply on party size is considered in more detail below. However, increased presence of adult males in parties with oestrous females is likely to be independent of food supply because such females act in their own right as a magnet for males.

Interestingly, it is not just the presence of more males that increases party size when oestrous females are present. There is also an increase in the percentage of females joining the party (after excluding the oestrous females themselves). Thus, presence of oestrous females is associated with larger party size generally. The maximum party size seen with one or more oestrous females present was 28 (mean 10.22), whereas the maximum party size seen with no oestrous female present was 22 (mean 5.88). Thus mean party size almost doubled when one or more oestrous females was present.

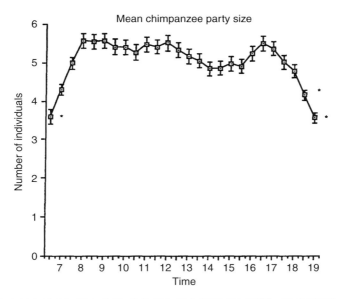

Fig. 5.4: Mean party size by time of day (courtesy of A. Plumptre, WCS, unpublished).

Time of day and party size

Mean party size was compared by time of day over a 15-month period using scan-based party size data (Fig. 5.4). As can be seen, party size increases from 6.30 a.m. until around 8 a.m., then levels off, dipping a little from mid-day until 2 p.m., then rises again until 4.30 p.m. after which it drops decisively until nest-making time at 7 p.m.

Seasonality

Temperature and rainfall records are collected every day at Sonso camp. Temperature at Budongo is fairly constant during the year, but rainfall varies with one major dry season, as was seen in Fig. 1.2.

For purposes of seasonality analysis, the major dry season (January–February) was compared with the remaining months (March–December). Tests were performed on

Table 5.3: Party size according to season.

	N	Mean	s.d.
Dry	173	8.50	5.95
Wet	1927	6.87	4.87

Table 5.4: Number of oestrous females in parties according to season.

	N	Mean	s.d.
Dry	173	1.07	1.21
Wet	1926	0.58	0.99

a subset of the data to see if there was any difference between party size or composition according to season. The results are shown in Table 5.3. The difference in party size was highly significant: party size was larger in the dry season.[35]

Why should this be? One of the main food sources in the dry season is the seeds of *Cynometra alexandri*. When this tree is fruiting, chimpanzees gather in large parties to feed on its seeds which are lipid-rich and provide one of the most nutritious vegetable foods the forest has to offer. Unlike some other forest tree species, *Cynometra* is highly seasonal. Only once in the 12 years of study has it not fruited between December and February and that was in January 1995, probably a side-effect of El Niño.

A second factor associated with increased party size is the presence of oestrous females (see above). Could the increase in party size in the dry season be a result of an increase in the number of oestrous females at that time? An analysis was done of party size comparing number of oestrous females in groups in the dry season with the wet season. The results are shown in Table 5.4. This result was, as before, highly significant.[36]

Thus the increase in party size seen in the dry season is likely to be associated with both the presence of *Cynometra* seeds and also with the larger number of adult females in oestrus at that time. Whether these two factors are causally related to each other is not known.

Influence of food supply on party size

The question of the effect of food supply on party size remains a central focus because we believe that food abundance and dispersion are important factors determining not only party size but also the evolution of the chimpanzee's fission–fusion social system. Evidence for a positive correlation has come from Gombe (Wrangham 1977), Kibale (Ghiglieri 1984; Isabirye-Basuta 1988; Wrangham *et al.* 1992; Wrangham 2000*a*) and

[35] The seasons were compared using a *t*-test (two-tailed). For purposes of the test, equal variances were assumed (Levene's test, $p < 0.000$). The result was highly significant ($t = 4.13$, df $= 2098$, $p < 0.000$).

[36] Levene's test indicated equal variances and the resulting *t*-test gave $t = 6.08$, $p < 0.000$.

Taï (Boesch 1996). However, the importance of food as a determinant of party size has also been questioned. Isabirye-Basuta (1988) found that this relationship did not hold when multiple sources of important foods were available. Boesch (1996) found that despite apparently similar levels of fruit production in each of two wet seasons, chimpanzee party size was large in one but small in the other. Stanford *et al.* (1994) found that parties of Gombe chimpanzees were largest during the dry season when food supply was restricted.

Newton-Fisher *et al.* (2000) explored this topic in detail. Over several years we made three independent studies at Budongo. In the first of these I took data from our long-term study of party size and composition, determined the mean size of feeding parties in 1-ha blocks of forest over a four-year period, and then compared this with the abundance of chimpanzee food using data from a study of the size and species of chimpanzee food trees in those blocks. There was no simple relationship. In the second study, Newton-Fisher compared party sizes obtained by scan-sampling over a 15-month period with the abundance of food on which they were observed feeding; again there was no simple correlation. In the third study, Plumptre observed chimpanzees over a 14-month period (following the period observed by Newton-Fisher) and obtained scan data on party size and also on food abundance; as before no simple correlation was found. The situation is very complex, but there was evidence from each of these studies that what matters in determining party size is not only the abundance of food, but its dispersion (how 'clumped' or otherwise it is), species and food type (whether leaves, ripe fruits, unripe fruits, etc.). We concluded that 'factors other than food supply were primarily responsible for the size of chimpanzee parties at Budongo' (Newton-Fisher *et al.* 2000: 623).

Why should this be? We posited a number of reasons. One is that it seems that the food supply at Budongo is particularly rich for chimpanzees (see Chapter 1) because of the logging history of the forest (Plumptre and Reynolds 1994). We concluded that 'with increasing food abundance the importance of feeding competition on party size decreases, eventually reaching a point where the relationship becomes negligible and other factors control the size of chimpanzee parties' (Newton-Fisher *et al.* 2000: 625). And in particular we posited that the presence of oestrous females might be an important factor determining the movement of males, and to some extent females as well, so that as food availability increases, after a certain point party size reflects the number of oestrous females present rather than of food.

Fawcett (2000) in a 16-month-long study focusing on female relationships and food supply at Budongo did find a positive correlation between food supply and party size. She wrote 'A frequent predictor of party size was the combined measure of abundance of fruit, flowers, buds and leaves, highlighting the importance of leaves, in addition to fruit, of the diet of the chimpanzees at Budongo' (p. 217). Fawcett also found that the number of females in the Sonso community exhibiting sexual swellings increased at times of high food availability. What she called 'sexual parties', i.e. parties with one or more females in oestrus in them, were larger when there were more oestrous females in the community. This moves us forward on the question of how food availability and presence of oestrous females affect party size: it seems the effects may be both independent and interactive. A rich food supply (including the presence of nutritious leaves) increases party size; and it increases the number of females in oestrus; the number of females in oestrus also increases party size. Thus, the near doubling in size of parties with one or more oestrous females (see earlier in this chapter) can be explained as an outcome of a two related causes, with food pushing sex and both pushing party size.

Yet another study of food abundance and patch size in relation to party size at Budongo was undertaken by Harris *et al.* (in progress) designed to test some of the above ideas. To overcome some of the difficulties presented by multiple species and food types we used the idea of 'food quality', with ripe fleshy fruits being of highest quality, unripe and hard fruits of medium quality, and leaves of lowest quality. In the analysis we related food quality, party size (recorded by scan-sampling with a 20-min interval), and presence of oestrous females, paying particular attention to changes in party size, i.e. whether individuals were joining the party or leaving it. The results showed that higher food quality was positively correlated with increases in party size over time and vice versa. Thus when food quality is lower, chimpanzees stay on the food for a shorter time and leave sooner than when food quality is high. Party size was greater when one or more oestrous females was present, but presence/absence of oestrous females did not have any effect on party duration or dynamics (movements to and from the group).

Nesting and nesting parties

Brownlow *et al.* (2001) collected data on the nesting behaviour of the Sonso community between October 1995 and December 1996. Chimpanzees were followed from dawn to dusk over periods of 24 and 48 h over this 15-month period; 104 day nests and 601 night nests were recorded.

A significant sex difference was found in night nest height, with adult males nesting lower than adult females (Table 5.5). Males nested lower than females, perhaps because they are better equipped to deal with nocturnal predators. Alternatively, the greater weight of males may lead them to nest on sturdier branches lower down trees than females, or their greater strength may lead them to choose somewhat sturdier branches, especially as these produce a better sprung bed than do the smaller branches further up

Table 5.5: Comparison of adult male and adult female nests.

	Males	Females	Significance (Mann–Whitney *U*)
Night nest height			
Mean	10.91 m ($N = 334$)	13.55 m ($N = 267$)	$p < 0.0001$
Range	2.5–32.5 m	2.5–32.5 m	
s.d.	5.83	5.77	
Day nest height			
Mean	14.75 m ($N = 40$)	15.78 m ($N = 64$)	n.s. ($p < 0.59$)
Range	0–27.5 m	7.5–27.5 m	
s.d.	5.96	5.51	

in the tree. In regard to the relationship between males and females, nesting close together, with the male lower than the female, may indicate a degree of sexual interest between the two, as evidenced by a number of occasions on which males have shaken branches at females in oestrus in whom they were interested, and later nested below them, nesting either before or after the female.

We explored whether presence of an oestrous female in the party had any effect on the height at which males nested; it did not.

Table 5.6: Day and night nests compared (both sexes, N = 104 day nests, 601 night nests).

	Day nests	Night nests	Significance (Mann–Whitney U)
Nest height			
Mean	15.4 m	12.1 m	$p < 0.0001$
Tree DBH			
Mean	43.0 cm	26.4 cm	$p < 0.0001$
% Re-used	16.3%	9.6%	$p = 0.04$
% Weak	60.0%	0.7%	$p < 0.0001$

The structure of day nests was compared with that of night nests. Day nests were structurally simpler than night nests, more quickly made and less dense and leafy. As a result they were less permanent and fell to pieces more rapidly than night nests. We concluded that day and night nests need to be more clearly distinguished than is often the case. They serve different functions: day nests for resting over short periods, night nests for sleeping over periods of 10–12 h. There was no sex difference in nest height for day nests, they were higher than night nests[37] and were made in larger trees. Both day and night nests were occasionally re-used (see Table 5.6).

The most commonly used tree for day and night nests was the ironwood tree, *Cynometra alexandri*. This is a species characterized by many small leaves creating a mass of foliage and doubtless making comfortable nests. A different species, *Uvariopsis congensis*, was favoured in a study of 143 nests in the northeast of Budongo forest, at Kaniyo-Pabidi (O'Hara and O'Hara 2001), with 74% of all nests being made in that species. *Uvariopsis* like *Cynometra* has many small leaves but is lower than *Cynometra*, being an understorey tree that often grows beneath *Cynometra* in Kaniyo-Pabidi (S. O'Hara, pers. comm.). Mean night nesting height was just below 12 m at Kaniyo-Pabidi, just above 12 m at Sonso (see Table 5.6).

Nest structure and nest types were distinguished by McVittie (1998). He climbed up to 65 nests and was able to distinguish six types. Monostatic nests, mostly made by males, were low (5–10 m) and made from a single small tree stem, often bowed over by the weight of the chimpanzee, while female monostatic nests were higher (15–20 m) and situated at the terminal tips of crown branches in smaller, unbent

---

[37] Nishida (pers. comm.) writes 'At Mahale, in my impression, day beds are built lower than night beds. I do not know why this is different between sites.'

trees. Polystatic nests, made by both sexes, used two or more branches to provide support. They were of two subtypes, 'tension' nests, made by pulling two or more saplings together and then interweaving their crown branches, or 'suspension' nests, made by bending a small branch down to rest upon a larger branch; the smaller branch provided most of the nesting material while the larger branch acted as the structural support.

In this study, males nested slightly later than females on average (mean 18.21 for males, 18.18 for females, non-significant) with a negative skew for both studies, some chimpanzees nesting as early as 17.00. Both McVittie's and Brownlow's studies found that over 30% of nests were constructed between 18.45 and 18.59, roughly corresponding to sunset.

Ground nests were found by Reynolds (1965), McVittie (1998) and Brownlow *et al.* (2001). They appear to be a regular, if not common occurrence in Budongo. The case described by McVittie concerned the adult female Kewaya (KY) who was in oestrus and the adult male Kikunku (KK). At 18.00 on 22 May 1998, KK was observed shaking small saplings to indicate his interest in KY. Both individuals were feeding on ground vegetation. At 18.42 KY built a ground nest for herself and at 18.50 KK built one for himself half a metre away from hers. The next morning both individuals were still in their ground nests at 06.30 but left shortly afterwards.

In an effort to get at the significance of nests for chimpanzees, Croxall (2002) studied nests of the Sonso chimpanzees in relation to their habitat, range and social groupings. She noted that chimpanzee nests are not concerned with giving birth or rearing offspring as they are in many species and do not represent 'homes' in any sense. However, they are places of security, fixed points in the landscape of their range, and, even if temporary, do represent markers of some kind within the territory.

We saw from Fig. 5.4 that party size declines at the end of the day as chimpanzees approach nesting time. From 6.30 p.m. as dusk falls they form parties that will nest together. We looked at night nesting party size and composition over a 14-month period (Brownlow *et al.* 2001); 147 night nest parties were sampled between 19 October 1995 and 6 December 1996. Mean night nest party size (including dependent offspring) was 4.5, with a range from 1 to 17. Party composition was analysed for the 147 night nest parties using a principal components analysis. Mother–offspring pairs was the commonest combination; another common one consisted of the alpha and beta male dyad (Duane and Vernon) and another was a disabled male with a juvenile male. Otherwise individuals nested with others who happened to be nearby at the end of the day, though there was a tendency for adult males to nest near (and lower than) adult females in oestrus, perhaps 'mate-guarding' them or at least putting themselves in line for an early pre-breakfast copulation, or even a nocturnal one (Singh 1999). One or more oestrous females was present in 34 of the night nest parties studied. The size of those 34 parties in which one or more oestrous females was present was compared with 20 parties in which non-oestrous females without infants were present. The presence of oestrous females had a significant effect, increasing the size of night nest parties, particularly the number of adult males in the nest party.

Consortships

A relatively unusual kind of party consists of a male–female consortship. This was first described for the Gombe (Kasakela) community and appeared to be an alternative mating strategy adopted by some males who 'sequestered' females and took them on safari away from the other members of the community for a greater or lesser time while the female was in oestrus. This appeared to be a mating strategy by certain males to avoid competition with other males (Tutin 1979; Goodall 1986).

At Sonso, consortships appear to be scarce and such data as we have comprise a few observations of pairs consisting mostly of single adolescent or young adult females in oestrus being found moving together with single adult males, seen together over a period of several days. Whether these pairs are preliminary courtships which might lead to subsequent mating when the female is fully mature is not clear and requires further study. For example, Kewaya, a subadult female, and Kikunku, an adult male, were seen together on several consecutive days, with Kikunku sapling shaking to Kewaya, a display which in the context of courtship seems to mean 'Come here!' or 'Follow me!'.

A different kind of consortship, more like the one described for Gombe, may however occur at Sonso. For two weeks, from 9 to 24 April 2002, the alpha male Duane disappeared and this coincided with the disappearance of the adult female Nambi. They reappeared together on 24 April. Sean O'Hara who was working on sexual strategies at the time wrote:

> We suspect, although cannot confirm, that the two were on a consortship. Duane had been seen the day before their disappearance trying to lead Nambi away from the group by sapling shaking. Nambi was reluctant and had no sexual swelling. At the time she returned (two weeks later) she again had no sexual swelling and it is suspected that she stayed away through the whole period of her tumescence, which Duane monopolized (if indeed they were on a consortship).

This is probably the first well-recorded possible consortship at Sonso, although Duane and Nambi, who had a swelling, were also absent together on one earlier occasion. On a third occasion Black was seen leading Kutu who had a swelling. And recently a very interesting case has been seen: in July 2003 we observed Black and Polly in a very tight relationship with each other. Polly was *not in oestrus* (completely lacking a swelling) but Black *was copulating with her*,[38] demanding that she accompany him by branch-slapping in the tree, and when he came down to the ground by knocking a large dangling liana against the tree trunk making a loud noise. At this Polly descended and Black followed her very closely as they disappeared rapidly on the forest floor. This incident is in contrast with the consortship situation more commonly encountered, e.g. at Gombe, where consortships are a mating strategy of low-ranking males; Black was at this time the second or third ranking male at Sonso.[39]

---

[38] This is very unusual and was the first time I (or others present) had seen a copulation where the female was not in oestrus. Black, an aggressive male, was behaving aggressively to Polly, who was a newcomer to Sonso. However, the copulations were not forced, in that Polly did not give fear screams or attempt to run away from Black.

[39] Thanks to Sean O'Hara for pointing out this difference.

The low frequency of observations of consort pairs may reflect a tendency on the part of such couples to keep quiet and thus avoid detection. It also undoubtedly reflects the difficulty of seeing them in the continuous forest of Budongo. As with patrolling, it is very hard to document consortships in the forest. So it is impossible to say how frequent consortships may be in the Sonso community.

Boesch and Boesch-Achermann (2000: 76–8) echo the difficulty of determining consortships in the forest habitat at Taï; from the cases they did observe they conclude that consortship as a strategy has 'low success' compared with other strategies for impregnating females. They state that at Taï, 50% of males and 56% of females never went on a consortship (over a 15-year period). At Kanyawara (Kibale Forest) also, consortships are recorded occasionally, based on simultaneous absence of certain males and cycling females; such mutual absences have lasted up to three months (R. Wrangham, pers. comm.).

Kinship

A common organizing principle in mammalian societies is kinship (Fig. 5.5). Matrilines were first described in wild Japananese macaques by Kawai (1958) and other members of the team of primatologists brought together by Kinji Imanishi.

Fig. 5.5: Nambi with her offspring. L–R: Nora, Nambi, Night, Musa (photo: S. O'Hara).

In chimpanzees matrilines are well known, as for example the matriline of Flo at Gombe (Goodall 1968), but because many if not all females emigrate at puberty into communities where they may not have kin,[40] the strongest bonds after puberty exist between mothers and their sons rather than mothers and their daughters. Before puberty sons and daughters alike are very close to their mothers. For the first five years of life they are almost always found together and in the very early years infants are never far from their mothers and are carried around by them, at first below their belly and later riding on their back. During infancy the offspring learns what to eat from its mother.

Assersohn (2000) found that infants frequently solicited food from their mothers who shared it with them. Of the food items fed on by mothers 52% was solicited by infants and shared with them. Mothers rarely gave food to infants actively; the initiative came from the infant. This process of mother–infant food-sharing peaked in the second year of the infant's life. In the first year of the infant's life the food had mostly already been chewed by the mother and the infant obtained it in the form of part of a food wadge from the mother's mouth or hand; later on infants obtained food independently. Begging for food was characteristic of older, not younger infants. A large proportion, 35%, of food obtained by infants in their first year came via the mother.

The above details show that feeding, one of the very basic and most important activities for any species, is learnt in the context of maternal kinship. Likewise, maternal kinship is the context in which infants learn about most other aspects of life, and begin their social development. By contrast, paternal kinship scarcely exists as a force in infant development. Males play with infants and are tolerant of them at Budongo. As related elsewhere in this book, the birth of Katia to Kewaya, observed by Zephyr Kiwede, was followed a few days later by the presence of adult males in the tree with the mother and her new infant in a wholly tolerant situation. Infanticide by within-community males seems not to cause fear among resident mothers at Sonso and has only been seen when males consider the infant to have been sired outside their community. The within-community infanticide by the resident females Passion and Pom recorded at Gombe (Goodall 1986) is something we have not encountered to date at Sonso.

Bennett (1996) made a study of kinship among the Sonso chimpanzees. In her words 'Males should perhaps be regarded as generalized fathers . . . They do not show much paternal care but are usually extremely tolerant of youngsters. Small infants are usually permitted to climb over adult males who are resting, to sit close beside them as they feed, and even to share their food.' She also notes that males are remarkably tolerant of youngsters who interfere during mating.

As Bennett points out, males have a protective role in relation to the entire community. They are not 'collective fathers' in that they compete to achieve reproductive success (see Chapter 6) and higher ranking males are most successful at siring offspring

---

[40] At Taï and Mahale, for example, almost all females emigrate at puberty; at Gombe in contrast daughters of high ranking mothers tend to remain in their natal community and have high rank there, as happens in macaques.

(see Appendix C). However, they share in mating with oestrous females. And in their territorial behaviour they act together on behalf of the community. At Sonso we have some evidence of patrolling by adult males, and of territorial defence (see below). So, while maternal kinship functions as a critically important nurturing and social learning device, paternal kinship functions to protect the community in its home range and to kill young infants of non-community males if their mothers bring them in, thus returning their mothers to reproductive condition.

Because males remain resident in the community, they are a close-knit group and have an extremely detailed working knowledge of their range, its food sources and dangers. They groom each other with high frequency and intensity (see next chapter). Their genetic relatedness is, however, less close than might be imagined because their mothers have come from a number of other communities and their sexual partners likewise. Thus there is a constant influx of genes from outside, ensuring genetic variety down the generations (see Appendix C).

The median birth interval for known births at Sonso is 66.0 ± 5.0 months (Emery Thompson *et al.* in press). The first recorded birth (Bahati) took place in 1994, so the kinship structure of this community lacks depth at the present time. Of our adult females, only one (Zimba) is *known* to be a grandparent (daughter Kewaya, grand-daughter Katia), others may be but we cannot be sure. The maternal kinship relationships we know of from births observed during our study period, and from our genetic studies, are shown in Table 5.7.

Table 5.7: Maternal kinship in the Sonso community. Offspring are designated by codes in the table; those in parentheses were born outside the community and immigrated with their mothers (see text). Birth years are estimated before 1995.

Year of birth / Mother	Before 1990	1990	1991	1992	1993	1994	1995	1996	1997	1998	1999	2000	2001	2002	2003
Banura	ZF	SH						BT					BE		
Clea														CT	
Flora						(FD)				(FK)					
Harriet						(HW)					HL				
Janie										JT					
Kalema						BH					KM				
Kewaya									KA						
Kigere				KD					KE						KI
Kutu					KT				KN						KS
Kwera							KZ						KR		
Melissa								(MR)							MN
Mukwano															MD
Nambi	AY			MS			NR								NT
Polly									(PS)						
Ruda		BO						RE							
Ruhara	NK	GT						RS					RM		
Sabrina							(SA)								
Wilma							(WS)								
Zana						ZL							ZD		
Zimba	KY	GZ						ZG							ZK

Ranges and ranging behaviour

The area over which the Sonso chimpanzees move has been studied by Newton-Fisher (1997, 2003) who focused on the males, and Fawcett (2000) who focused on the females. According to Newton-Fisher the community's core area amounts to 6.89 km² of forest, which is small and indicates a food-rich area of forest, but this does not include the wider area they use towards the outer borders of the Sonso territory. Fawcett reports that the range utilized by the females falls entirely within the range of the males, female range being about 80% of male range. This is very similar to the situation in Mahale (Hasegawa 1990) and is in keeping with the idea that females are less inclined than males to venture out towards the margins of the community's territory, and that males do so and include patrolling of community borders in their ranging activities (see also Chapman and Wrangham 1993).

Fawcett observed patrolling behaviour by parties composed mainly of males and oestrous females. 'During a patrol the chimpanzees would silently leave a feeding tree and move purposefully towards a boundary area, stopping occasionally as if to listen and/or wait for other chimpanzees' (Fawcett 2000: 173). In August 1999 at 07.30, Dempsey (pers. comm.) observed the four senior Sonso males, Duane, Jambo, Maani and Nkojo, to the north of camp. While observing them calls and drumming were heard coming from the Waibira community's range, some distance away to the north. At this the males moved silently in a southerly direction, away from the calls. It may therefore be that they had been patrolling to the north but on hearing calls from the Waibira direction they became afraid (at that time the Waibira community males had been making sorties into the Sonso community's range) and moved away. Once they were well inside the Sonso range they paused and sat quietly for 13 min. Then, at 08.36, more calls and drumming were heard from the north and this time they responded loudly with 'aggressive sounding calls and drumming'. They called again 2 min later after which they moved on and joined a feeding party on some *Broussonetia* trees.

Individual females were found to have extensively overlapping ranges, which were approximately half the size of male ranges. Range overlap was greater than at Gombe and more like the situation at Mahale (Hasegawa 1990: 111). Male ranges overlapped with each other even more than those of females, and they also overlapped with female ranges, so that the community forms an integrated bisexual whole. Females in oestrus increased the size of their ranges to the same size as males. Females not in oestrus were inclined to remain in their core areas to a greater extent than other females, especially mothers with infants, and males. Also, female core areas overlapped less than male core areas, in line with the ideas of Wrangham (1979); such core areas were no more than 15% of the community range. Why range overlaps should be greater at Sonso than in some other areas (but not Taï, where male and female ranges are congruent) may be due to the extensive food supply of the Sonso chimpanzees which reduces feeding competition; Fawcett (2000) draws attention to the importance of the exotic *Broussonetia* in the food supply at Sonso, and Hasegawa (1990) writes that food supply may underlie the overlap of ranges at Mahale.

In support of this, Fawcett did not find evidence that females competed for core areas in the centre of the range. In contrast with the situation at Gombe (Goodall 1986), older, higher ranking females at Sonso did not have core areas in the middle of the community's range, with lower ranking females on the periphery. Nor did adolescent or young adult immigrant females establish core areas on the periphery, they moved into the centre of the range with the resident adult males and females, sometimes using maximal sexual swellings as a 'social passport' (Boesch and Boesch-Achermann 2000).

Newton-Fisher (2000) concurs that while both males and females at Sonso have core areas or 'preferred' areas of activity, these areas are not exclusive and overlap extensively. He found that males spent most (80%) of their time in their core areas, moving out for occasional territorial behaviour or for peripheral food sources. In the case of females, the same was true but they moved out when in oestrus for purposes of joining parties with adult males. To test whether good food supply was responsible for males having core areas Newton-Fisher compared them with surrounding areas and found no difference.[41] Nor did they appear to be related to male rank. He suggests that they may have territorial functions as zones of safety, and social functions as areas where alliance partners can be located.

Inter-community movements of adult females

Reference to certain adult females as 'immigrants' makes the assumption that they have come from another community. An alternative explanation is that some or all of these adult females are not immigrants but are 'peripheral' females who have not hitherto made their presence known to observers because of extreme shyness, so that when they move into the centre of the range and we see them for the first time it is a case of 'coming out' within the community rather than entering the community from another one. Harriet, for instance, was first seen with her juvenile Hawa in 1996, since when she has always been a rarely seen individual who pops up from time to time. In 1996 the Project was still young and it is unclear whether she was an immigrant or a previously unrecognized individual. It is also possible some females come and go between communities. Mukwano, a well integrated and easily seen female, twice disappeared and we concluded she had emigrated or left the Sonso community. The first time this happened, when she was a subadult, we did not see her for 20 months; the second time when she was an adult with a son, Monday, for a shorter period during which she lost her infant son in unknown circumstances (perhaps as a result of infanticide), returning without him.

The arrival of Melissa with her juvenile son Mark, Sabrina with her juvenile daughter Sally, and Wilma with her juvenile son Willis, who arrived in quick succession[42] in October–November 2001, took all observers by surprise. Of these three new adult females, only one, Sabrina, is suspected to have been seen before, for a short time one

---

[41] This is not to say that core areas elsewhere are not related to food supply, which may indeed be the case, as for example in the case of female core areas at Gombe.

[42] Melissa was first seen on 15 October 2001 and Wilma on 16 October 2001 and Sabrina on 3 October 2001.

Table 5.8: Named immigrants into the Sonso community, 1995–2003.

Name (code)	Date of immigration	Age–sex category	In oestrus?	With offspring? (age–sex categ.)
Mama (MM)	6/95	AF	Yes	Muhara (I2F)
Janie (JN)	9/95	SAF	Small swelling	No
Harriet (HT)	9/96	AF	No	Hawa (I1M)
Clea (CL)	8/97	SAF	No	No
Emma (EM)	9/97	Juv/SAF	No	No
Melissa (ML)	10/01	AF	Yes, full	Mark (JM)
Sabrina (SB)	10/01	AF	Yes, full	Sally (SA)
Wilma (WL)	10/01	AF	No	Willis (JM)
Juliet (JT)	1/02	SAF	Yes, full	No
Polly (PL)	1/03	AF	No	Pascal (JM)
Flora (FL)	5/03	AF	No	Fred (SAM)
				Frank (JM)

Key: I2F = infant 2 female; I1M = infant 1 male; other age–sex codes as throughout.

year previously when she was in oestrus. Now they suddenly appeared, confidently and without fear of observers, in the heart of the Sonso community. In addition, they associated closely with each other (M. Emery Thompson *et al.* in preparation). A fourth adult female, Flora, who arrived at the same time[43] with two sons, a subadult, Fred, and a juvenile, Frank, appeared, in contrast, to be frightened of human observers. A fifth new adult female, Polly, has joined Sonso with her juvenile son Pascal; she too was initially frightened of observers and, though probably first seen in early 2002, was not described until 2003.[44] All five of these adult females have settled in to the Sonso community, and, because the whole community was well known to all observers by the time the first newcomers arrived, we can be 99% confident that these are immigrant adult females (Table 5.8). At least six other shy, new parous females have been noted by observers as of 2004, though we are still waiting to see if they will permanently integrate into Sonso.

Emery Thompson *et al.* (in preparation) comment on the fact that whereas some immigrant females are fearful of community members and of human observers, others, even when they arrive with offspring, are able to enter the Sonso community without problems with residents or humans. A possible explanation is that some of these females may be natal females returning after many years absence. This cannot be discounted; our recognition of *all* community members dates back only to 1995. These confident females rapidly become well established and show no problems of habituation to observers; it is as if they knew the members of the Sonso community already. As seen above, in recent years several have arrived together.

Could it be that we have actually missed seeing these females, that they are not in fact new arrivals but had hitherto successfully evaded detection by our field assistants and students? This is extremely improbable according to Emery Thompson (in preparation)

[43] Flora was first seen on 10 November 2001 but not named until May 2003.
[44] Polly was named in July 2003.

and others[45] who have looked into this carefully. We were finding and naming new *adult* females until 1996 (Harriet was the last), after which there were no new *adult* females found until the three who arrived in 2001. Added to this, recognition of the new females was made easy by the fact that both Wilma and Flora were missing hands, while Melissa was quickly and easily identified by her resemblance to the adult male Jambo. Interestingly also, none of the new females, neither the more habituated ones who were not scared of observers, Melissa, Wilma and Sabrina, nor the more skittish ones who seemed to be less habituated, Flora, Polly and Juliet, moved around on the periphery of the Sonso range; they were first seen near the middle of the range. This is in contrast with the situation in Gombe and Kibale where only the most well-established females occupy the middle of the range. As regards the behaviour of our adult male residents, Sabrina's first oestrus in October 2001 caused great excitement among our males; the swellings of the other females caused no more than normal levels of excitement; and no aggression against the incoming females was seen, either from the adult males or the resident adult females.

One of the recent immigrant females, Sabrina, with her daughter Sally, did not immediately settle in Sonso, but disappeared after 10 months, in August 2002, and was seen only twice during the following year, after which she returned. These movement patterns by females are not normal either in Sonso or in other communities such as Taï (Boesch and Boesch-Achermann 2000), Mahale (Nishida 1990) or Kanyawara (R. Wrangham, pers. comm.)

At other sites, notably Gombe and Mahale, adult females have joined a new community when the adult males of their own community were attacked by males from the new one. We obtained no evidence of such attacks at the time these females appeared. Having considered various options, those who observed these events have concluded that this influx of females and young may have been the outcome of the disintegration of a neighbouring community, perhaps as a result of human activity such as intensive pitsawing, or for other reasons.[46]

Inter-community fighting

The first report of inter-community fighting in chimpanzees came from Goodall's Gombe studies (Goodall 1986). Over a period of several years, males of the Kasakela community killed males of the breakaway Kahama community and took over the Kahama territory and the Kahama females. In similar fashion, at Mahale, prolonged conflict between the K and the M communities appears to have been responsible for the annihilation of the K community, its males were probably killed and its females moved to join the M community. Both the above were in Tanzania. In Uganda inter-community conflict has

[45] Besides Melissa Emery Thompson, Sean and Catherine O'Hara and Lucy Bates have, with me, explored this possibility thoroughly in the course of numerous exchanges.

[46] I am grateful to Lucy Bates, Melissa Emery Thompson, and Sean and Catherine O'Hara for providing relevant data and for lengthy discussions about this issue. The idea that the females came from a disintegrating community was Melissa Emery Thompson's and was subsequently accepted by the group.

been reported at Kibale Forest. The subject has been reviewed by Wilson and Wrangham (2003).

At Budongo such fighting is rare, but we have observed hostile interactions between the Sonso community and the Nyakafunjo community living to the west and south of the Sonso range.

Fawcett (2000: 32) reports:

> Only one direct community encounter was witnessed. This occurred within the eastern boundary of the Nature Reserve. [The Nature Reserve is to the west of the Sonso range.] The Sonso community were feeding in *Cynometra*, they replied to some nearby calls (we initially assumed that these were from another party of Sonso chimpanzees) by rapidly climbing down and moving towards the east. The Sonso community chimpanzees were relocated about 200 m east, again feeding on *Cynometra*. As Geresomu Muhumuza [field assistant] and I stood beneath the tree trying to identify all individuals, two unknown males charged through the undergrowth, directly past us. The response of the only Sonso male on the ground, Kikunku (KK), was to turn and run east. Upon seeing us the unknown chimpanzees ran back immediately... The Sonso males continued to move slowly north east of the river and the other community moved towards the south. It is probable that the presence of human observers affected the outcome of the encounter as when we were spotted by the charging males, they stopped and ran back to the south where other community members were spread out in the trees.

Together with field assistant Kakura James, I was in the forest on that occasion, some 300 m to the east of the boundary where Katie Fawcett and Geresomu Muhumuza were witnessing the above events. In particular I was watching an adolescent female Mukwano (MK). She appeared very frightened at the noise coming from the fracas to the west. She abruptly climbed down the tree in which she had been feeding, a few steps at a time, stopping each time and looking towards the noise, her arms and legs rigid and her whole figure a portrait of apprehension. Finally she got to the ground and ran rapidly towards the east, the centre of the Sonso range.

In general, however, the lives of the Sonso chimpanzees appear, during the years we have been observing them, to have been peaceful ones. There have been no examples of inter-community killing as far as we know. Possibly some of our adult males who went missing were killed by other communities. Over a four-year period from 1997 to 2000, during which there were occasional inroads into the range of the Sonso community by the Waibira community to the north, we lost four of our adult males (see Table 2.1): Chris, last seen on 9 August 1997, Kikunku, last seen on 6 July 1998, Vernon, last seen on 28 June 1999, and Muga, last seen on 23 March 2000. The loss of these males was a severe blow to our community. The number of adult males was reduced from 15 to 11 over this period. It could be that they were killed in snares or traps, or died of diseases. However, during this same period the number of adult females who disappeared was two, Mama and Sara, who had immigrated and may have moved to another community. No adult females are *known* to have died during this period. Further, we did not find any of the males' bodies, suggesting they died some distance from camp, possibly on the fringes of the range where inter-community fighting is known to take place.

The antagonism between the Waibira and the Sonso communities declined in 2000 when an ever-increasing number of illegal pitsawyers began felling mahoganies in Budongo. They built a large camp on a site along the old logging road, where a concession had been given for some legal felling. This site happened to fall between the Sonso and the Waibira communities' ranges. The logging camp served as a collection point for timber from all over the forest and at one point over 100 men were working there, with trucks moving up and down the road several times a day taking mahogany to Kampala. During this time we rarely heard the Waibira community and they ceased making incursions into the Sonso range.

This interaction between logging activities and chimpanzee ranging behaviour is described by White and Tutin (2001) for the Lopé site in Gabon. There, large-scale logging had severely disruptive effects on chimpanzee ranges and there was a steep and long-lasting decline in the number of chimpanzees seen. As they point out, where logging is on a relatively small scale chimpanzees can move to other parts of their range, but when it becomes large scale they no longer have this option and may be driven into the ranges of neighbouring communities with consequent fighting and deaths. At present we do not know what happened in the case of the Waibira community but the logging operations stopped after a year or so when all the valuable timber had been removed and at the present time there is certainly a community living in the Waibira range and it has encroached on the Sonso range once again recently.

Inter-community aggression is always rare, but maybe it is more frequent if resources are scarce or population numbers are high. Neither is the case in Budongo. Resources seem to be plentiful and, alas, deaths from snare and trap injuries are the most likely factor keeping the population to its current size. We shall return to the problems associated with hunting by the local population in Chapter 9.

In the next chapter we turn to the social behaviour and relationships of the Sonso chimpanzees and explore in more detail the ways in which they interact with one another.

6. *Social behaviour and relationships*

7.15 Into forest with Tinka John. We move east to block 8D. Duane walks past us unconcerned. Nambi feeding on *C. mildbraedii*. Other chimps around on ground and in trees. Musa seen, now a big juvenile! Others: Gonza, Shida, Muhara, Kadogo, Gashom & Bwoba. Move east across sawmill trail. 7.57 Find Duane grooming Vernon. Duane chases Janie (in estrus 4). Clea — oestrus 2 — plus inf seen, not id. Gashom, Zesta, Bwoya, Zefa (getting big, subadult), Nick. Duane waves a sapling to get Janie's attention. He does this many times, moves and does it again. He vocalizes, she replies, both give short harsh barks. We are in 9C now. 8.16 Muga (following Gashom). 8.18 Tinka (26 September 1998).

In the previous chapter we looked at the social organization of the Sonso chimpanzees. The social organization of any species is the outcome of social relationships, and those relationships are the outcome of social behaviour. These three levels of society have been distinguished and described by Hinde (1976) and provide a way of looking at society: in the large (social organization), at one level closer to actuality (social relationships) or at the level of what we see taking place (social behaviour). There is a further level to which social behaviour can be taken, namely the level of social or cognitive psychology. This level has been well explored by Chadwick-Jones (1998) and promises to be at the forefront of future developments in the understanding of primate social behaviour. One of the fundamental tenets of this approach is that individuals, in their interactions with each other, are involved in an exchange of benefits, and this process of social exchange is cognitively mediated. We shall return to this approach at various points in this chapter.

Communication

To be social, behaviour has to be communicative. Each species has a system of communicative behaviours. A detailed description of the communicative gestures, postures, facial expressions and vocalizations of chimpanzees is to be found in Goodall (1986); this description fits the Sonso chimpanzees well. There would therefore be no point in my describing the social behaviours[47] of the Sonso chimpanzees here, and I shall only

---

[47] This does not refer to cultural peculiarities. Chimpanzee cultures are described in Whiten *et al.* (1999, 2001). To give an example, hand-clasp grooming, an everyday occurrence in the chimpanzees of the Mahale Mountains in Tanzania, has never been seen in the Sonso community. Sonso culture was described in Chapter 4.

refer to communicative signals as and when they arise in the course of describing other things.

From the point of view of cognitive psychology, communication is not so much an exchange of emotionally mediated calls, gestures and postures as a series of actions, characteristic of the species and understood by other members of the species, that convey particular meanings in the social context. Each individual exchanges communicative signals with others and in so doing expresses meanings such as sexual interest or aggressive intent or a wish to groom or be groomed (Chadwick-Jones 1998). De Waal (1982) gave an excellent account of the complicated machinations of three adult male chimpanzees in their struggle for dominance in the captive colony at Arnhem Zoo; this account shifted emphasis from the communicative signals and their functions to the underlying tactics and strategies of these three individuals in their interactions with each other and with the group's females over a lengthy period of time. What has become clear in the last three decades of chimpanzee research is that these animals have evolved a high degree of what can be called social intelligence, involving appeasement, deception and counter-deception, alliance formation, reconciliation after conflicts and sympathetic consolation for victims of aggression (De Waal 1982; Whiten and Byrne 1988). The result is what I have termed a 'thinking society' (Reynolds 1986). Harré (1984) has described the differences between such a society, based on the intentions, cognitive plans and strategies of its members, and the polar opposite kind of society based on largely automatic cause-and-effect processes. The danger of anthropomorphism resulting from the use of terms that have their origin in the description of human thoughts and interactions has been well analysed by Asquith (1984) who points out the inevitability of anthropomorphism in the description of non-human primate behaviour, especially in the case of our closest relative the chimpanzee. With the publication of Goodall's definitive monograph on the Gombe chimpanzees (Goodall 1986), as well as her other works and the many videos coming out of Gombe, it has become customary for primatologists to use words such as 'reconciliation' for particular kinds of chimpanzee social behaviour without fear of scientific inaccuracy. Quiatt (1984) has described the 'devious intentions' of monkeys and apes and discussed what is involved in such a characterization of behaviour. Quiatt and Reynolds (1993) have written at book length on the cognitive basis of primate, and especially chimpanzee, society and linked it in with human social evolution.

Grooming and other affiliative behaviours

Grooming, a strongly affiliative behaviour, occurs between mothers and their offspring (see Fig. 6.1) and it is in that context that the infant learns to groom. Later, as a semi-independent, and then as an independent individual, grooming is transferred to other members of the community, siblings in particular. Between males, which stay in the community (whereas females mostly move out at adolescence or early adulthood), grooming continues right through life. Indeed, the closest and most intimate grooming we see at Sonso occurs in close-contact huddles between adult males (Spini 1998).

Fig. 6.1:   Nora grooming her mother Nambi (photo: N. Newton-Fisher).

Of 709 consecutive instances of grooming recorded in 2001, when there were 10 adult males and 17 adult females, the highest number of approaches for grooming was made by Maani, the second highest ranking male, with 8.9%. Jambo, also a high ranking male, scored 5.8% and Duane, the alpha male, scored 3.7%. The highest number of approaches for grooming by a female was made by Nambi, the alpha female, with 2.0%. Males thus initiated grooming more often than females.

The highest score for receiving grooming was achieved by Duane, the alpha male, with 10.9% of all cases. Black, third ranking male, scored 6.9% and Maani, second ranking male, scored 6.5%. The highest scoring female was Nambi, alpha female, with 2.0%. Males thus not only initiated (see above) but also received more grooming than females.

Of these 709 cases of grooming, 306 were mutual, i.e. during the grooming session the chimpanzees involved groomed each other. The highest frequency for mutual grooming was achieved by Maani (second ranking male) with 32.0% of all mutual cases. Duane (alpha male) was involved in 25.3% of cases. Black (third ranking male) in 23.3% of cases. Tinka (omega male) was involved in 4.8% of cases. Nambi (alpha female) was involved in 4.7% of cases, of which 68% were with Duane. Most other cases of mutual grooming involved males, sometimes in groups of 3, 4, or even 5. Mutual grooming is thus a very male activity.

Adult males are highly selective about whom they groom. Newton-Fisher (1997), in the course of his study of the Sonso males, made a special analysis of 81 occasions of grooming by adult males. In those instances when 3 or more males were present, individuals showed a preference for a particular partner by selecting him and not the other male or males present. Distance to the selected individual was not the main criterion, and the selecting individual nearly always moved to the selected partner to groom him, rather than being approached by him. In cases where a lower status male groomed one of higher status, he approached and started grooming more quickly than when the statuses were the other way (i.e. the subordinate was keener to groom than the dominant).

The grooming relationships found by Newton-Fisher reflected the association patterns of these males. They positioned themselves nearer some males than others, and Newton-Fisher called these their 'proximity partners'. The positioning and the grooming reflected and strengthened the fact that some males were more closely bonded than others, but this 'bonding' was not constant, and kinship may not be the explanation. The clue was found in the fact that males position themselves closer to higher status males and further away from lower status males, and this goes for grooming too. 'Proximity may therefore reflect social tactics, as close associates are likely to be potential allies, or important competitors, and higher status males more useful as either allies, or as maintainers of social harmony (Harcourt 1989, de Waal 1982)' (quoted from Newton-Fisher 1997). This opens up a new dimension to grooming which takes it beyond the 'affiliative' category and gives us an insight into chimpanzee political relationships, particularly between males. The grooming patterns observed by Newton-Fisher among the 15 adult and subadult males he studied are shown in Fig. 6.2.

Subordinate chimpanzee males, in their grooming interactions, adopt careful spatial positioning near senior males, and are quick to groom higher ranking males once there is a chance, as against the slowness to respond by the latter.

The Sonso males, who are the stable core of the community, groom more, and groom each other more, than do females. (By contrast, in female-bonded societies such as macaques, females do more grooming than males.) An analysis of 3108 grooming

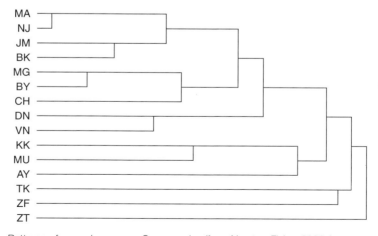

Fig. 6.2: Patterns of grooming among Sonso males (from Newton-Fisher 2002*a*).

Table 6.1: Grooming sessions between Sonso adults, both sexes, 1995–2000.

	Groomer AM	Groomer AF	Total
Groomee AM	1927	362	2289
Groomee AF	534	285	819
Total	2461	647	3108

Key: AM = adult male; AF = adult female.

sessions recorded at Sonso over a five-year period (1995–2000) shows the number of occasions on which males and females groomed each other (see Table 6.1; the analysis is based on adults only so as to exclude mother–offspring pairs). In each year, in the community, there were more adult females than adult males (see Table 2.1).

As can be seen, males groomed one another more than six times as often as females groomed each other, and they groomed females more than females did. Females groomed males five times less often than males did. Females, in fact, groomed less than a quarter as much as males. However, they should not be thought of as rarely interacting, as seems to be the case at some sites (Gombe: Goodall 1986; Kibale: Wrangham *et al.* 1992; Mahale: Nishida 1989); as we saw in the previous chapter, females do often form all-female groups consisting of mothers, infants and childless females (Bates 2001; Munn 2003).

Arnold and Whiten (2003) also studied grooming among the Sonso chimpanzees. They found that adult males had an average of 13 grooming partners of either sex of which 7.5 were other adult males. Adolescent males typically had 7 grooming partners, but adult females averaged only 2 grooming partners, of which only 0.17 were other adult females. Both sexes groomed males most frequently. After correcting for the number of available partners in each age–sex class the same pattern emerged. Grooming tended to be directed up the hierarchy, with individuals ranking close to each other grooming more than those less close. Higher ranking males had more grooming partners than lower ranking ones.

Looking at the above from the viewpoint of cognitive psychology (Chadwick-Jones 1998), it looks as if males are more interested in, or stand to gain more from, grooming exchanges than females do. Why should this be so? In some chimpanzee populations where hunting is common, grooming can be seen as a mechanism for male bonding that then expresses itself in co-operative hunting. But at Sonso hunting, while it occurs, is infrequent. A more important reason for co-operation among males is territorial defence, for the Sonso range is juxtaposed on three sides by other communities, the males of which appear to be hostile to the Sonso males. The bonds established by grooming may be important in achieving good co-ordination between Sonso males in the event of territorial disputes. Females do not have any clear-cut co-operative role to match that of males.

However, as we have seen, males do not just groom any other males, they have specific grooming partners. Here the psychological explanation is likely to be linked to the power struggles within Sonso society. Grooming, as we noted, goes up the hierarchy: individual males use grooming to appease and to please their superiors. Alliance partners, who stand to gain from co-operating with each other against one or more third parties, groom each other while they are allies, confirming their trust in one another. Their relationships are, however, 'fickle' (Newton-Fisher 1997), and when it no longer suits two males to ally but instead one decides to dominate the other, their grooming partnership falls apart.

We have used grooming as an indication of affiliation and this is indeed a good indicator. However, grooming is not the only way in which affiliation can be expressed. There is evidence from Taï that certain dyads of females had higher association indices (based on time spent together) than any of those between males, and in these cases the high rate of association was not accompanied by a high rate of grooming each other. Three pairs of adult females spent respectively 66%, 71% and 79% of their time together (Boesch and Boesch-Achermann 2000). The females were of similar age and therefore not thought to be sisters. The authors describe them as 'friends'; they shared food and supported each other in conflict situations. One such friendship lasted for four years, the other two lasted for five years each and all ended only when one of the partners died. The authors suggest that these female–female friendships may have been associated with high inter-female competition at Taï (p. 107); there were 25 adult females in this community, and the female friendship pairs had high status and a high rate of success in food competition, as well as in helping their sons rise in the male hierarchy. They had, it seems, transcended the tendency for females to be less sociable than males in their own, and their sons', interests. Boesch and Boesch-Achermann propose that the status of females, higher in Taï society than in other chimpanzee communities studied, is based on 'stronger intra-group competition, a highly biased sex ratio, and large party size' (p. 263). This would fit with the Budongo situation where the food supply is plentiful and feeding competition is apparently not great for much of the year (see Chapter 4).

Play

Play of many kinds is seen at Sonso, mostly between juveniles when they happen to be together in a party. Juveniles like to climb lianas, one after the other, and dangle from thin branches together, and can play-chase for long stretches of time in such simple ways. A proper account of play is not included here, but a single recent example is worth recounting, in which the play, rather unusually, involved water:

> On 8th February 2004 during a focal follow of Janie, her daughter Janet (aged almost four and a half years) was seen leaf sponging in the river Sonso. 5 mins later she was observed splashing with one hand in the water, from the bank of the river. Janie crossed the river on a pole bridge and Janet followed, but stopped on the bridge. Nora (a late stage juvenile aged eight years) joined her on the bridge and they both put one hand in the water and splashed, moving the water towards one another. Nora then left to play with Beti (a female aged seven and a half years). Janet followed her

briefly but returned to resume splashing in the river, this time leaning over and splashing vigorously with both hands. Janet returned to do this twice more before joining Nora and Beti playing in the trees. (Events Book, recorded by Katie Slocombe and Raimond Ogen.)

Sex and reproduction

Darwin (1871) suggested that in addition to natural selection (the process of differential survival based on inter-individual competition) there was also sexual selection based on the ability of individuals to attract mates and produce offspring. Female chimpanzees show large sexual swellings around the time of ovulation; normally these signal fecundity and attract the attentions of males. There are two situations in which swellings do not signal ovulation: first, pre-reproductive females show sexual swellings during adolescence, and second, non-reproductive females who are pregnant or lactating also in some cases show sexual swellings. The reasons for these 'dishonest signals' have been much debated and it may be that such swellings in pre-ovulatory females give them some early experience of sex with resident subadult males (adult males largely ignore them, however 'nubile' they may look to a human observer[48]) or help them gain acceptance if they transfer to other communities; in such cases swellings serve as a 'social passport' (Boesch and Boesch-Achermann 2000: 58). In the case of pregnant or lactating females they are pseudo-sexual signals that may 'trick' males into tolerating them at times when they are not fertile (see Wallis 1992*a* and Wallis and Goodall 1993 for discussions of these and related phenomena).

Turning to males, a notable anatomical feature is the large size of their testes. Compared with the other apes, or humans, chimpanzee testes are large. Clutton-Brock and Harvey (1977) showed that relative testis size (i.e. testis : body size) in primates varies according to social organization. Species living in multi-male societies (including chimpanzees) have the highest ratio, one-male group species have an intermediate ratio, and monogamous or pair-bonding species have the lowest ratio.[49] This has been interpreted as signifying that sperm competition is important in chimpanzees and other species living in multi-male societies. Large testes are advantageous when frequent mating with available females is the norm. The production of large amounts of sperm is advantageous to males when several males are mating sequentially with each female around her time of ovulation; the sperm of different males then compete within the female's reproductive tract (Dixson 1997).

Seraphin (2000) examined the question of whether increased production of testosterone would be found in higher ranking male chimpanzees but did not find a straight-line correlation between testosterone (extracted from faeces) and rank. Perhaps this is related to the fact that most competition for females occurs between

---

[48] In a study of 245 copulations at Sonso, only 15 were with young, nulliparous females (S. O'Hara, pers. comm.).

[49] Humans fall into the one-male group cluster, indicating that we evolved as a one-male group 'polygynous' or 'harem' species.

males of intermediate rank, whereas the alpha male is able to monopolize females when it suits him to do so.

Considering the data for the year 2001, 149 copulations were recorded. For each observed copulation we recorded the approaching chimpanzee.[50] Females approached more often than males, the highest number of approaches being made by Nambi (alpha female, 12.8% of all approaches), Gonza (11.4%), Melissa and Clea (8.7% each) and Kwera (7.4%). Black and Nick scored highest for males, with 6% each. Tinka, the omega male, made just one approach (0.7%). Interestingly, Duane, the alpha male, made only two approaches (1.3%), and no male scored more than 6% on approaches.

Comparing this with the number of copulations, we find that Gonza, a young female just becoming adult, received the highest number of copulations with 20.8%, followed by the alpha female Nambi (14.8%), Melissa (12.8%) and Clea (10.1%). We thus see that the females who made the most approaches also received the highest number of copulations.

In the case of males, the highest copulation frequency was for Duane with 15.4%, followed by Nick with 14.1%, Black with 11.4% and Zefa with 8.7%. These males did not often approach females for copulation, preferring to be approached by them. Interestingly, as pointed out by Newton-Fisher (2004), Vernon, the second highest ranking male after Duane in terms of respect shown to him by other males (pant-grunt frequency), had throughout his life a low copulation frequency quite out of keeping with his rank. In addition, as our genetic studies show (Appendix C) he failed to sire any offspring. This is surprising and raises the question why he did not benefit reproductively from his close relationship with the alpha male Duane. At present we have no answer to this question.

The circumstances surrounding copulation in chimpanzees can be illuminative of chimpanzee relationships. Here is an example from my field notes of an unwilling female:

> Friday 4th October 1996.
> Zimba in oestrus 4, Bwoya, Magosi, Duane, Kikunku all in attendance. They rush about. Chris & Zesta arrive; they fight in the trees. Much noise and excitement. The 4 adult males groom each other (not Zesta and Chris) on the ground, for a long time, totally unconcerned by us but looking up into the canopy from time to time to where Zimba is sitting. Then noise again and they climb up but can't reach Zimba who screams and keeps up in the thin branches so when they move near the branch sways and down she leaps to a new branch or tree. She doesn't want to mate it seems and they are getting frustrated! All morning this goes on.

In the above case, Zimba did not want to mate. Often the reverse is the case but even then things are complicated. For example, there was an occasion when Nambi, the alpha female, was in oestrus; her sexual swelling had been increasing in size for three days and she was now approaching maximum swelling and ovulation. During the preceding days she had been accompanied by a number of subadult and adult males; they had been

---

[50] That is, the one making a physical approach; in some cases the approach was preceded by a signal from the male, e.g. a branch-wave. Records are not sufficiently detailed to record all occasions when this occurred.

following her, staying near her and feeding in her company. When she moved, they moved. Black, Maani and Nkojo had all mated with her. The alpha male, Duane, however, had not been with them. He had come visiting a few times and examined Nambi's swelling, its size and its odour, but he had not been seen to copulate with her. On day 3 he sat on the ground, observing Nambi who was in the canopy. On day 4, however, he climbed up into the tree with the other males and Nambi. The three males gave way to him as he moved towards her. Once or twice they came close; he chased them away fiercely, threatening to attack if they didn't flee even if it meant hair-raising acrobatics in the treetops. They fled screaming and eventually took to sitting on the ground beneath the tree, eyeing Nambi from below.

Duane was now clearly interested in Nambi but not yet copulating with her. He sniffed her swelling from time to time but still they did not copulate. She for her part was not just passive. During the previous days she had been making life hard for the three adult males by resting up a lot of the time on small branches at the periphery of feeding trees, so each time they moved towards her the branch swayed and made things difficult. But she couldn't do this all the time, she had to feed and move around in the treetops, so they succeeded in mating with her. She now moved to a sturdier branch with Duane and positioned herself making it easy for him to mate with her and eventually he did.

After staying with her for two hours, Duane needed to feed. He moved away from Nambi and immediately there was a scramble from below as Black, Maani and Nkojo climbed up into the treetops to attempt to mate with Nambi; Maani succeeded first as she was receptive, Black succeeded next after a chase, and finally it was Nkojo's chance but he had left it too late. Duane returned, having fed, and monopolized Nambi again. He stayed with her for most of the day, moving with her on the forest floor and sometimes waving saplings to draw her attention when he wanted to move on or to mate. The two stayed together in this way the next day as well, but the day following she was already showing the first signs of detumescence and he left her; the other males could now mate with her and they did so from time to time when she made herself accessible. Over the next three days her swelling declined further and she ceased to be of sexual interest to the males, or to show any interest in them.

Copulation can take place on the ground, in the understorey or up in the treetops where we see it most often. On rare occasions the pre-copulatory stage is noisy, when a female emits screams loud and long as if she were terrified, while a male postures close to her demanding sex; this can be the end of a long period during which the male has been showing an erection and eyeing a female with a full swelling. O'Hara (in progress) is looking at the ways in which female sexuality influences male behaviour and is exploring the consequences of male aggression.

Often a female will give small high-pitched squeaky screams during copulation, and sometimes break off after mating with a scream and run a short distance away rapidly. But on other occasions this does not happen, she just remains with her male until they copulate again.

Males often mount a female several times but, unless the male is interrupted or the female runs away, after one or more mounts with pelvic thrusts he ejaculates. Hasegawa and Hiraiwa-Hasegawa (1990) working at Mahale define 'copulation' as one or more mountings taking place over a maximum period of 10 min. If there is an interval of more than 10 min between mountings, they consider that a second copulation is taking place. Sometimes a vaginal plug of coagulated seminal fluid can be seen at the vaginal entrance after copulation, or dislodged when the next copulation takes place.

A study of female mating patterns in the Sonso community was made by Oliver (2002). She collected data from June to December 2000, and combined her own observations with data collected by BFP field staff over the previous five years. All except one of the copulations observed took place in a group setting; the single exception was a case of copulation with just the pair concerned present. She compared the patterns of sexual behaviour of the Sonso females with those reported in several long-term field studies at Mahale and Gombe.

Two main kinds of copulatory relationship were distinguished: 'opportunistic', in which a female copulated with a second male within an hour of copulating with the first male, and 'possessive' or 'restrictive', in which a female did not copulate with a second male for at least an hour after copulating with the first, despite a second male being present in the party. Of the copulations she analysed 84% were opportunistic, while 15.8% were possessive. These proportions are not dissimilar to those reported for Gombe (73% opportunistic, 27% restrictive; Tutin 1979) or Mahale (73% : 27% in one community, 93% : 6% in another community). Copulation frequency of females was measured by Hashimoto and Furuichi (in press) at Kalinzu Forest; the mean for 329 copulations was 2.7 copulations per hour. One female copulated 6.8 times per hour over a 343 min period. Promiscuity was very marked in all except one female, for example the female Ha copulated with 9 of 11 males who were observed in the same tree during a 52 min period. For males, the greater prevalence of opportunistic mating over possessive mating reflects the fact that they may be unable to prevent other males from mating with any given oestrous female even if they want to. Vaginal plugs, whatever their function may be, do not prevent this.

Unlike at Gombe and Mahale, female age in Oliver's study did not correlate with the proportion of possessive copulations although the alpha male did focus on older females and possess them (see below). Hasegawa and Hiraiwa-Hasegawa (1983) and Tutin (1979) report that in Mahale and Gombe respectively possessive matings were more frequent in the case of older, resident females than in younger, immigrant females. Also, there was no difference in the proportions of promiscuous matings by younger, immigrant females and older, resident females at Sonso (promiscuity was measured by the number of different mating partners a female had during a given observation session). However, a significant positive relationship was found between male rank and copulation *rates* with older females. Thus, at Budongo, high ranking males copulated more frequently with older females but did not possess them more often.

Male alpha rank was correlated with the number of possessive copulations, and the majority of possessive copulations were found in the alpha male (57%). Apart from the alpha male, however, the correlation was not statistically significant, and it seems that possessiveness was a successful strategy of the alpha male only, as was found at Gombe (Tutin 1975) and Mahale (Hasegawa and Hiraiwa-Hasegawa 1983).

Immigrant females were not shy of mixing with the community males; in contrast, they spent more time with the adult males than resident females did. This was the case when they were not in oestrus (when they were in oestrus the difference between residents and immigrants disappeared). This may be because the immigrant females were establishing themselves in the community prior to their first birth, gaining familiarity with the males and hence reducing the risk of later infanticide if and when they gave birth (see Chapter 7 for details of infanticide).

Oliver (2002) concludes that the commonest tactic used by the Sonso females is to stimulate competition among the males rather than to actively seek out the high status, high quality, males. This fits with my own observations. The presence of a female in full oestrus, with a large pink swelling, has an electrifying effect on the adult males who gather around her, often displaying erections, looking in her direction, moving and choosing to feed near her, or sitting on the forest floor watching her every move. From my own observations I find it hard to be sure whether males other than the dominant male would like to be able to 'possess' such a female. In fact, when several males including the alpha male are present, only the alpha can do so. In his absence, higher ranking males can intimidate younger or lower ranking individuals from soliciting an oestrous female. An alternative strategy for possession, i.e. 'consortship', may occur (see Chapter 5). There is a little circumstantial evidence for this at Sonso[51] but the data are very scant. To some extent it appears that the Sonso males other than the alpha male accept that they cannot have an exclusive relationship with a fully oestrous female and so they move in to mate with her as opportunities arise. For her part she is mostly receptive at any time other than when she is being mate-guarded by the alpha male, perhaps for a mixture of reasons: reducing the possibility of immediate and future aggression from them, sexual desire and improving the likelihood of fertilization.

Copulation and time of day

Unlike the situation at Gombe (Wallis 2002a), Sonso chimpanzees do not concentrate copulations first thing in the morning. Data indicating this were at first thought to be due to differences in the time that observations were started between the two sites, i.e. observers at Sonso were missing early morning copulations because they were not present to see them (Wallis 2002a). However, newer data, including systematic de-nesting observations, indicate that this is a true difference. Data, including sexual and feeding behaviour, were recorded by Zephyr Kiwede, senior field assistant at BFP, and Melissa Emery Thompson. The results are shown in Table 6.2. Sonso chimpanzees generally

---

[51] For example, over a 12-day period in April 1997, Ruhara (adult female in full oestrus) was seen occasionally but the dominant male, Duane, was not seen at all. She was not observed to copulate with any adult male. On 15 November 1997 she gave birth to an infant, Rose. Was she copulating exclusively with Duane during her April oestrous period? Such sleuthing, based on our records of party composition and behaviour, is indicative but does not amount to proof. A second case of possible consortship involved Black and Polly in July 2003.

Table 6.2: Copulations by time of day, September 2001–December 2003. Based on approx. 2000 hours of observation, 600 hours with oestrous females present (data compiled by M. Emery Thompson).

Time of day	No. of copulations observed	No. of copulations by high status males	No. of copulations by low status males	No. of scans	No. of e-scans*
0600–0700	0	0	0	29	5
0700–0800	11	2	9	411	91
0800–0900	62	21	41	1455	432
0900–1000	76	17	59	1524	490
1000–1100	65	23	42	1374	479
1100–1200	47	18	29	1019	339
1200–1300	36	5	31	676	241
1300–1400	10	2	8	320	88
1400–1500	12	0	12	336	65
1500–1600	8	2	6	421	91
1600–1700	5	1	4	362	75
1700–1800	3	0	3	179	40
1800–1900	0	0	0	55	11
Total	335	91	244	8161	2447

* e-scans = scans with one or more oestrous female present.

wake just after dawn at 7 a.m., yet no successful copulations were observed prior to 7.30 a.m. Instead, the chimpanzees were most likely to spend their earliest hours feeding, concentrating their copulations in the mid-morning hours.

As can be seen from Table 6.2, few data were collected for the first period (6.00–7.00 a.m.) when light was poor and most chimpanzees were not yet moving. Copulations were seen with increasing frequency up to the 9.00–10.00 period. There is no evidence that copulations were concentrated at first waking, indeed they are concentrated around the mid-morning time. There were only five observations of parties in the first hour with oestrous females present, but no copulations took place.

To explore the effects of male status on copulation rates, rates were calculated for each of the males over the day. The results are shown in Fig. 6.3 which distinguishes three high status males, Duane (DN), Maani (MA) and Black (BK), from those of lower status.

Interestingly, the three highest status males copulated less frequently than the lower ranking males and this was consistent throughout the day. Also, the pattern of copulations differs for these two groups. When the high ranking males are reaching their peak rate in the late morning, the lower ranking males' rate declines, and then it picks up at midday and into the early afternoon when the rate of the dominant males declines. This is consistent with the idea that the high ranking males take the earlier opportunity to inseminate the females who are in full oestrus, leaving lower ranking males to rely on sperm competition to achieve reproductive success.

We also investigated the relationship between copulation and feeding. Do males copulate more when they are less busy feeding, and vice versa? The situation turned out to be a little more complicated than this. The study was confined to groups with oestrous

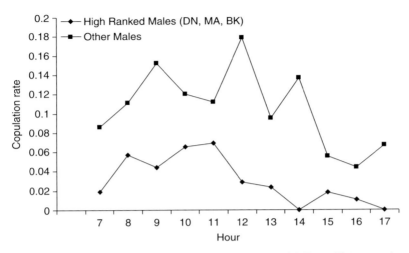

Fig. 6.3: Copulation rates per hour for high and low ranking males (M. Emery Thompson, in prep.).

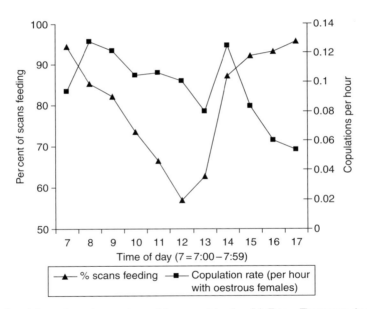

Fig. 6.4: Copulation rate and percentage of time spent feeding (M. Emery Thompson, in prep.).

females present. Before 9 a.m., there was an inverse relationship between the copulation rate and the percentage of time spent feeding. After 9 a.m. the rate of both copulating and feeding declined, reaching a low point in the heat of the midday period. Both rates picked up sharply at the start of the afternoon, but while feeding time continued to increase until nesting, copulation rates declined in the latter part of the day. This is shown in Fig. 6.4.

Seasonality of oestrous cycles

At Budongo, females do not 'come into oestrus' (or show oestrous cycling) equally all year round. This has been one of the more surprising things we have discovered in the Sonso community, and the extent of seasonality was first documented by Janette Wallis, who had worked at Gombe (Kasakela community) before coming to Budongo and so was able to make a comparative study of sexual behaviour and oestrous cycling at the two sites. The following is taken largely from her work on the subject (Wallis and Reynolds 1999; Wallis 2002*a*).

Gombe, in Tanzania, is 4° 40′ south of the equator while Budongo is 1° 35′ north of the equator. Though not a huge distance apart, they show mirror image seasonality and this applies also to sexual cycles. Seasonality of sexual behaviour is more marked at Budongo than at Gombe, with very few sexual swellings or copulations at Sonso from May to August over the period studied (1993–1998). Onset of sexual cycles by post-partum females occurs in the late dry season at Gombe, Mahale (Nishida 1990) and Budongo, and adolescent females also tend to show their first swellings during this season. At Budongo the dry season is from December to February. At Gombe the main dry season is from May to October. When the rainfall distribution for the two sites is plotted from January to December they look very different; when the same data are realigned to start with the first month of the rainy season, the similarity between them becomes apparent, although the dry season at Gombe is longer, six months as against three months at Budongo.

Wallis compared reproductive events at the two sites by reviewing long-term data from Gombe from 1964 to 1994 (6419 hours of observation) and from Budongo for 1990–2000 (3515 hours of observation). She looked at the presence of oestrous females and the occurrence of sexual behaviour. At both sites, presence of oestrous females varied seasonally. Seasonal variation was greater at Budongo than at Gombe, but both sites showed that fewer females came into oestrus during the later part of the wet season (see Fig. 6.5).

The same pattern occurs for resumption of postpartum cycling which most frequently occurs during the dry season at both sites, and for copulatory activity, which declines during the wet season and rises again towards the end of the wet season at both sites.

How are these patterns to be explained? Wallis argues that, rather than being a wet season/dry season pattern, the differences in levels of sexual behaviour and oestrous cycling coincide with periods of environmental change. This would help to explain the fact that at Taï the situation appears to be the opposite from that at Gombe and Budongo, with sexual swellings at their minimum during the dry months of June to August (Boesch and Boesch-Achermann 2000; Anderson *et al.* 2002), rising during the wetter time of year. What factors could account for such a reversal between the East African and West African chimpanzees?

In an attempt to find an explanation, Wallis is inclined to favour the effects of secondary plant compounds, especially plant hormones, in the diet of the chimpanzees. This takes us into largely uncharted waters. In the forests of East and West Africa,

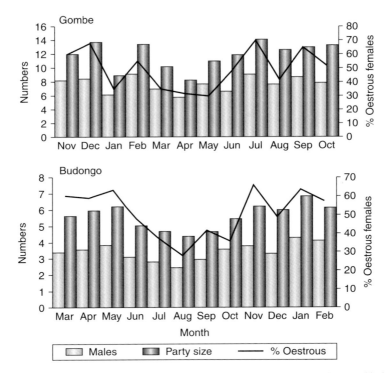

Fig. 6.5: Seasonality of oestrous females at Gombe and Budongo. Data are aligned with the graphs beginning at the onset of the wet season. Bars represent mean numbers per month. Solid lines represent the percentage of females in oestrus (from Wallis 2002a).

different tree species and other plant species shed their leaves, and then produce new leaves, flowers and fruits at different times. Chimpanzees eat the flowers of many tree species, and flowers contain phytoestrogens which can affect oestradiol action and the sexual cycle. Other plant compounds such as phenolic acids can inhibit gonadotropin release, thus blocking ovulation. Further, the action of phytochemicals may not be seen until three or four months after ingestion. We are a very long way from understanding how such processes work, but it is possible that they underlie the seasonality of reproductive and sexual behaviour found at the various sites.

Status

Status is about access to scarce resources. At various points in this book I have referred to the alpha male of the Sonso community (it used to be Magosi, is currently Duane, and he will be superceded). The dominant male is supremely confident and relaxed, but he can be harsh depending on circumstances. He has a glossy coat of hair and a powerful gait. As we move down the hierarchy males have rather less machismo, then lower still they have missing hands or feet, or may be like our omega male, Tinka, who has injured hands and a constant itchy skin complaint.

Since the work of de Waal (1982) published under the title '*Chimpanzee Politics*', we now know much about the relationships between males in a chimpanzee community. Coalitions, scheming for dominance, buttering up would-be allies, bluffing and finally all-out attack are some of the ingredients of male–male relationships. Status is a near-total preoccupation for males. And yet, in a community such as the one at Sonso, these males are in some cases related to one another, can be tremendously friendly and mutual with one another, and move around together in seemingly perfect harmony. When they turn aggressive the results can be devastating. Small fights, often over access to females, are easily contained though they may sound as if someone is being killed. But when real trouble breaks out there seem to be no limits to the aggressiveness of chimpanzees.

Jane Goodall was the first to describe really terrible fighting. Some of the males and females of the Kasakela community at Gombe had split off from their fellows and moved to the south to form a new community called the Kahama community. Not long afterwards the Kasakela males, acting together, sought them out one by one and savagely killed them. They killed a female too. They seemed to be avenging the crime of the breakaway individuals who had set up an exclusive territory to the south (Goodall 1986).

The ranks of the males in the Sonso community, calculated on the basis of their agonistic dominance plus conferred respect, are shown in Table 6.3. Within the community, the rank order between the higher ranking ('political') males seems to be enough to control male–male relationships most of the time.[52] Lower ranking males defer to higher ranking males, groom them and give them the pant-grunt greeting vocalization that signifies subordinacy (Goodall 1986; Hayaki 1990; Takahata 1990*a*) or 'conferred respect' (Newton-Fisher 1997). Not all the males are 'political', however; as Newton-Fisher (1997, 1999*d*) has shown, some males prefer to keep out of the political arena and forgo status-seeking for the quiet life. Plotting the status of males according to pant-grunts received against rank determined from agonistic interactions (see Fig. 6.6), we see the expected positive correlation, but the 15 males fall into three groups. In the 'top' group are the alpha and beta males, DN and VN. They formed an alliance pair and were very close during the study period (from October 1994 to December 1995). Although DN was dominant and could displace VN, VN was never heard to give DN pant-grunts: it appears their relationship was too close for that. (All other males did give DN pant-grunts.) The remaining males fell into two groups. The 'bottom' group consisted of four males, all very subordinate, two of whom were adolescents (AY and ZF), the third a young male (ZT), the fourth was TK, the omega male. The middle group consisted of nine males of intermediate rank, the top three of whom (BK, MG and MA) were in muted but not vigorous competition with one another as seen by their relatively high

---

[52] Not all the time. In Chapter 8 I describe a fatal attack on one of the Sonso males by his fellows. This is highly unusual, however.

Table 6.3: The males of the Sonso community in mid-1995, together with their status (from Newton-Fisher 1997).

Name	Code	Rank
Duane	DN	1
Vernon	VN	2
Black	BK	3
Magosi	MG	4
Maani	MA	5
Chris	CH	6
Jambo	JM	7
Bwoya	BY	8
Kikunku	KK	9
Nkojo	NJ	10
Muga	MU	11
Zefa	ZF	12
Tinka	TK	12
Zesta	ZT	14
Andy	AY	15

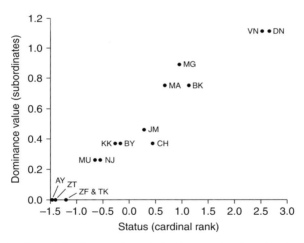

Fig. 6.6: Cardinal rank vs. status from pant-grunts for Sonso males (from Newton-Fisher 1997).

scores and their linear distribution. These were the 'political' animals, the rivals among whom competition for rank was greatest. The remaining six males of the intermediate group appeared to be less interested in achieving higher status and were less competitive during the period under study.

Aggressive behaviour is not common at Sonso. In 2001, only 28 instances of aggression were recorded in the course of project data collection. The highest scoring actors were Nick (5 instances), Duane (4), Zefa (3) and Maani (2). The most frequent recipients of aggression were the adult male Tinka (3), the subadult female Gonza (3), and the adult female Kwera (3). The aggression in all these cases consisted of threatening and chasing, and in the case of Gonza and Kwera this was in the context of sexual coercion. In the case of Tinka, the actors were Nick, Zefa, two higher ranking males, and Zimba, a higher ranking female; in these cases no sexual element was involved.

Because the number of instances in any one year is small, a second analysis of the incidence of aggression was done over a longer time period, from January 1995 to December 2001 (seven years). The highest scoring actors were Duane (45 instances), Andy and Nick (adolescent males) (22 each), Black (19) and, interestingly, the adult female Kewaya (18), who is an aggressive female who also drums on tree buttresses, both of which are considered male-like behaviours. In all other respects Kewaya is typically female.[53] The most frequent recipients were the immigrant adult female Clea (27 instances) who appears to have been unpopular for reasons I am not clear about, Nick (16) and Gashom (14), both adolescent males, and again Kewaya, the rather aggressive adult female.

Status change

Status was not constant over time. During the 15 months of this study, MG ceased to be the alpha male and was replaced by DN in an almost bloodless *coup* (MG ended up with a bloody finger and, even more visibly, a very wounded pride). At the start of the study MG was together with DN and VN at the top but this was an unstable period; the situation settled down during 1995 when DN became the undisputed alpha, as MG had been for many years during the period before Newton-Fisher's study began. The situation shown in Fig. 6.6 above is the one that prevailed after DN had achieved alpha status. Two further changes in status during 1995 concerned BK and BY, the former rose from mid to high status, while the latter fell from mid to low status. BK's rising status at this time was to cause problems later for DN and also, in particular, for low ranking ZT; we shall return to these events later.

Tactical association and alliance partners

Much of the manoeuvring of males in their quest for high status can be understood in terms of 'tactical association'. Males who are engaged in trying to rise in status have two strategies they can pursue. The first is aggression, but this has disadvantages because it is risky and can antagonize opponents who can find ways of thwarting the progress of aggressive individuals. The second is both more subtle and more normal in male chimpanzees: to form coalitions with alliance partners. We have already noted the alliance that existed between VN and DN while DN was ousting MG from alpha status in early 1995. That alliance weakened as DN consolidated his position at the top, but it re-emerged later in response to challenges.

Within their fission–fusion social system, the Sonso males associate with one another on the basis of choices: choice of whom to move with, feed with, sit near and, especially,

---

[53] A second adult female who has been seen buttress-drumming is Mukwano.

whom to groom. Association can be measured by proximity and by behaviours; the most striking behaviour of associating males is grooming, which can be intense, prolonged and mutual. One form of association can be called 'tactical' because, as the name suggests, there is more to the association than just interpersonal affiliation (or, to be more anthropomorphic, friendship). Behind tactical association is a goal the individual wants to achieve. Males forming tactical associations are generally called alliance partners. Together they can achieve a goal that neither could achieve alone. That goal is tied up with gaining higher status, or holding on to high status. DN and VN had been alliance partners for some time before DN moved against MG; DN's strategy had been there all along. VN did not have such a strategy, he showed no signs of trying to topple MG, and neither did he display against or have any conflict with DN during the time DN was gaining power. The alliance had no clear advantages for VN and presumably it was mainly driven by DN. We cannot know when DN developed the intention[54] to defeat MG and become the alpha male, but we do know that DN was forming an alliance with VN during 1994 and that he moved against MG at the end of 1994 and in early 1995, finally defeating him in April of 1995.

Newton-Fisher (1997, 1999*d*) made a detailed study of the strengths and fluctuations of association between the males of the Sonso community during 1995. He divided the associations (represented as clusters) into the four quarters of the year (Fig. 6.7).

We can see that in the first quarter MG was associating closely with DN and VN (DN was interacting with MG frequently, aiming to depose him, and VN was accompanying DN). Most of the males seemed to recognize the DN–VN alliance as dominant to MG during the first quarter but MG, who, judging from the genetic evidence,[55] had been alpha male for some time, had yet to accept the situation. In the second quarter MG was not with DN and VN at all, instead he was interacting with a different set of males (BK, JM, ZF and ZT) lower down the hierarchy. By the third quarter he had started associating with DN and VN again and by the final quarter his rate of association was somewhat higher with them. I was at Sonso in April and again in September of that year, and I recall seeing MG in April just after his fall from power, when he spent much time sitting looking crestfallen in day nests, with the juvenile male BB nearby as a companion, nursing a bloody finger (probably bitten by DN though we did not see a fight). That was in the second quarter, the time he mixed with the subordinate males only. By September I was very pleased to see that MG had recovered his sense of pride and was strong again, moving and interacting confidently with the dominant males DN, VN, MA and BY, calling to them with vigorous pant-hoots and feeding close to them. He had been displaced with a minimum of bloodshed and not long after had been accepted back into the dominant clique, wholly subordinate to DN but nevertheless, I felt, a respected member

---

[54] I use the language of intentionality purposely. Chimpanzee social life is cognitively organized and individuals have intentions towards one another. I concur with Takahata when he writes 'Male chimpanzees of a unit-group cannot recognize their paternal kinship. Thus, lacking established kinship connections, they intentionally form complicated relationships in which they simultaneously associate and compete with one another through social interactions' (Takahata 1990*b*).

[55] See Appendix C.

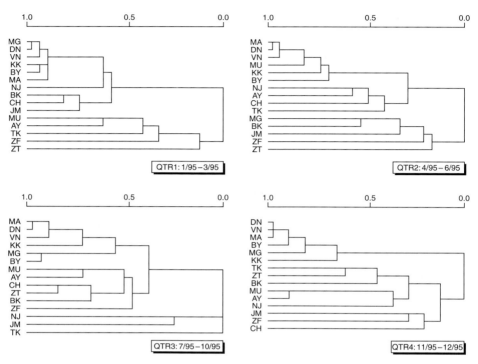

Fig. 6.7: Dendrograms showing the results of cluster analysis of male–male dyadic associations for each quarter of 1995 (from Newton-Fisher 1997).

of the Sonso community. (And so it should have been, for as it turned out when we later did our genetic analysis of paternity (see Appendix C), he was the father of many of the next generation.)

How to defeat an aggressive challenger

MG lost his alpha position in 1995 (he had been the alpha male since the start of our studies, and probably for many years before) as a result of a direct challenge by DN. We did not have anyone studying in detail how MG held off the challenge of DN, and possibly challenges by other males, before he lost power, but he did associate fairly closely with BK. From 1998 to 1999 there was a second study of our adult males by Arnold (2001) and this was followed by a third, shorter study in mid-1999 by Dempsey (2000). These studies shed light on how DN defeated a challenge by an aggressive male, BK, and also how he safeguarded his position afterwards. These details make a fascinating story and show how complex and, in a way, brilliant, adult male chimpanzees can be in the political arena around them.

At the time of these studies, DN was the alpha male but there was rivalry for the beta position, held by VN. The gamma male, BK, had been challenging him for some months (Arnold 2001). BK was an aggressive male, not involved in any alliances and he competed with VN by threats and displacements. But he could not succeed because VN had DN, the alpha male, as a very strong alliance partner (they had been alliance partners since before DN took over from MG). DN did not tolerate challenges to VN from BK, and BK had to back down each time because of the intervention of DN on VN's side. In the earlier part of 1999, BK started to settle down and ceased to actively challenge VN (Arnold 2001). Then VN disappeared. He was last seen by our field assistants on 29 June 1999, after which he was never seen again.[56] This precipitated BK into the beta position and the top four males in mid-1999 were: DN → BK → MA → JM. After them came four other adult males who were less power-minded, and finally two adolescent males, AY and ZF, who were becoming self-assertive but not yet serious contenders for male rank.

Dempsey focused on the beta male, BK, to see if he might try to usurp DN now that DN's alliance partner had disappeared. What happened was different. MA, the no. 3 male, associated closely with DN, and proceeded to dominate BK. This seems to indicate a firmer grasp of chimpanzee politics by MA than BK. Whereas BK was a rather aggressive male, not good at forming alliances but inclined to go it alone, MA was more 'laid back', and in fact over the 12 years we have known him he has always been a calm individual. At this time, however, he did threaten BK, always supported by the smaller male ZF. For example:

> On Sept 3 1999 at 12.30 p.m. MA and ZF arrived in an area containing BK, AY, KL and BB. On arriving MA displayed and was joined by ZF. He displayed again at 12.33 directly below BK. At 12.35 MA shook branches and displayed again, forcing BK to climb higher up the tree. At 12.40 BK presented his backside to MA in a gesture of subordination. They then groomed one another, alternately and mutually, for another 35 minutes (Dempsey 2000: 28).

Over the course of September MA gained dominance over BK until there was no doubt at all of his position as no. 2, with BK as no. 3. Indeed, in due course of time, MA replaced VN as DN's alliance partner.

Regarding ZF's close partnership with MA, was this a case of MA 'using' ZF to bolster his displays against BK? The evidence from the grooming relationship between the two indicates that this was not the case: ZF groomed MA but MA was not observed to groom ZF. From this it appears that ZF was the active partner in this alliance, and was ingratiating himself with MA ('sycophancy' is the word used by Dempsey) while MA was making no great effort to reciprocate or keep ZF on board. Instead, MA was grooming his rival BK as well as dominating him — perhaps because, astute male that he was, he understood that BK was aggressive and feared that without grooming he might

---

[56] We don't know the reason for his disappearance. He may have been killed while crop-raiding, or perhaps he died in a snare or trap, but his body was never found. At this time the Sonso community was receiving incursions into its territory from the Waibira community to the north, and it is also possible that VN was killed by the males of the Waibira community (of whom the Sonso community males seemed to be very afraid) but evidence for this is lacking.

alienate him, leading to challenges in the future. However, MA may have been flattered (if we can use such a word, and maybe we can) by the attentions of the younger male ZF. We have a similar case from Gombe in Goblin's relationship to Figan (Goodall 1986: 431); in such cases the younger male gains status more rapidly by associating with dominant males than he would otherwise do. ZF did rise in rank after the events of 1999 but he had a setback in September 2000 when he got all four fingers of his right hand caught in a snare; for some time he was preoccupied with trying to remove the snare which was fixed tightly round his fingers. He did manage to remove it (with the loss of his two central fingers) and has since become a vigorous medium-ranking young male with the potential to rise higher in the future.

MA, meanwhile, has remained the beta male in the Sonso community, while DN remains alpha and BK remains gamma. Since the events of 1999 described above, however, there was a period in 2000 when MA seemed to be using his special relationship with DN to achieve a status sufficiently high to begin to worry DN. As a result, for a time, DN and MA's alliance fell apart and DN began to increase his association with the fourth ranking male, JM. This was sufficient to alert MA to the dangers of pushing ahead too far too fast and he ceased doing so, remaining very much the beta male. Had he not done so, there is a danger that BK would have dominated him, because MA was at that time not enjoying the support of DN.

These details give us a good insight into the ever-fluctuating status competition between chimpanzee males. This competitiveness is, however, very much tempered by affiliative tendencies expressed in grooming, and by co-operative actions in respect of both hunting and territorial defence.

Undoubtedly it is the close ties between the adult males that moderate the amount of physical violence they show in their competition with one another. This is evident from their tendency to groom each other frequently and for long periods which helps to prevent their relationships from sliding into hostility. But beneath the closeness is a competitive dimension which determines their all-important access to oestrous females, as we showed earlier in this chapter.

Reconciliation

Arnold (2001) demonstrated reconciliation for the first time in wild chimpanzees at Sonso. Although reconciliation was not common, she found it mostly occurred between pairs of individuals that had strong affiliative relationships. This supports the idea that when a pair of individuals value their relationship highly, they are quick to reconcile after a tiff. It might therefore be expected that males would reconcile with each other more readily than with females, because males have very close relationships, support their alliance partners, and act co-operatively in a number of contexts such as territorial defence and hunting. However, Arnold found this not to be the case at Sonso. Arnold and Whiten (2001) discuss post-conflict behaviour in the Sonso community, showing that in contrast to captive chimpanzees, individuals at Sonso did not reconcile with each other often (the mean conciliatory tendency was 12.3% of conflicts in Budongo as opposed to 35%–41% in captivity). This low frequency of reconciliation was found in relation to male–male, male–female, and female–female conflicts. In fact, aggressive incidents are not as frequent in the Sonso community (and other wild chimpanzee communities) as they often are in captive situations. When reconciliation did take place it fitted the 'valuable relationship hypothesis' under which individuals who spend more time together in affiliative contexts are expected to reconcile more often. Surprisingly, however, oestrous females were not more likely to achieve reconciliation with males than anoestrous ones. In relation to status, reconciliation was almost always initiated by the lower ranking individual after a conflict. Of all successful attempts at reconciliation 54% were preceded by some form of submissive signal, most commonly the pant-grunt, which is a universal signal of subordinacy in chimpanzees. Gestures used before reconciling included holding out a hand or foot or presenting the rump, but mouth to mouth kissing in this context (de Waal and von Roosmalen 1979) was never seen, and nor was any form of consolation, which the above authors have reported seeing in captive chimpanzees. Summarizing, Arnold and Whiten conclude that only one dimension of relationship significantly influences the likelihood of reconciliation: compatibility. Thus a history of friendly contacts constitutes a relationship that is sufficiently valuable to be worth restoring after a conflict.

Finally, on this subject, we have the finding by Seraphin (2000) of a positive correlation between male rank and cortisol production (the cortisol was derived from faeces). We can interpret this to mean that higher ranking males have higher stress levels than lower ranking males, in which case the fluctuating strategies of rivals and the need to keep a constant watch on rivals may constitute, as Seraphin suggests, a cost on high rank. A study of the Kanyawara community of the Kibale Forest by Muller and Wrangham (2004) also found this correlation, which they attributed to elevated energetic expenditure rather than to psychosocial factors. As they point out, aggressive displays, characteristic of dominant males, are energetically expensive.

Status among females

With the qualification that females may influence males in a variety of ways, there is no doubt that at Sonso females are subordinate to males. They are highly sociable and often congregate together or in mixed parties of males and females (Fawcett 2000; Bates 2001; Munn 2003). Between themselves there is not so much a rank order as a division between 'central' females and 'peripheral' females. This refers partly to spatial location and partly to social position. If we compare the ranging patterns of central and peripheral females, some females, such as Ruda or Ruhara, are rarely seen in the central area of the community; they seem to keep to themselves and prefer to be in small groups of females and juveniles or with lower ranking males away from the action. By contrast the central females such as Nambi and Zimba are very often found with the dominant males of the community in the centre of the range.

We have already noted (Chapter 5) that at Taï females form stronger bonds than at other sites studied. At all sites, however, there is some evidence for dominance-striving by females: at Gombe (Goodall 1986), at Mahale (Nishida 1989) and at Kibale (Wrangham *et al.* 1992). De Waal (1982) described the female dominance hierarchy as being more influenced by subordinacy (via pant-grunts) than by dominance (via aggression). Pusey *et al.* (1997), analysing 35 years of data from Gombe, found that high ranking females had higher infant survival and shorter birth intervals than low ranking females. The reason, they suggest, is that high rank gives females improved access to areas of high food quality. As suggested in Chapter 2, food may be more of a limiting factor at Gombe than at Budongo. For example, the same authors found that immigrant females are sometimes attacked by resident females, even coalitions of resident females; this has occasionally been seen at Sonso, where immigrant females are sometimes bold and integrate well as soon as they arrive, but at other times are shy and evasive and may be attacked by resident females. For example, a prolonged attack was observed on the immigrant Juliet, a subadult nulliparous female, by Nambi and Ruhara during January 2003 (the aggression lasted 15 min on 29 January). At Mahale the situation is similar: Nishida (1989) found that younger, nulliparous and recently arrived females tended to pant-grunt to older, multi-parous and longer resident females indicating some degree of fear of them.

Female association patterns at Sonso, and status among the Sonso females, have been studied by Fawcett (2000). The main behaviours she used to record dominance were aggression, chasing, displacing and pant-grunting. She recorded 102 dominance interactions between females. Although it would be risky to claim there was a hierarchy on the basis of this sample size, interactions between the same pair of females always went the same way. The alpha female was Nambi (NB), who was dominant over the largest number of females. Two other females, Kwera (KW) and Kewaya (KY), were second and third in dominance over other females; as related earlier these two females were very close associates as adolescents and had their first oestrous periods at the same time. A recently immigrated female, Clea (CL), was subordinate to the largest number of females and did not dominate any other female. Thus one factor emerging out of the Budongo data (as elsewhere) is that long residency is linked to high status. By the same token, Fawcett found that three immigrant females, Emma (EM), Mama (MM) and Harriet (HT), were more loosely associated with the community than females who had been resident for longer.

Five pairs of females did have close relationships,[57] echoing the situation at Taï. There was also some evidence of female alliances, albeit short-lived ones, e.g. when Zimba and Kewaya jointly chased away an unidentified immigrant female, and when Ruhara and Sara formed a temporary alliance to stop Kalema entering a feeding tree; the first of these pairs is a mother–offspring pair[58] so that kinship is involved. Kewaya had two close associates. Nambi, the alpha female, had a close associate, Kutu, a lower ranking female.

---

[57] Closeness was measured by having significantly higher than the mean dyadic association index between females.

[58] This was indicated by behavioural data and has subsequently been confirmed by genetic analysis.

Summarizing, the existence of female dominance interactions confirmed that some degree of female–female competition exists, and it is hierarchical, though Fawcett could not construct a full-scale hierarchy (as was possible for Newton-Fisher in his study of the males) because so many female–female pairs were not seen involved in any dominance interactions at all. Length of residence and age were important factors and they interacted with each other: all adult females were dominant over all subadult females. However, within the adult females age was not necessarily correlated with dominance as some younger females such as Kwera and Kewaya, by virtue of having been in residence for very many years, were dominant over some older females who had immigrated.

Bethell (1998, 2003) looked at rank relationships in both males and females in a novel way, by studying social monitoring, or 'vigilance' which she documented by counting the number of times individuals looked at each other in parties. She found that both males and females monitored their social environment but in different ways. For lower ranking individuals of either sex, vigilance was directed to higher ranking individuals. Lower ranking males spent time looking at higher ranking males, and lower ranking females spent time looking at both higher ranking females and males. The need for vigilance in lower ranking individuals meant that they were only secondarily able to focus on potential mates and attend to reproductive strategies. In contrast, high ranking individuals could concentrate on reproductive strategies without fear of interventions such as threats or attacks. This was borne out by the fact that Duane, the alpha male, spent nearly a quarter of the time he was being observed looking at females, more than any other male. Dominant females, on the other hand, spent significantly more time looking at food sources than lower ranking females.

Vocalizations

The basic set of vocalizations of chimpanzees was described by Reynolds and Reynolds (1965). Goodall (1986) listed 34 calls, grouped into 13 different emotions or feelings with which they are associated (Table 6.4). These emotions or feelings are expressed in social situations and the different calls give information to others in the group, either about the context (e.g. the 'Wraaa' sound indicates fear and directs the attention of others to anything strange in the vicinity), or the rank order (the 'pant-grunt' indicates subordinacy), or enjoyment of good food (the 'pant-hoots' of chimpanzees, which often occur in choruses of many individuals calling together, can indicate the presence of good food in a single tree or an area of forest). There are many variations of these calls, which form a 'graded' system (Marler 1969) rather than a set of discrete calls. For an excellent description of the calls of chimpanzees the reader is referred to Goodall (1986: chapter 6); most or all of the calls described there are also to be heard in Budongo, with some significant variations (see below).

Pant-hoots

One kind of call given by chimpanzees is a loud and complex call given by individuals or by several animals joining in one after another to make up an impressive chorus.

Table 6.4: Chimpanzee calls and the emotions or feelings with which they are mostly associated (from Goodall 1986: 127).

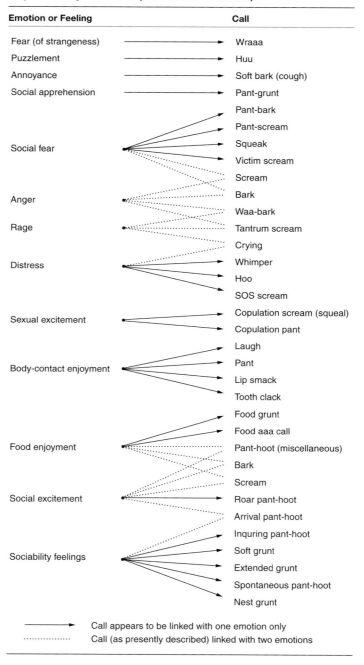

Emotion or Feeling	Call
Fear (of strangeness)	Wraaa
Puzzlement	Huu
Annoyance	Soft bark (cough)
Social apprehension	Pant-grunt
	Pant-bark
	Pant-scream
Social fear	Squeak
	Victim scream
	Scream
Anger	Bark
	Waa-bark
Rage	Tantrum scream
	Crying
Distress	Whimper
	Hoo
	SOS scream
Sexual excitement	Copulation scream (squeal)
	Copulation pant
	Laugh
Body-contact enjoyment	Pant
	Lip smack
	Tooth clack
	Food grunt
	Food aaa call
Food enjoyment	Pant-hoot (miscellaneous)
	Bark
	Scream
Social excitement	Roar pant-hoot
	Arrival pant-hoot
	Inquiring pant-hoot
	Soft grunt
Sociability feelings	Extended grunt
	Spontaneous pant-hoot
	Nest grunt

Call appears to be linked with one emotion only

Call (as presently described) linked with two emotions

There are four phases of the pant-hoot: the introduction, build-up, climax and letdown (Fig. 6.8), and these differ in 'length' and 'shape' in different chimpanzee populations.

Mitani *et al.* (1992) added a new dimension to our knowledge of chimpanzee vocalizations when they described chimpanzee 'dialects' — minor but distinct variations in the structure of the same call-types in different parts of their range.

Males at Mahale were found to have a faster rate of delivery, shorter duration of build-up elements and a higher pitched climax than Gombe chimpanzees. Males at Kibale (Kanyawara community) in Uganda had longer elements in the introduction and a shorter build-up than at Mahale. Kibale males also delivered acoustically less variable pant-hoots than those at Mahale. Playbacks of the calls made it possible to determine where they were from with a greater than chance frequency.

An additional finding was that of Mitani and Brandt (1994) in which they suggested from an analysis of pant-hoot choruses that chimpanzees could modify their calls to sound more similar to those they were calling with. They also found that the more time males spent with each other, the more similar their calls were. This fits with the 'action-based' model of vocal learning in birds (Marler 1990). In human terms it can explain how individuals come to adjust their speech to that of others in the group, thus, for example, taking on a local accent. This, together with the existence of dialects in chimpanzees, indicates that over and above the basic call structure of chimpanzees there are learned components in the calls. Clark Arcadi (1996) suggested that modification of pant-hoots in a community leads to them coming to resemble those of the alpha male.

Clark and Wrangham (1994) extended our understanding when they distinguished between two wholly different aspects of meaning in pant-hoots. On the one hand they are associated with good quality food, and it was (and is) thought that they attract fellow members of the community to rich food sources where competition is not so great as to

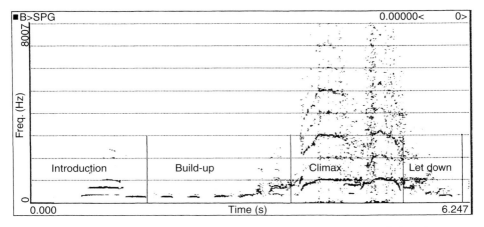

Fig. 6.8: A representative pant-hoot sequence from Budongo showing the four distinct phases (from Notman 2003).

reduce their own ability to find as much food as they want.[59] On the other hand they found that the frequency of arrivals of other chimpanzees supposedly 'called' to feeding trees by pant-hoots was no greater than the frequency of arrivals when no pant-hoots had been given. They noted that pant-hoots were often given by males when they were near, but out of sight of, their alliance partners, and concluded that these calls were equally or more likely to be given by adult males when locating or ascertaining the presence of alliance partners.

At Sonso chimpanzee vocalizations have been studied by Wong (1995) and Notman (1996, 2003, in press). Both Wong's study and Notman's first study were comparative. Wong compared Budongo pant-hoots with those described for Gombe, while Notman compared Budongo pant-hoots with those of Kibale.

Wong, in her 1995 study, examined the pant-hoot calls of two populations of chimpanzees to see whether there were differences (possibly 'dialects'). The calls from Gombe she used for the comparison study were made by Charlotte Uhlenbroek in 1993.[60] 32 calls from seven Sonso individuals[61] were compared with 34 calls from five Gombe individuals. The alpha male contributed largely to these samples in each case. Wong focused on the build-up and climax phases of the pant-hoot in her analysis, following the methods used by Mitani *et al.* (1992) and Mitani and Brandt (1994).

The differences between the build-up and climax phases of the pant-hoots of Budongo and Gombe chimpanzees are summarized in Table 6.5. The results of Wong's analysis indicated that the rate of delivery of the build-up elements is the only parameter that differs significantly between Gombe and Budongo chimpanzees. Budongo chimpanzees have a slower rate of delivery than Gombe chimpanzees. The differences in duration of build-up elements and frequency of the climax elements found by Mitani *et al.* (1992) when they compared Gombe with Mahale were not found in Wong's comparison of Gombe and Budongo.

In addition Wong looked at the question of whether, as mentioned above, individuals may be influenced in their call types by those around them, giving rise to accents and dialects. 'This' she wrote 'seems a plausible explanation for the difference in call structure between Budongo and Gombe chimpanzees.' She also noted that there was no significant variation within populations, suggesting that there may have been 'convergence' among individuals of the same population.

As to the function of regional accents, Wong followed Goodall's (1986) suggestion that distinct community accents might have arisen for reliable and easy recognition of

---

[59] This 'meaning' was rejected when I first suggested it in 1965. I was told in the friendliest possible way that the existence of competition between individuals would eliminate any possibility that pant-hoots could be telling others of the location of good food: individuals would be expected to keep quiet about such things and enjoy them in peace! It was only later, in the 1970s, when the degree of co-operation between males (who do most of the pant-hooting) was discovered, together with the advent of sociobiology, that zoologists were able to take on board that pant-hoots might indeed be co-operative calls. Up to that point these ideas were regarded as unscientific.

[60] Wong acknowledges Uhlenbroek's generosity in making her recordings available for the comparison.

[61] Most calls were from the alpha male Duane, and she established that there was no statistically significant difference between repeated pant-hoots by this male, i.e. his pant-hoots (and probably those of others) had a distinct pattern, repeated each time he called.

Table 6.5: Mean values with standard deviations (in parentheses) of two acoustic features (build-up and climax) of pant-hoots in five individuals each of the Budongo and Gombe populations, and mean values for the two populations (from Wong 1995: 44).

Parameter measured	Population											
	Budongo						Gombe					
	DN	MG	CH	MA	TK	X̄	FD	GB	BE	FR	PF	X̄
Build-up												
Rate of delivery	2.62	2.49	—	—	1.51	2.21	2.80	3.17	3.46	3.16	—	3.15
(elements/second)	(1.3)	(1.2)	—	—	(1.3)		(0.9)	(0.9)	(0.4)	(0.8)	—	
Duration (seconds)	0.41	0.42	0.48	—	0.54	0.46	0.36	0.22	0.25	0.24	—	0.27
	(0.2)	(0.1)	(0.3)	—	(0.3)		(0.3)	(0.0)	(0.0)	(0.0)	—	
Fundamental	635	589	649	—	774	662	498	486	445	691	—	530
frequency (Hz)	(347)	(250)	(178)	—	(665)		(242)	(256)	(7.4)	(15)	—	
Climax												
Duration (seconds)	0.79	0.71	0.94	0.48	0.58	0.70	0.59	0.76	0.76	0.66	0.98	0.75
	(0.3)	(0.1)	(0.1)	(0.1)	(0.2)		(0.1)	(0.1)	(0.1)	(0.1)	(0.1)	
Minimum frequency	718	656	583	592	578	625	643	809	539	730	577	660
(Hz)	(143)	(119)	(183)	(188)	(66)		(165)	(194)	(142)	(280)	(19)	
Maximum frequency	1308	1268	1040	1230	1250	1219	1163	1425	1100	1388	1056	1226
(Hz)	(247)	(253)	(287)	(105)	(66)		(257)	(272)	(413)	(218)	(291)	
Frequency range	565	613	456	633	672	588	520	616	561	659	479	567
(Hz)	(174)	(149)	(107)	(76)	(133)		(159)	(138)	(271)	(63)	(310)	
Average fundamental	1001	969	831	874	958	927	941	1115	819	1110	769	951
frequency (Hz)	(217)	(203)	(186)	(141)	(155)		(233)	(231)	(332)	(148)	(96)	

other groups in order to avoid agonistic interactions and I strongly concur with her on this. Having observed the reaction of members of the Sonso community to calls coming from the Waibira community to the north, it is clear that they not only know they are from foreign neighbours but show fear of them and when they hear them they move away fast to the south. This is less the case with their western neighbours, the Nyakafunjo community, whom they appear to fear less, and with whom we have seen them involved in border confrontations from time to time.

A year after Wong's study, Notman (1996) continued this work with his comparison between the Sonso chimpanzees and those of the Kibale Forest, some 150 miles to the southwest of Budongo. Being closer to Budongo than Gombe we might expect differences between these two populations to be slight, and the existence of Bugoma Forest and numerous forest outliers and patches of forest between Kibale and Budongo means that, at least until recently, there is likely to have been some degree of contact between chimpanzees of the two forests. Following the 'action-based' model used by Wong, Notman added some subadult males to his sample to see if their vocalizations were less standardized than those of adults, as might be expected if they were in the process of conforming to the local dialect.

Notman analysed 17 usable calls from Budongo and also used 10 calls from Wong's study. For the calls from Kibale he used 39 calls from Clark Arcadi's sample. Both samples were heavily based on the alpha male of the two communities as were the samples in Wong's study.

Initial visual and audio inspection of the results indicated that there were two main differences between Budongo and Kibale pant-hoots: the build-up phase was more clearly defined at Budongo than at Kibale, and Kibale calls were higher pitched than those at Budongo. The numerical differences between the two populations are shown in Table 6.6 and results of statistical comparisons are shown in Table 6.7.

Table 6.6: A comparison of the sums of mean values of the acoustic features of adult male pant-hoots from Budongo and Kibale (standard deviations in parentheses); from Notman (1996).

Parameters measured	Population	
	Budongo	Kibale
Build-up		
Rate of delivery (p/s)	1.5852	1.2315
	(0.5685)	(0.4579)
Duration (sec)	0.5535	0.5987
	(0.2566)	(0.1804)
Fundamental frequency (Hz)	566.9	910.1
	(251.6)	(249.2)
Climax		
Duration (sec)	0.7458	0.7474
	(0.2245)	(0.1553)
Minimum frequency (Hz)	621.9	777.0
	(178.2)	(118.2)
Maximum frequency (Hz)	1117.9	1297.4
	(295.3)	(84.6)
Frequency range (Hz)	495.9	521.3
	(170.6)	(92.9)
Ave. fund. frequency (Hz)	958.3	1065.8
	(208.6)	(157.0)

Table 6.7: Results of analysis of variance between the adult male pant-hoots of the two populations, and of five Budongo individuals and three from Kibale (from Notman 1996).

Parameters measured	Population (F value) 2-way ANOVA	Individual	
		Budongo (F) 1-way ANOVA	Kibale (F) 1-way ANOVA
Build-up			
Rate of delivery	6.587*	4.5872**	13.032**
Duration	0.618	1.2148	3.0140
Fundamental frequency	26.475**	2.1946	6.7399**
Climax			
Duration	0.975	0.5481	2.2295
Min. frequency	15.658**	4.3683**	0.2013
Max. frequency	11.054**	5.4260**	1.2739
Frequency range	0.523	2.0130	1.0578
Average fund. frequency	2.431	1.4203	0.6934

$^* = p < 0.05$; $^{**} = p < 0.01$

The results of this study were that significant inter-population differences were found to exist in the rate of delivery and fundamental frequency of the calls, with Budongo chimpanzees having a faster build-up phase and a lower fundamental frequency (lower 'pitch') in both build-up and climax phases than Kibale chimpanzees. Bringing in Wong's comparisons also, she found that Budongo chimpanzees had a slower rate of delivery of the build-up phase than Gombe chimps. That puts the Budongo chimpanzees between Kibale with the slowest build-up or even no build-up phase at all (Clark Arcadi 1996), and Gombe with the fastest build-up.

The fundamental frequency (pitch) of the build-up phase in Budongo was found by Notman to be significantly lower than that at Kibale. Wong did not find a significant difference between Budongo and Gombe. Mitani *et al.* (1999) found that males at Mahale had a higher pitched climax than Gombe chimpanzees. Thus these four-way comparisons are not able to be put into a linear sequence as with the build-up phase (above) but we don't have a proper comparison between Budongo and Mahale. It may be therefore that the Budongo chimpanzees are at the low end for pitch in their pant-hoots, with Gombe and Kibale in the middle, and Mahale at the high end. This remains to be ascertained.

Since the work of Notman and Wong (above), Marshall *et al.* (1999) in a study of the vocalizations of captive chimpanzees in two facilities in the USA found that there were differences between them. Although the individuals in each place had come from a variety of sources, their vocalizations in each shared common features which must have been learned. In addition, a novel pant-hoot variant introduced by one male to one of the groups spread to five other males in the same group suggesting that males modify their calls through learning.

One thing is clear from the acoustic analyses we have looked at: the variations found do not follow a geographical pattern. It is not the case, for example, that as you move from south to north the variation changes in a regular way. Budongo is not between Kibale and Gombe geographically. In this respect the variations in pant-hoots are like some other variations found, e.g. in hand-clasp grooming (see Whiten *et al.* 1999). The variations found so far are probably not ecological in origin. The evidence indicates that they are learned differences, cultural in origin, and transmitted by learning from generation to generation. And, if pant-hoot structure changes according to the calls of the alpha male, it must change quite often, for changes in leadership mostly occur at a frequency of less than 10 years, so that the process of learning is a constant one.

Notman (2003) made a second, more detailed study of pant-hoots in the Sonso community. He explored whether pant-hoots were a single type of call with a great deal of variation, or whether distinct kinds of pant-hoots could be distinguished, and he looked for referential aspects (meanings) of different variations of pant-hoots. He discovered four different kinds of pant-hoots (although there were gradations between them) as follows:

1. Roar pant-hoots. These were most likely to elicit a response and were primarily given during travel on the ground. They may be concerned with locating and joining other parties.

2. Slow roars. These were mainly found in the context of arrival at a particular place, often when the food was abundant; correspondingly they often attracted other individuals to join within 15 min of calling.

3. Wail pant-hoots. These were not likely to attract others to join the group. They were given mainly in the trees when feeding peacefully or resting.

4. Pant roars. These were produced mainly during periods of display or arousal while travelling or arriving. Pant roars contain no climax phase.

The four pant-hoot types are illustrated in Fig. 6.9.

Notman also concluded that listening chimpanzees can infer context from acoustic cues accompanying the calls. Thus, if drumming accompanies pant-hoots then it is clear that the caller is on the ground, travelling; pant-hoots with drumming were more often responded to than other calls indicating that the caller-drummer may be interested in the whereabouts of other chimpanzees and inviting them to respond.

Besides these general characteristics of pant-hoots, Notman (pers. comm.) considers that individual variations may be communicatively as important as referential or contextual meaning. They indicate *who* is calling. Thus listeners can tell, for example, that Duane is calling, that he is on the ground, that he is travelling in their direction, and they can reply to indicate their whereabouts and indicate that food is available on their tree. Notman, like Clark and Wrangham (1994), found no evidence that the type of food or its quality were indicated by variations in pant-hoots given when chimpanzees were in a tree, but he did find that individuals arriving at a food tree were more likely to call if the food on the tree was abundant ($> 50\%$ of its total capacity) than if it was scarce. The individuality of pant-hoots may be of primary importance (just as, in a setting where everyone knows each other, when a person enters the room we look up and note who they are), with the context being secondary. By contrast, in the context of alarm calls, the meaning may be primary, as indicated below.

Human observers (our field assistants in Budongo for example) can identify many of the members of the Sonso community by the sound of their calls; it therefore seems certain that the chimpanzees can do the same, and studies of chimpanzees in captivity by Bauer and Philip (1983) have proved that they can. But we don't know whether one chimpanzee can inform other chimpanzees about the presence or absence or activities of a third chimpanzee, either present or not present. When chimpanzees engage in a chorus of pant-hoots at a food tree, are they also identifying themselves and are the chimpanzees some way away, who carefully listen to these calls, identifying the callers? It seems very likely they are. Evidence for this comes from the response of the Sonso chimpanzees to pant-hoot choruses coming over the forest (a loud chorus can travel up to 2 km) from another community. There is tremendous interest, and the reaction of females may be to flee, while males may flee, be more alert and nervous, or move towards the calls. As Wilson and Wrangham (2003) write:

> The long distance over which pant-hoots are audible enables chimpanzees to advertise their presence and numerical strength to rival communities (Clark 1993, Ghiglieri 1984) and to assess the numerical strength of rivals from a safe distance (Boesch and Boesch-Achermann 2000).

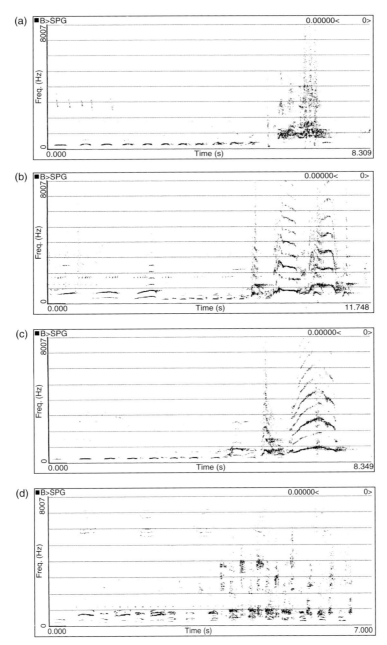

Fig. 6.9: Four pant-hoot variants identified in Budongo chimpanzees (a) Roar pant-hoot; (b) slow roar; (c) wail pant-hoot; (d) pant roar (from Notman 2000).

In the case of drumming, are individuals (males mainly but also a few females) identifying themselves when they drum on the buttresses of trees, using their hands or feet to produce a characteristic series of 'boom-boom-booms'? The drumming patterns of individual chimpanzees are in some cases distinct and can be distinguished by human beings, so evidently individual chimpanzees have their trademark drumming patterns. Besides our adult males we have two adult females at Sonso who occasionally accompany their roar pant-hoots with drumming, each in her own distinctive way: Mukwano and Kewaya. At Gombe, the adult females Fifi, Gremlin and Patti all do likewise, giving roar pant-hoots accompanied by drumming occasionally, but never as often as males (B. Wallauer, pers. comm.).

Barks

A recent study of chimpanzee vocalizations focuses not on pant-hoots but on barks. Crockford and Boesch (2003) used discriminant function analysis to see whether the barks (alarm calls) of chimpanzees at Taï were context specific in the way that, for example, the alarm calls of vervet monkeys at Amboseli have been shown to be (Seyfarth *et al.* 1980). They found some evidence that they were. Their analysis was able to distinguish two acoustically graded bark subtypes in the context of (a) hunting and (b) presence of snakes. Second, they found some evidence for context-specific combinations of barks with other call types or drumming. Chimpanzees produced context-specific calls not only in the alarm context but also when travelling and hunting; the authors speculate that these may convey specific contextual information to listeners, but more study is needed to confirm this by analysing the responses of listeners. This work and the method used may mark a beginning of unravelling the semantics or 'meanings' of chimpanzee calls. However, even if barks contain specific semantic messages that would not imply that the same is true of pant-hoots, which could be functioning in a different way.

Screams

Studies of vocalizations are continuing at Budongo (Slocombe and Zuberbühler, 2005). This study focuses on a hitherto neglected but common type of call given in agonistic contexts: the scream. In this study, 14 different individuals contributed scream bouts to the data, as both victims and aggressors; in addition 19 other individuals contributed scream bouts but only in the role of victims. Most scream bouts consisted of three or more screams, but the first three screams formed the basis of the analysis. Victim screams were given by chimpanzees of all ages, ranks and sex. In contrast, aggressor screams were mainly given by low ranking males, females or juveniles, whereas high ranking males were typically silent when engaging in aggressive acts.

Slocombe and Zuberbühler found regular differences between victim screams and aggressor screams in terms of the acoustic structure of the calls. Victim screams were flatter and more symmetrically curved whereas aggressor screams were characterized by a distinctive down-sweep after mid-call. The difference is shown in Fig. 6.10.

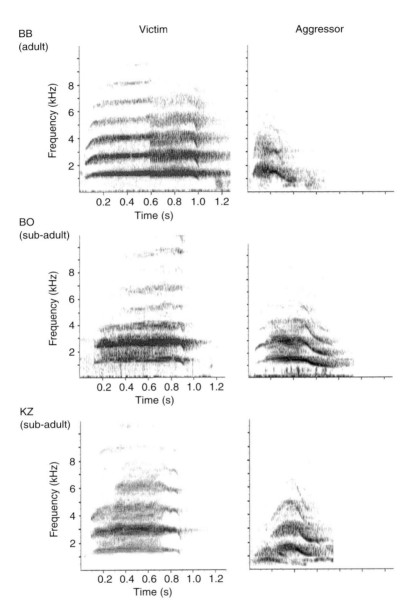

Fig. 6.10: Spectrograms of scream vocalizations given by three randomly chosen males. Illustrated are screams given by these individuals during agonistic interactions where they acted as victims and aggressors (from Slocombe and Zuberbühler, 2005). Copyright © 2005 by the American Psychological Association. Reprinted with permission.

The authors conclude that listening chimpanzees could use the acoustic information in these calls to infer the role of the individuals involved in a conflict. Listeners who move to intervene in agonistic encounters may base their decisions on this acoustic information. It is possible that these agonistic screams are functioning as referential calls, i.e. they contain specific information that can be of use to other individuals listening. However, the authors acknowledge that playback experiments, testing the hypothesis that listeners take advantage of the information available in the calls, need to be conducted before these calls can be labelled as functionally referential.

7. *Infanticide*

Sunday morning, 7 a.m. Colobus called in the night, again at dawn, chimps too. Baboon barks from time to time — a big male just inside the forest. People now getting up and going to the choo. Geresomu taking the weather records. A beautiful background medley of bird songs, some very laid back, unhurried, long notes in sweet cadences. Sunlight breaking on the treetops, moving down towards camp. Mozzies floating around . . . Tomorrow Budongo university meets, Mr Bean in the evening. Tuesday General meeting. Weds Makerere students arrive. Little Arambo, Mary's daughter, wanders around starkers with braids and beads in her hair . . . Now I want to finish my toast and honey with a cup of tea and Lariam tablet . . . Evace is filling the shower tank, the day's round is beginning (26 September 1999).

First observation of infanticide

The first observation of infant killing by chimpanzees was actually made in the Budongo Forest by Suzuki (1971). He described and photographed an adult male, Ropoka, a member of the 'Picnic Site' community, a large community of chimpanzees living in the Busingiro area of Budongo, eating an infant chimpanzee. He described this as 'cannibalism'. He had not seen the killing of the infant, nor had he any knowledge of the provenance of the infant or its relationship to Ropoka. The observations were made on 13 November 1967:

> The author first found Ropoka, the largest male in the group, perched on a branch of a *Cordia millennii* tree with the prey, the baby chimpanzee, in his right hand . . . . Mkubwa, another male, emerged from a nearby tree, groomed Ropoka and stretched out his right hand to grab hold of the baby, which was dangling from Ropoka's hand. Ropoka snatched the baby right back. Mkubwa then elaborately groomed Ropoka for about three minutes. This done, Mkubwa grabbed hold of the body, this time with his left foot, and hung it upside down from the tree and patted it.
>
> Soon Ropoka got the cadaver back again . . . .
>
> The baby chimpanzee was in a horrible state. Its right leg was gone and bled from the groin and the head. It still had a navel cord five to six centimeters long — an altogether newborn baby. What now ensued is recorded in the author's field notes as follows:
> 10:27 Ropoka tears off flesh from the baby's right thigh with his mouth and 'mumbles' it.

10:37 Ropoka, still mumbling, moves away from his perch, holding the baby in his right hand, and goes away from Mkubwa.

10:41 Ropoka tears off some leaves of *Cordia millennii* within his reach and eats them.

10:48 Mkubwa comes again close to Ropoka to touch occasionally the baby on its back with his hand.

Ropoka now seemed to eat the baby's flesh on its right groin. Surprisingly the baby at this stage was still alive and emitted faintly audible cries.

All at once, chimpanzees around Ropoka became restless. Mkono and Centi climbed down the tree as Laini, a medium sized male, climbed up. At the same time, Mzee climbed up the tree and crying loudly, its hair standing on end on the head and shoulders, approached Ropoka.

Now all the chimpanzees in the neighborhood joined in the crying which in turn touched off cries from all other members of the group separated in three sites in the forest....

Two hours after the author first found them, Ropoka, holding the baby chimpanzee in his mouth, came down the tree and went away... Others followed him and all soon disappeared (Suzuki 1971: 31–4).

Frankie and I had watched the chimpanzees of the Busingiro community for the best part of a year in 1962 and we had never seen anything like this, so we were exceedingly puzzled when Suzuki sent us a reprint of his paper in 1971. At that time we did not know that chimpanzees live in mutually exclusive communities, and that the males do not move from one community to the next but in fact defend their territories fiercely. Once that fact became known, the rationale for male chimpanzees to practise infanticide became clearer, if and only if the infant victim had been conceived outside their own community. But first, they must be clear about this fact, so the question becomes: do chimpanzees (and other species practising infanticide) know about the relationship between copulation and pregnancy and birth? Do they know about the length of time between copulation and birth? (The length of gestation in chimpanzees is roughly seven and a half months.) Or do they know that a female immigrant into their community who arrives with an infant is not familiar and has thus conceived with a 'foreign' male? And is this the trigger for their attack on the mother and the killing of her infant?

Infanticide by Sonso males

The first evidence we obtained of an infanticide arose in an unexpected way (Bakuneeta *et al.* 1993). On 23 September 1991, Chris Bakuneeta, the Co-director of the Project, was washing dung through a sieve as part of his Ph.D. study of the feeding ecology of the Sonso chimpanzees. He was looking for seeds that would later be identified. He noticed in one faecal sample a large bolus of thin black hairs (approximately 500 hairs, together with a small piece of bone and cartilage). We considered that this was not the result of grooming, which can lead to the ingestion of a few hairs but not 500. The hairs appeared to be from a chimpanzee raising the question of infant-eating. We did not think they were monkey hairs, also sometimes found in chimpanzee faeces, because of their black colour and length.

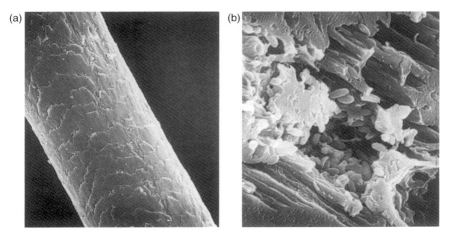

Fig. 7.1: (a) Hair scale patterns and (b) ultrastructure of the medullary cavity of a hair from an infant chimpanzee (from Bakuneeta *et al*. 1993).

The hairs were collected and dried. A sample was sent to H. Inagaki in Japan, and identified by him as belonging to a chimpanzee infant below the age of 3 years. Scale patterns on the hairs were wavy, and the medulla was rich in granules with the appearance of ants' eggs (see Fig. 7.1). These characteristics corresponded with those of chimpanzee hairs previously studied by him (Inagaki and Tsukahara 1993). Our first evidence that infant eating occurred in the Sonso community was thus circumstantial.

The first events directly indicating that an infanticidal event may have happened at Sonso were observed and recorded in the Events Book by Zephyr Kiwede on 27–28 December 1993.

> On 27 December at 9.40 a.m. Kalema (KL), a subadult female was observed carrying an infant estimated to be about a month old on her belly. This infant was not hers. The adult male Tinka (TK) and other adult males chased her, she screamed. The [then] alpha male Magosi (MG) reached her and displayed by shaking branches. There was much screaming and beating of tree buttresses by other males. MG pulled KL to the ground and appeared to be trying to get the infant from her. KL screamed. MG bit her on the face, drawing blood. The infant screamed for the first time. MG was then seen carrying the infant and running, chased by KL. KL climbed a tree and licked the blood on her face. A subadult male Zesta (ZT) remained with KL, the other chimpanzees left.
>
> Next day, 28 December, at 9.00 a.m. the adult male Kikunku (KK) was seen carrying the same infant which was screaming. At 9.30 KK left with the infant. It was never seen again.

Newton-Fisher (1999*c*) reported on two further cases.
Observed by field assistant Geresomu Muhumuza, on 2 February 1995:

> At 7.58 a.m. Geresomu heard some pant grunts and pant hoots and moved towards them. At 8.10 he found Magosi (MG) sitting on the trail. At 9.00 a.m., he saw the alpha male Duane (DN), sitting close

to his alliance partner Vernon (VN), in the trees, feeding on the cadaver of a young chimpanzee. At 9.12 VN began feeding on the carcass which DN had discarded. By 9.15 both males had finished feeding. Other than the head and a few bones, the infant was eaten in its entirety.

The head was recovered intact and preserved in formalin at camp (where it still is) (see Fig. 7.2).

Observed by Newton-Fisher on 25 September 1995:

> It became clear by 13.53 that Black (BK) had discovered a female chimpanzee, and he began to display ... [the alpha male DN moved towards the female]. Both the Sonso females and the unknown female were screaming. [Other males moved towards her.] It became clear that the female was clasping a small infant to her chest. In response to DN's charges the female huddled and turned away, shielding the infant from attempted grabs. She also presented to DN, inviting copulation, despite showing no sign of anogenital swelling. DN ignored the invitation. In the following 5 minutes, DN, VN, and the adult males Chris (CH) and Maani (MA), all showing piloerection, made repeated attempts to snatch the infant, and the female continued to scream, shielding the infant with her body ... At 14.06 DN seized the infant and leapt away through the branches ... Before beginning to feed at 14.08, DN killed the infant with a bite to the head. [He continued to eat the infant until 14.43 when he passed the cadaver to Kewaya (KY), a young adult female in estrus. She fed on it for a further 3 minutes before dropping it.] One and a half legs, one arm, and some of the intestines had been eaten. The infant was a very young male. Body length was approximately 10 inches (25 cm) (Newton-Fisher 1999*c*).

Fig. 7.2:   Head of infant killed on 2 February 1995 soon after it was recovered (photo: V.R.).

Newton-Fisher adds that both the above cases occurred in the same general area near the western border of the Sonso community. In the case he observed, the female whose infant was eaten was not a current resident member of the Sonso community and most likely was a peripheral female from a neighbouring community to the west, the Nature Reserve or the Nyakafunjo community. In this case there was no evidence of harm being done to the mother. It may be that the Sonso males knew her, perhaps she was originally from Sonso. In that case she may have emigrated and sired the infant in the neighbouring community. Why she had returned to Sonso with the infant is not known; the same question can be asked of all such cases where females move to a new community with a new infant. The fact the infant was a male is also of interest as the majority of cases of infanticide occur to male infants (future competitors of the resident males but also potential allies for hunting and territorial defence).

A fourth case that we considered was an infanticide was observed on 1 August 1997. The following is taken from notes in the BFP Events Book made by a doctoral student, Clea Assersohn:

10.45 High levels of excitement and vocalizing within a group of 21 individuals. At 11.54 an adult female, Kwera (KW), with her infant Kwezi (KZ) on her back, was seen carrying an infant carcass. Males were excitedly displaying and moving. At 12.00 KW was seen without the carcass. At 12.06 the adult male Bwoya (BY) had the carcass. He sat with the carcass between his legs, not attending to it, then at 12.12 made a day nest and took the carcass into it. At 13.02 he left the nest with the carcass still whole. At 13.11 and 13.16 he groomed the carcass, and at 13.26 held it up and inspected it, then at 13.32 made another nest and took the carcass into it. At 14.50 the carcass fell to the ground and BY left the nest. (Sex of this infant was not determined.)

The following notes were made on the carcass when it was inspected on the ground: it was 27 cm long, a small baby, perhaps a week or two weeks old. It had deep lacerations through both legs to the bone. There were tooth marks on its back. Cause of death could not be ascertained but the injuries were too severe to have been sustained simply through being carried in the mouth. None of the Sonso resident females had given birth recently and it was therefore concluded that this infant was that of a non-community female.

For three other cases we have only scanty evidence. On 12 August 1997 the young adult female Sara (SR) was seen with an (unsexed) infant not thought to be hers. The infant was screaming. Other chimpanzees were very excited, trying to grab the infant from her. Tinka (TK), a low ranking male, bit SR on the back apparently harassing her to get her to release the infant. SR and other chimpanzees were screaming a lot. We do not know the circumstances surrounding the death of this infant, nor whose it was.

On 13 December 1999 the adult male Jambo (JM) was seen carrying a recently born male infant around until it starved to death, after which he continued to carry it for a further two days. The incident is described by Notman and Munn (2003). Several Sonso females were around near JM and the infant but none appeared distressed or anxious. The infant was producing high-pitched squeals and whimpering; JM responded by 'cuddling' it to his chest, to which it was clinging. He carried it on his chest throughout the day, and when he rested it attempted to suckle. At 18.46 he nested with the infant. On 14 December JM continued to carry the infant but now it was weaker and he had to support it. No other Sonso individuals showed any interest in JM or the infant. It was still alive in the evening and JM nested with it again. On 15 December the infant was dead, but JM was still carrying it. By the following day he was no longer carrying it. There was no evidence he ate it. It is not known whether the infant was born in the Sonso community or outside. Nor are JM's motives known; the authors suggest he may have been initially motivated by interest in the infant as a 'possession' and later lost interest as the infant weakened and died.

It was suspected that this infant might belong to Mukwano (MK), a young adult female of the Sonso community who had not been seen for a while; when she reappeared she had a recently damaged wrist which might have happened during the snatching.[62] If this was indeed MK's offspring then it would have been a case of infanticide of a resident female, as has been seen at other chimpanzee study locations (see below). MK is the Sonso female who is best known for her disappearances for longish periods after which she reappears; she may move between communities and this may also be a background factor to the events described.

In February 2000 the adult female Zimba (ZM) who had an infant male Zig (ZG), was seen carrying a dead infant (not her own and not sexed). She dropped it and then Musa (MS) (son of Nambi (NB)) carried the carcass for a while and groomed it. How the infant died is not known, nor whether it came from another community or was a Sonso infant.

Infanticide within the community

Within-group infanticide and infant-eating also occur in chimpanzees. At Sonso we have the case of the death of the infant suspected to belong to the resident female, Mukwano, described above.

In March 2004, while this book was being written, Katie Slocombe, a research student, and field assistant Raymond Ogen, heard prolonged screaming and moved towards it, to find Zimba holding a dead infant male aged about one week old. Nambi took the infant from Zimba and held on to it for over an hour, inspecting it and biting its feet gently. Wilma and Mukwano were permitted to come close to the infant but other individuals[63] were not. At this time Katie and Raymond found Flora lying 10 m away in dense undergrowth with a fresh bleeding gash on her left arm. Flora tried to approach the infant but Nambi would not let her come closer than 2 m. After an hour Nambi and the other females present dropped the infant on the ground and moved away. In this case it seems that the mother may have been Flora, in which case this was apparently an

---

[62] Though we have no evidence for this; the injury might have been caused by a snare but it was not a typical snare injury.

[63] These included Gashom, Bob, Zana, Kalema and Banura.

intra-community killing, perhaps by Zimba. Unfortunately it was not possible to preserve a sample from this infant for genetic analysis.

Two better documented cases were reported on by Hamai *et al*. (1992). These occurred in the M Group of the Mahale Mountains chimpanzees. In both cases the infants were male, 5–6 months of age, and in good health when killed. Their mothers had immigrated into M Group 4–5 years previously as nulliparous females, i.e. they were not new immigrants. In both these cases the infants were killed by the alpha male and there was considerable competition between the adult males of the community for a share of the meat.

These two were among seven cases at Mahale reviewed by Hamai *et al*. In all cases the infants were male. Most were completely eaten, with competition for the meat by both adult males and adult females. The head was always eaten, in contrast to the situation in Budongo where the head was not eaten in the two cases where the carcass was recovered.

The authors noted that only infants of immigrant females were killed. Four of the seven mothers were young and primiparous. After considering a number of reasons why they should have been killed, the authors concluded that

> The common feature through the seven cases of within-group infanticide is that mothers of victimized infants mated more with older adolescent and other immature males rather than with fully adult males before they gave birth to victims.... We suggest, therefore, that one function of infanticide might be to...coerce her [the mother] into more restrictive mating relationships with adult males and high-ranking males.

In other words, females who mate with the lower ranking or less mature males in the community run the risk that their infants will be killed by the more mature and higher ranking males. To the extent this is true, it shows that the males in the community, even though they live their lives together without emigrating and consequently know each other well, are highly competitive when it comes to mating and reproduction. We have already noted this fact when we considered mating strategies (Chapter 6). It also makes genetic sense for, surprisingly, males are not as highly related to one another within a community as might be expected, the genetic variation coming about as a result of the new genes brought in by immigrants (see Appendix C on the genetics of the Sonso community). And finally this competition between intra-group males helps to explain an anomaly we examine in the next chapter in which I describe the killing of a low ranking, young, virile male in the Sonso community by his older, senior peers.

The role of females in cases of infanticide

Information on an active role for females in infant killing is sparse compared with that on males. At Budongo, we have two cases of females found carrying infants that may be infanticide victims: the cases of Kalema (December 1993) and Zimba (February 2000) described above. We do not know at what stage they became involved, whether before, during or after the death of the infant, and we do not know what role they played in the infant deaths.

A more fully documented case of infant killing in which females were involved, already referred to above, was as follows:

> Observers were with NB, MS, KU, ML and WL at a *Cynometra* tree off grid to the SE at 13.50. At 13.53 they started to climb down and a few minutes later a large screaming bout started out of sight to the south. We followed the females towards the screaming. After 7 mins of continuous scream-ing, we reached the screaming party of chimps. ZF was seen displaying in a tree and many females and juveniles were screaming up in the trees. A group of chimps was seen moving into a vine tan-gle, observers followed and found blood on the ground. ZM [adult female] was first seen carrying the dead infant but shortly after NB [adult female] was seen charging at her. A brief fight ensued out of sight and NB was then seen with the infant. NR [juvenile female, daughter of NB] then took the infant and KT [adolescent male, son of KU] and NR dragged and pulled it off each other as if play fighting. NB then took the infant and sat with it. NB then threatened KT who tried to approach; KT ran screaming to FL [adult female]. FL was lying in thick undergrowth with a fresh bleeding gash 2–3 inches long on her left upper arm. The infant was a male of approximately one week old. It seems likely FL was the mother.

Various other individuals then came close to NB and the infant to inspect it. After a while FL approached sat 1.5 m from NB and the infant. FL attempted to come closer but NB threatened her (bark and lunge) and FL screamed and did not try again. Later it was noticed that BN (adult female) had a cut on her leg; she may have been involved in the fracas but is not thought to have been the mother. The infant was recovered and preserved in formalin in the camp museum. (Above notes taken from the BFP Events Book. Observers were Katie Slocombe and Raimon Ogen, date 12 March 2004.)

The most well-known cases of infant killing by females are the cases of Passion and her late-adolescent daughter Pom of the Kasakela community at Gombe Stream National Park in Tanzania, described by Goodall (1968, 1977). At first considered pathological, these have been reinterpreted as cases of female–female competition, but they are truly extraordinary; in most cases it is thought that females would not run the risks associated with such attacks in view of the rather marginal food benefits they accrue from eliminating other female competitors. Evidently these Gombe females were exceptional. In October and November 1976 they fiercely attacked two mothers, Gilka and Melissa, also from the Kasakela community, killing and eating their three-week old infants, one a male, the other a female. As Goodall writes: 'Passion and Pom attacked the mothers of their victims only in order to acquire the infants as meat. Once they had possession of the babies, no further aggression was directed towards the mothers. Indeed, when Melissa approached the killers, who were then eating her infant, Passion reached out and embraced her' (1986: 284).

Another case is that of a prolonged attack on a mother and her two-year-old daughter resulting in the death of the infant in the Kanyawara community living in the Kibale

National Park, Uganda (Clark Arcadi and Wrangham 1999). This attack took place on 14 December 1996 and was seen by the first author and field assistants, who were within 7 m of the events. The mother, MU, was a border-area resident from the northern part of the range, who, like Mukwano at Sonso, was not totally committed to living in the community but disappeared from time to time, returning later. Whether she mated with males from other communities is not known.

On the day in question, MU and her infant MB were attacked by an adult male SY and an adult female LP carrying a six-week-old infant, both more central members of the Kanyawara community. The female initially attempted to intevene on the victim's behalf, chasing SY away and interposing herself as a kind of 'peacemaker' between SY and MU. However, later LP joined in the attack after receiving aggression from SY. Subsequently, MU was found without her infant which is presumed to have died as a result of the attack.

The authors conclude: 'Our interpretation of this is that LP and her young infant risked serious attack from SY by not forming a coalition with him.' In this case, therefore, it appears to have been coercion by the male that led LP to attack MU and the infant MB. In the Gombe case of Passion and Pom the attack by the females was not coerced. There appear to be many circumstances in which females attack infants; and for males too there are a number of explanations for infanticide, including the killing of infant males sired from outside the community, and the killing of infant males sired by junior males inside the community. At least in the case of males there does seem to be a regularity: the killing of males, potential rivals in the future. In the case of females (for whom we have far fewer cases) the motivations seem to include fear of retribution, and the desire for meat.

8. Intra-community killing — the case of Zesta

Into forest at 7.00 (up at 5.30). Meet Zimba (estr 2) & Mukwano on *F. sur*, also Gonza. This is the tree they were not feeding on previously. I'm with Zephyr who's on his 3rd follows day. He's gone off to look for 'his' chimp i.e. his focal so I'm sitting on a log watching these 3. There's a redtail monkey group also here. Tinka arrives. I move to *F. mucuso* tree in fruit on the logging track. Duane + Jambo + Kigere in estr. 4. Copulation with cop call by Kigere. Nambi, Musa, Muga, Kadogo. We are on the road, people are passing. Chimps are noisy today, lots of calling between 3 gps of which we've seen 2 (15 March 1997, 7.32 a.m.).

Inter-community killing between adults

During the 'war' at Gombe, the males from the larger Kasakela community brutally attacked members of the smaller Kahama community, killing all the males and a female (Goodall 1986). At Mahale, all the males of the smaller K community of chimpanzees were apparently killed by males from the larger M community (Nishida *et al.* 1985). Two cases of lethal inter-community killing have been described for the Kanyawara community in Kibale Forest (Muller 2002).

Boesch and Boesch-Achermann (2000) have described a number of different strategies for such attacks. Manson and Wrangham (1991) have argued that one of the main factors underlying them is the numerical superiority of the attackers, proposing the 'imbalance of power' hypothesis. One-on-one attacks across borders are rare, in most cases a number of males have been on patrol, have encountered a lone individual from a neighbouring community, and have launched an attack. The attacks consist of prolonged biting while the victim is held down, and they continue after the victim is incapacitated and unable to run away. The victim is either killed during the onslaught or dies later of his or her injuries.

Zesta

We have had one case of *intra-community* killing by adult males of one of their fellows in our Sonso community. This is not unique but it is very rare. Nishida (1996) described a case where the alpha male, Ntologi, of the M community at Mahale, was killed by the other males of his community in a successful bid to remove him from power. Goodall (1992) and Nishida *et al.* (1995) have described severe within-group attacks by males on other males, the former case being on an alpha male, the latter, as in the present case, on a young adult male.

This case, which was not lethal, nevertheless has some features in common with the case of Zesta. The incident, involving the males of Mahale's M group, was seen in 1991. The victim was Jilba, a young adult male, who was attacked by five adult males, including the alpha and beta males Kalunde and Shike, and by two adult females. Jilba was attacked after he displayed near a group of males and females eating a recently captured colobus monkey. The attack was fierce, including much biting, but lasted only one minute before Jilba escaped. In explanation, Nishida *et al.* point out that Jilba had never pant-grunted to the alpha male Kalunde and had a tense relationship with him. They write:

> The proximate cause of the aggression may have been as follows. Frustration regarding Jilba had accumulated in many adult chimpanzees. On the day of the attack reported here, tensions were amplified by the excitement of meat-eating. This excitement may have found an outlet in aggression towards Jilba, which was triggered by Kalunde's violent chase. It is also possible that lower-ranked males such as Bembe took advantage of the higher-ranked males' aggression in order to raise their own dominance status (Nishida *et al.* 1995: 210).

As we shall see, this last point is relevant in the interpretation of the attack on Zesta.

A lethal intra-community attack was filmed in the Ngogo community at Kibale, in which a large number of older males of this community severely injured a younger adult male with much noise and screaming. This attack seems to have features in common with the attack on Zesta at Budongo.

The start of the attack on Zesta was not seen, although the accompanying loud, high-pitched screaming was heard by our students and staff at Sonso. They arrived at the scene when Zesta was already badly injured. What happened is described here in as much detail as is available, together with the best interpretation possible using our knowledge of the individuals concerned and the social context both before and at the time of the event.

Let me say at the outset that it was a complete and utter surprise to us all when it happened. Up to that time we had seen some cases of aggression between Sonso adult males, based on competition, mainly over status, sometimes over oestrous females, but injuries had been light or non-existent. We had seen the takeover of the community by Duane, usurping the former alpha male Magosi with no more physical damage than a bite to the finger. Later, when Duane was established, he occasionally safeguarded his position with threats or even mild attacks. But he was tolerant: Magosi, whom he had usurped for alpha status, was once again admitted to the caucus of dominant males and was not ostracized. It seemed as if Duane and the rest of our adult males were well able to control their aggressive tendencies.

Fig. 8.1: Zesta (photo: J. Lindsell).

Background to Zesta

Zesta (ZT) (Fig. 8.1) was first named on 17 February 1993 at which time he was a subadult male. He is recorded as moving from subadult to adult status in December 1996. Thus by November 1998, when he was attacked, he had been an adult, with fully descended testes, for a couple of years.[64] Our definition of an adult male is as follows: 'Testicular development complete. Face fully black. Now dominates all females and challenges other males.' This last element of the definition is noteworthy: young adult males are starting on the road to dominant status and thus do from time to time challenge adult males, which can lead to them being threatened or attacked by the more dominant ones.

Newton-Fisher (2002*a*) studied the grooming relationships of the males of the Sonso community from August 1994 to December 1995 when ZT was still a subadult. His analysis of grooming partnerships (see Chapter 6, Fig. 6.2) shows that ZT was an outsider among the males, but this is normal for a subadult male and was probably related to ZT's age.

Three years later, in 1998, ZT had become more sociable. Checking his grooming and copulation records for the month preceding the attack, the following picture emerges

---

[64] Later inspection of Zesta's skeleton showed that his third molars were fully erupted, and his humeral heads were fully fused, indicating fully adult status (Newton-Fisher, pers. comm.).

Table 8.1: Grooming and copulation partners of Zesta, 1 October–4 November 1998.

Activity	Partners involved (age–sex group)	No. of times groomed by ZT	Mutual grooming	No. of times grooming ZT
Grooming	MG (AM)	1	1	0
	BB (JM)	7	2	7
	BY (AM)	2	2	0
	JN (SAF)	5	2	2
	MA (AM)	5	1	5
	NJ (AM)	1	0	0
	MK (SAF)	1	0	0
	SR (SAF)	0	0	1
	BO (JM)	0	0	1
	TK (AM)	1	1	2
	AY (SAM)	0	0	1
Copulation	JN (SAF)	7	—	—

(see Table 8.1). The data in Table 8.1 tell us the following about Zesta in the month before he was attacked:

1. He was sociable.

2. He both groomed and was groomed by several adult males.

3. These did not include the three top ranking males, DN, VN and BK.

4. His favourite copulation partner was JN.

5. His favourite grooming partner was BB, a juvenile male.

We can thus see a not unusual picture for a young adult male, whose social inter-actions were with both some of the lower and higher status adult males in the community and also with youngsters. What is perhaps most obvious from this one month is that he was sexually interested in JN, who was at that time still considered a subadult female because although she was sexually active, she was nulliparous (her first infant was not born until a year later, in October 1998).

The killing of Zesta

Here are relevant facts from Fawcett and Muhumuza (2000):

At 0640 hr on 4 November 1998, 'wraa' calls (fear: Goodall 1986) were heard southeast of camp. At 0720 in the core area of the Sonso range, the chimpanzees were located. Wraa calls were still being emitted. Visibility was poor owing to low lighting conditions and dense undergrowth. There was a flattened area of vegetation approximately 5–7 m^2 and an adult male chimpanzee, lying in a prone position, tentatively identified as Zesta. At approximately 0730 Magosi, an old adult male (ex-alpha) charged at and then circled Zesta. He was followed shortly afterwards by Black, another prime adult male (gamma rank, Newton-Fisher 1997). Black, while displaying, ran at Zesta, leapt on his body, then violently and repeatedly shook Zesta. We retreated to the edge of the group of chimpanzees, approximately 10 m from Zesta. At 0840 the chimpanzees were quiet and beginning to disperse; we were able to approach Zesta and obtain a positive identification. His breathing was laboured and shallow and he died shortly afterwards. Most of the other community members had now left. Ruda, an adult female, with her infant and juvenile son, and Sara, a subadult female, remained. They approached Zesta. Sara 'whimpered' (Goodall 1986) as she gently shook and pulled Zesta. By 1030 Zesta was left alone. In total we observed six adult males, one adult female and infant, two subadult males, four subadult females and two juvenile males at the scene of the incident (not including Zesta). Of these, Janie, a late subadult female, was observed to be present in maximal tumescence. By the time of nesting all community males had been seen. Duane, the alpha male, Andy, a subadult male, and Black, were all observed to have sustained superficial injuries. A fecal sample collected from Duane during the afternoon of the attack was found to contain pieces of flesh, thus further implicating his involvement in the attack (fecal samples were not collected from other individuals). The flesh (approximately 2 cm^2) was identified as that of a chimpanzee by the hairs attached. We suggest, from the nature of Zesta's wounds, that the swallowing of flesh was probably accidental, rather than intended cannibalism. Two sections of muscle (4 cm^2 and 2 cm^2) were also found in the flattened area where the fight had occurred.

The next day, Ruda and her family returned to visit Zesta's body.

The severity of the attack on Zesta (see Fig. 8.2) is akin to the severity of inter-community attacks on single males, as described for example by Muller (2002: 118–9).

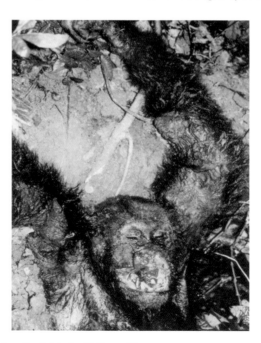

Fig. 8.2:   Zesta after the attack (photo: K. Fawcett).

Zesta's injuries

One of our visitors at camp at the time, Diccon Westworth, who went to the site of the attack and observed the death of ZT, was a qualified veterinarian; he was able to make a detailed examination of the cadaver *in situ*. As such a report is very rare I include it in full here:

POSTMORTEM
SUPERFICIAL EXAMINATION AND PALPATION OF ZESTA
(ADULT MALE IN THE SONSO COMMUNITY OF CHIMPANZEES)

4th November, 1998

Subject: Zesta (adult male)
Time of death: 09.52 approx.
Time of postmortem: 14.30
Time of initial attack: 06.40 approx.
Duration of attack: 50 minutes approx.
Attack known to have been carried out by Sonso community members.
The postmortem was carried out in the field at the site of death, supplemented with photographic documentation.

Prior to the postmortem examination it was ensured that no chimpanzees were within a 200 m radius of the body. It was decided that a minimally invasive examination would take place to ensure the cadaver was left undisturbed. It was suspected that the cadaver would be important over the following week for the adjustment of community members.

On initial inspection the cadaver was lying supine in rigor mortis approximately 15 m from the area of initial attack. Within this area two sections of muscle (both approximately 3 cm in diameter) were found and blood was present on under-story vegetation.

Both the thorax and abdomen were tapped with a 19 gauge needle and 20 cc syringe for the presence of blood or air. No blood was aspirated from the abdomen on either side of the thorax. Air was easily aspirated from the right thorax. Approximately 5 ml of clear frothy acellular fluid was aspirated from the left thorax between the third and fourth ribs.

Head and Neck

Superficial abrasions and lacerations to both ears, face and cranium were present. A large full thickness 8 cm tear to the upper lid on the right side extended from the level of the upper canine to the entry of the right nares.

A superficial triangular defect was present over the bridge of the nose consisting of a 2 cm tear from the bridge to the lateral canthus of the left eye and across the mid-line. Multiple tears to the upper pinnae of the left ear were present, resulting in a 2 cm defect. The right ear was macerated with multiple lacerations all over the pinnae. There was a superficial 3 cm laceration over the occipital area of the right side of the cranium.

Thorax

There was an 8 cm diameter defect over the right upper thorax. The superficial and deep pectorals were transacted with a 4 cm defect and there was a small puncture hole into the thorax through the inter-costal muscles between costa 3 and 4. In addition there was a superficial 3 cm laceration over the xiphoid.

Right Upper and Lower Arm

There was a large laceration extending to the bone in the region of the right axilla exposing a 10 cm defect. The muscles in this area were severely macerated. The latissimus dorsi was lacerated by a 5 cm deep gash. Another 15 cm long, 10 cm wide defect over the caudo-medial aspect of the upper right arm exposing the caudal humerus over its proximal two-thirds. The triceps muscles were macerated and the brachial artery was exposed over the entire length of the upper arm. Another defect of 5 cm was present over the cranial surface of the upper arm, exposing a mid-belly thickness tear into the biceps and 1 cm deep tear to brachialis.

At the proximal third on the anterior surface of the anti-brachium there was a deep laceration resulting in a 5 cm diameter defect. The superficial flexor muscles in this area were macerated with multiple punctures down to the radius.

Right Hand

The medial and palmar surface of the carpo-metacarpal junction of the thumb had multiple deep lacerations transecting the opponens muscles exposing the proximal two-thirds of the first metacarpal. In addition there were deep punctures and lacerations over the entire surface of the hand.

Left Upper and Lower Arm

In the region of the left shoulder there was a 7 cm laceration over the dorsal surface of the deltoids extending down to the humerus. The deltoid muscles had been transacted over two-thirds of their width. There was a 8 cm laceration through the proximal caudal upper arm. The triceps had been partially severed to a 3 cm depth. There were three other large deep puncture wounds over the caudal medial surface of the mid-third upper arm.

There was a 11 cm long 4 cm wide defect over the medial aspect of the left elbow joint. The flexor and extensor muscles of the medial anti-brachium had been severely macerated exposing the elbow joint and the proximal ulnar. Adjacent to this over the medial surface of the anti-brachium was a defect 15 cm long and 10 cm wide extending to the distal third of the lower arm. The superficial flexor muscles in this region had multiple lacerations and extended laterally to include radio-brachialis, which was completely macerated exposing the deep flexor muscles of the anti-brachium.

Left Hand

There were multiple lacerations on the palmar surface of M1. The opponens muscles were transacted to the bone exposing M1 over its proximal half. There were also many other deep and severe punctures and lacerations to the palmar and dorsal surfaces of all five digits.

Right Upper and Lower Leg

There was a 4 cm laceration to the right groin over the sartorius with multiple deep punctures. Over the mid-cranial surface of the thigh there were two long lacerations resulting in a defect approximately 6 cm in diameter. The rectus femoris and sartorius was almost completely transacted. Over the mid-cranial surface of the tibia there was a 3 cm superficial laceration.

Right Foot

There were superficial lacerations over the palmar surface. There was also a deep puncture wound at the P1/P2 junction of the fifth digit.

Left Upper and Lower Leg

There were two superficial lacerations over the cranial surface of the mid thigh over sartorius. There was a 4 cm laceration over the cranial surface of the lower leg also. On the caudal surface of the lower leg extending over the proximal two-thirds of the lower leg there was a deep defect 18 cm long and 10 cm wide. This exposed two-thirds of the caudal tibia. All the superficial and deep flexor muscles in this are were severely macerated.

Left Foot

There were multiple lacerations to the palmar surface of the heel and first digit.

As can be judged from the above, Zesta was most savagely attacked: his face was torn apart and most of his body was bitten, ripped and lacerated. This attack was certainly the work of several chimpanzees acting together.

Zesta's cadaver was retrieved from the forest and the skeleton was beautifully cleaned and prepared by our students, notably Donna Sheppard. Later, the skeleton was mounted by Robin May and Wilma van Riel, two volunteers to our project, and is now located in the BFP's museum (Fig. 8.3). One of the phalanges of the left hand had been scraped by

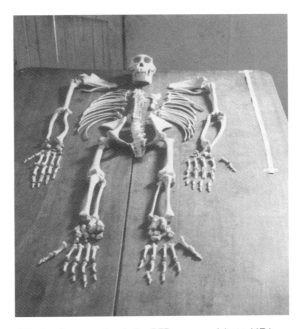

Fig. 8.3:   Skeleton of Zesta after mounting in the BFP museum (photo: V.R.).

a tooth during the attack, the right humerus shows five tooth scrapes and there are bite scrapes on the bones of the right hand. These were serious bites indeed.

Interpretation

How can these highly unusual events be explained? We can follow the points made by Fawcett and Muhumuza (2000) in their interpretation.

First, it was clear from the extent of the injuries to Zesta (ZT) that several chimpanzees were involved in the attack, and they were members of the Sonso community. This meshed with the fact that three of the Sonso males had new wounds including the alpha male Duane (DN) and the gamma male Black (BK). (The beta male (VN) was not present, as far as we know.) The third male to show wounds was Andy (AY) and it is not clear how this came about. He was smaller and younger than ZT and may perhaps have tried to defend him, or become accidentally involved in the initial fighting. The fact that this attack occurred in the core of the Sonso community range also indicates that it was a 'domestic' fight and did not involve extra-community chimpanzees.

ZT was the youngest and lowest ranking of the adult males, he had never attempted to usurp power, and so this was clearly not an attack by senior males on a rival, not a political attack or a result of a power struggle.

Food was plentiful at the time, and it was not considered that feeding competition had any part to play in the attack.

Mating competition provides a possible avenue of explanation. At the time of the attack there was only one female in oestrus in the community, Janie (JN), and she was in maximal tumescence. At this time, there were 12 mature males in the community, all competing for this one female. Other than JN, none of the 13 adult females was in oestrus, and all had infants. There was thus intense competition for access to JN. From previous records of copulation and grooming events, ZT was the most frequent copulation and grooming partner of JN. As Fawcett and Muhumuza write:

> It is hypothesized that a preferential relationship between Zesta and Janie, and the presence of Janie in maximal tumescence, may have both instigated and escalated the violence of the attack. We suggest that Zesta, a young adult male, may not have observed his position within the male hierarchy, and thus became the target of aggression.

They point out that at Mahale, the ostracism and eventual attack on a young adult male (non-lethal, the male escaped) was also interpreted to be the result of inappropriate social behaviour (Nishida *et al.* 1995), as described earlier in this chapter. Comparing the two cases, we can note the underlying excitement of meat-eating in the Mahale case, suggesting that what underlay the case of ZT may have been sexual excitement centred on JN. Why the case of ZT escalated whereas Jilba was able to escape may be related to the presence of BK in ZT's case. BK was a very aggressive male. He was gamma male in the hierarchy but was at that time trying hard to rise to beta place. He had for some time been unsuccessfully competing with the beta male Vernon (VN), and had attempted to overcome VN by aggressive means but been defeated by an alliance between DN and VN (see Chapter 6). In the month before the attack on ZT, however, VN had a leg wound, possibly inflicted by BK. Thus BK was an aspirant beta male who may have been trying to enhance his status as beta male by joining DN in the attack on ZT. Coupled with this, we know that BK was sexually interested in JN, for he had been observed copulating with her three times the previous day, 3 November, so that an element of sexual competition may be important in his case, as well as that of DN.

VN was apparently absent or took no part in the attack on ZT, either because of his leg wound, or because of the presence of BK. BK was, however, very much in evidence and may have been involved in the initial attack on ZT during which he was injured; he alone was still attacking the seriously injured ZT after the students arrived at 0730. Whether DN or BK or both made the initial attack on ZT is not known, but each would have had his own motives: BK's sexual and to consolidate his political ambitions, DN's sexual and to emphasize his alpha status. Whether, in the absence of BK, ZT would have *killed* for obstructing access to JN is doubtful. DN was (and at the time of writing still is) a very composed alpha male who normally controls his aggressiveness. BK on the other hand has always been prone to aggressive outbursts and it is in keeping with his character to have initiated or escalated the attack on ZT with the end result that this young adult male, who had hitherto been a well integrated member of the community, lost his life.

A tail-ender to the story: 11 months after ZT was killed, JN gave birth to a daughter, Janet; genetic analysis (see Appendix C) shows that the father was DN. Thus ZT, for all his efforts, did not impregnate JN; if he had she would have given birth seven and a half months later. Nor did BK sire JN's subsequent offspring. It appears that some three or four oestrous cycles after ZT's death, JN conceived by the alpha male DN.

9. *The problem of snares*

Up at 5 so have time to write. Hyraxes in full swing, stars out. Some rain y'day but still unusually dry for April. Everything around camp has memories for me — even things not there like the lack of smell from the water drain by the tank of House 1 — it smelt bad for years, now it's o.k. The fridge is working! Thanks to Jeremy. (Hasn't been working for a year or so). Cooker fine. There's a second car port here for vehicle no. 2 — a nice Toyota pickup, white like no. 1 but petrol not diesel. Alas its battery is flat so won't start. Yesterday we met with the TCs to discuss the snaring problems . . . we decided there has to be a moratorium on the snare removal programme (8 April 2000).

At any one time around one-quarter of our Sonso chimpanzees are suffering from the effects of having been caught in snares. They have missing hands, feet, or are crippled and have to struggle to climb trees and feed themselves. We do not know if they are in pain, and if they are, how much. We do know, however, that from the moment of that encounter with a snare, their lives are depleted and their future to a greater or lesser extent blighted.

At present, following two very comprehensive surveys, we know that there are some 584 chimpanzees (excluding dependant infants aged 4 years or less) in the Budongo Forest (Plumptre *et al.* 2003: 24). We also know that of the 49 non-infant chimpanzees in the Sonso community (again excluding dependant infants aged 4 years or less), the following snare injuries have occurred:

1. Tinka — both hands deformed

2. Zefa — right hand damaged

3. Bwoba — left hand damaged — recovered

4. Gashom — right hand damaged — recovered

5. Kalema — right hand severely damaged

6. Kigere — right foot lost

7. Zana — both hands severely damaged

8. Mukwano — right foot damaged

9. Kewaya — right hand damaged

10. Shida — left hand lost

11. Kana — snare on right ankle

12. Nora — snare on right wrist

13. Muga — right hand lost

14. Wilma — right hand lost

15. Beti — right hand snared — recovered

16. Zig — snare on right wrist

If 16/49 (32.6%) of non-infant Sonso chimpanzees are suffering from snare injuries, then, in so far as the Sonso range is representative of the whole forest with its population of 584 chimpanzees, there may be as many as 190 injured chimpanzees in the whole forest. There may be more, because in the Sonso range (see below) we are removing snares from the forest every day. On the other hand, there may be less because the Sonso range is closer to human habitation than some of the ranges of other communities in the forest. If these factors balance out, then we may be looking at approximately the right figure. It is a sobering thought that in one forest of just 136 square miles there could be 190 injured chimpanzees, 1.4 for each square mile of forest. And we have to remember that not only is Budongo Forest a Reserve, i.e. a Protected Area, but we are talking about the range of the Sonso chimpanzees which receive all the protection the BFP has to offer.

The background to snaring

The people of Western Uganda, where the Budongo Forest and other forests along the escarpment of the Rift Valley are located, are for the most part subsistence farmers with a very low income. We shall return to look at the human population and its problems in the next chapter, but for the moment we need to recall that for these people meat is a luxury. They are subsistence agriculturalists, growing a variety of crops (cassava is the poor man's staple in the area) and most days they do not have any meat. If families have any domestic animals they have a goat or a few chickens, but very many do not have any at all. If they want to eat meat, they have to buy it in the market. Beef, goat, chicken and occasionally pork are all locally available but at a price: beef costs 2000 Uganda shillings a kilo (about 66 UK pence) but many families do not have more than 100 or 200 shillings to spend in the course of a week.

Given these facts, it is not hard to imagine the allure of walking into the forest and placing snares on the trails that criss-cross the forest floor and returning a day or two later to see what has been caught. Snares used to be made of tough thin lianas and trailing herbs,[65] but for many years now the majority (85%) are made of wire: wire from inside car tyres, or from bicycle brake cables, or from telephone lines, or from electric cable. Sources of wire are markets, other people, and old, non-functional sawmills. Nylon cord from locally available nylon sacks accounts for the remaining 15% and can do as much damage as wire but is not liked as much because it decays in the moist conditions of the forest (Tumusiime 2002). The BFP has a large collection of confiscated snares. The hunters who put them in the forest are not specialized or highly skilled; most up-country Ugandans can do it. In a study of snaring in Budongo Forest, Tumusiime (2002) found that 23% of respondents to a questionnaire admitted to having ever made snares; 97% of snares were set either in the forest or in their cultivated plots; the remaining 3% were sold to hunters.

Snare types

There are two main types of wire snare:[66] one is attached to a sapling at the edge of the animal trail and hangs in a noose shape just above the ground placed so that a duiker will put its head into the snare as it walks along, and the snare will tighten around its neck as it walks forward. The second, more effective type is sprung: the noose is as above but the end of the snare is attached to a sapling that has been bent over and held in place with a toggle, so that when the wire is pulled by the animal the sapling springs back into place tightening the noose around the animal's neck or, often, its foot. Both types are found in Budongo. Chimpanzees apparently have difficulty spotting these snares or don't know what to look out for, and they put a hand or foot into the noose and then panic and pull tighter and tighter; if, eventually, they succeed in getting away the noose is embedded tightly around fingers, wrist or ankle, with consequences to be described below. Six Budongo shaares are shown in Fig. 9.1.

Traps

Besides snares there are traps. Traps are normally set for a different reason than snares, and in a different location. Whereas snares are set inside the forest to catch duikers, pigs, porcupines, rats or guinea fowl, traps are set around the edges of farmers' fields to catch crop-raiding animals. The commonest form of trap is the leg-hold trap: a pair of toothed metal jaws is held open by a metal spring mechanism; when the animal steps between the open jaws it releases the spring and the trap snaps shut gripping the arm,

---

[65] The climber most commonly used for snares is called *Mutega nende* meaning 'catching blue duiker'. The species is extremely tough, particularly when dry. A chimpanzee would have no difficulty in chewing through it, thus this species can be seen as 'chimp friendly'. Personally I would like to see all snares made of this material in order to save the hands and feet of so many chimapanzees. All hunting is illegal in the Forest Reserves and BFP cannot promote illegal activities; however, we are exploring this possibility with Uganda Wildlife Authority at the present time.

[66] Wire for snares is these days obtained (illegally) from the Kinyara Sugar Works, located near the Budongo Forest (Tumusiime 2002).

wrist, hand, leg, ankle or foot of the victim (see Fig. 9.2.). The trap is attached to a metal chain which is tethered to the ground. Traps are sometimes quite unnecessarily large — big enough to seriously injure a buffalo or a human being. Chimpanzees do engage in crop-raiding from time to time and as a result get caught in these traps, usually with disastrous consequences (see below).

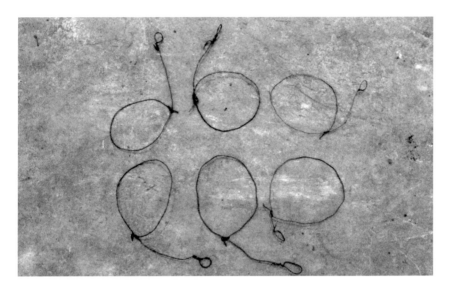

Fig. 9.1: Typical Budongo snares made of wire (photo: V.R.).

Fig. 9.2: A leg-hold trap of the kind found in fields around Budongo Forest (photo: V.R.).

Injuries from snares and traps

Waller and Reynolds (2001) investigated the injuries of the Sonso chimpanzees. The injured limbs fell into two broad morphological categories: claw-hands and wrists, and missing digits and limb segments.

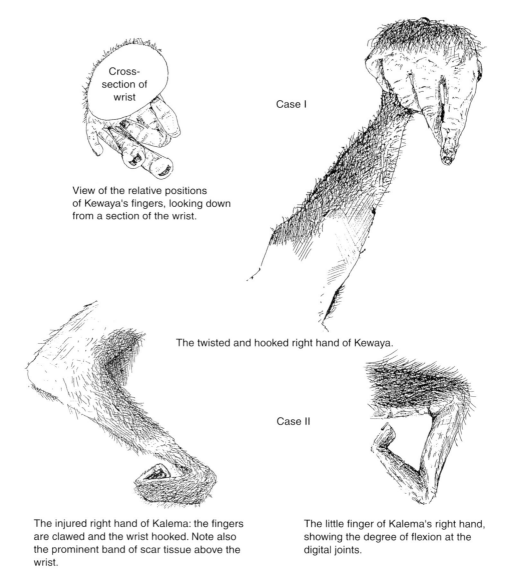

Cross-section of wrist

Case I

View of the relative positions of Kewaya's fingers, looking down from a section of the wrist.

The twisted and hooked right hand of Kewaya.

Case II

The injured right hand of Kalema: the fingers are clawed and the wrist hooked. Note also the prominent band of scar tissue above the wrist.

The little finger of Kalema's right hand, showing the degree of flexion at the digital joints.

Fig. 9.3:   Cases of claw-hand and hooked-wrist (from Waller and Reynolds 2001, line drawings by J. Waller).

1. Claw-hand and wrist. Six individuals of those studied showed extreme flexion of the wrist and digits, with accompanying hair loss and scar tissue (see Fig. 9.3). In each case the individual had lost the use of the hand. The explanation found was tendon damage, which had left the hand in a viable state (i.e. it had not dropped off or become gangrenous) but without functionality. This tendon damage was the result of the individual trying to extricate itself from a snare by pulling against the top of the wrist, severing tendons or damaging them beyond repair in the process. As in similar human injuries, this resulted in claw-hand. No voluntary movement was possible in all these cases. In Case I, there is evidence that the individual (Kewaya, now an adult female with an infant, Katia — the name means 'wire') has, in addition to damage to tendons, dislocated the bones of the carpal region with severe damage to muscles and nerves; her hand flops about in an uncontrolled way as she moves. In Case II of Fig. 9.3 (Kalema, now an adult female and mother of Bahati and Kumi — the name means 'lame') the wrist and fingers are clawed, immobile and wasted, the thumb is adducted and immobile, and the hand is partially adducted at the wrist. The result is lack of any functionality. Other cases include that of Tinka, our oldest adult male. Both his hands are injured in much the same way, but he has some voluntary movement in the right hand, none in the left. The wrist and joints of the digits are flexed and incapable of voluntary movement, and the thumb is immobile. Tinka has a real problem moving quadrupedally because he cannot knuckle walk in the normal way. As a result he puts his hands down on their dorsal side, with the result that the skin is bare and often subject to infections and sometimes develops red raw patches. Indeed, poor Tinka has a chronic skin infection all over his body and is often found scratching himself. He manages to climb trees with some difficulty because he is not only injured but old; I am always surprised when I return to Sonso and find him still going strong. He is the lowest ranking of all our males, partly owing to age but mostly owing, I suspect, to his injuries (see below).

2. Missing limb segments. At the time of the study, two individuals had missing hands (Muga and Kikunku, both adult males), and one (Kigere, an adult female, mother of Kadogo and Keti) has a missing foot. These may well be the results of stepping on a trap because in each case the lesion is clear-cut, almost surgical. The resulting wound appears, in each case, to have healed up perfectly. In the case of Kikunku, the stump has healed but the joint is enveloped in scar tissue accompanied by a complete absence of hair on the distal two-thirds of the forearm.

A full description of these injuries is given in Waller and Reynolds (2001), together, for comparison, with a description and illustration of Banura's congenital deformity (club foot) which has an altogether different appearance and aetiology from that of the snare injuries.

Regarding how chimpanzees deal with snares once they are caught, it seems from a number of cases that the individual has actually chewed its way through the wire. If the wire is multi-stranded, then each strand has to be individually bitten through. Once escaped from the snare site, the commonest situation is for the animal to trail the remaining wire, still tightly attached to the fingers or wrist, so that the first thing we notice is the loss of use of the affected hand and the strand of wire itself. At this stage the animal is usually paying attention to the wire, licking the affected wounds and occasionally using its teeth to try to remove the snare. This may go on for days or weeks. In due course, either because the skin falls off or for other reasons, the snare disappears and is not seen any more. Several of our younger animals have succeeded in removing snares and their hands have healed up completely. In the case of adults this is less likely and the damage is permanent. Evidently, once the snare is off, the power of regeneration is greater in young individuals than in older ones.

On 26 March 2004 at 2.28 p.m. two of our field assistants, Monday and Zephyr, found the infant male Zig (ZG) caught in a snare in block GC:

> He screamed loudly and eventually all the chimps present, especially the females, joined him and also started screaming and calling in a strange sound. Females present were KW, KL, ZM (ZG's mother), NB, KU, CL and JL. The senior males DN, MA, ZF, BK, together with the males GS, TK, BO and MS all approached while displaying all round the area. DN came nearer to ZG perhaps to assist but ZG did not allow him to come close. ZM also became very aggressive to any individual who came near ZG. ZG kept on struggling, biting and pulling the wire snare, to the extent that his mouth started bleeding. He struggled on until 3.02 p.m. at which time he managed to pull the wire from the sapling it was tied to. The wire remained on his right hand (from the Events Book, observers Monday Gideon and Zephyr Kiwede).

We have not encountered a case where one chimpanzee has removed the snare from another. However, Boesch and Boesch-Achermann (2000) describe such a case: a young female, Vera, in the Taï community had her left wrist caught in a snare. She managed to break the cable but the snare remained deeply embedded in her wrist. 'Rapidly, Schubert, the beta male, approached, and while she held her arm towards him, he removed the cable with his canine' (p. 251).

Deaths

Since the Budongo Forest Project began in 1990, two individuals have been found dead in the forest near camp as a direct result of getting caught in snares. This proves that, despite the many cases where chimpanzees have extricated themselves from snare situations, suffering serious injuries in the process, it is not always possible to escape and death may result. It is impossible, given the small sample size, to obtain an accurate estimate of how many chimpanzees lose their lives to snares in the whole forest each year, but on the assumption of a 5% fatality rate over 10 years (0.5%/yr) there would be 2–3 chimpanzees dying from snare injuries each year in Budongo.

Deaths from snares

On 26 May 1997 I was informed that the Project's trail cutters had found a rather decomposed chimpanzee in the forest which they thought had been caught in a snare. It was some distance from camp, in Block HH, but nevertheless in the range of the Sonso chimpanzees. The men returned to the place with a container and brought all the remains they could find back to camp. I put together as much of the carcass as remained. Mostly it was skeletal, but some skin and hair remained, no muscle or internal organs. The skeleton was intact with the exception of the right humerus and the left leg below the femur. We assumed that the body had been partly eaten by predators which had removed the missing bones. The teeth indicated it was an adult and the size of the skeleton indicated it might be a female.

One large mass of hairs contained the remains of the snare: a circular loop of multi-strand wire, with a maximum diameter of just 3.9 cm. The additional length of wire which had been attached to a sapling at the site was missing and had possibly been

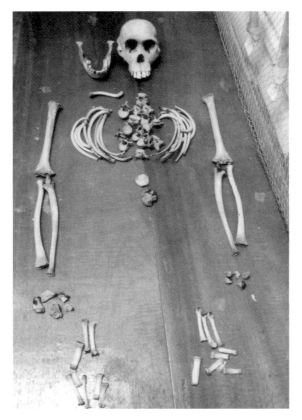

Fig. 9.4: Willis, killed by a snare (photo: V.R.).

removed by the hunter. This unfortunate individual, whom we were not able to identify, had been unable to free itself from the snare and had died at the site of the snare, either from septicaemia or starvation or predation.

The second case was similar. On 11 March 2002 Ofen and Dominic, our two snare removers, returned from the forest with the remains of a chimpanzee who had died, in their opinion as a result of being caught in a snare. The remains were washed and they are shown in Fig. 9.4. The pelvis and lower skeleton were missing, presumed eaten by predators. The men had been unable to find the snare itself but were convinced that this had been the cause of death because the area where the cadaver was found was very much trampled down, and was around a sapling which showed, from the abrasions on the bark, that a snare had been attached to it. Probably a hunter had returned to the site after the chimpanzee had died and retrieved his snare so he could use it again, and also to avoid any danger that he would be identified by it, for by now we were heavily engaged in anti-snare patrols and educating the villagers against putting snares in places where chimpanzees might be caught. We had told them that if a chimpanzee were to be caught in a snare, they must come to the Project immediately and report the fact, and we would then investigate all methods of releasing it, and would not take action against the person responsible. But no hunter has, to date, ever come forward to us to admit he has caught a chimpanzee in a snare, partly because all hunting is forbidden by law, and partly because they do not believe we would not take action against them.

Examination of the skull, dentition and skeletal remains, together with consultation with others,[67] led us to conclude that this individual was probably Willis, son of Wilma. He was a late stage juvenile aged around 8–9 years who went missing at this time.

Deaths from traps

Besides these snare deaths, there have been two cases of deaths from leg-hold traps, both in cultivated areas to the south of the main block of the Budongo Forest, one in Kasokwa Forest and one nearby in Kasongoire Forest. We do not have many details about the Kasongoire case, but we have details about the Kasokwa case because as an offshoot from the BFP we have been studying the small Kasokwa community (12 members) of chimpanzees since September 1999, employing a field assistant to make daily observations of them (see Chapter 11). The incident at Kasokwa was written up by Julie Munn and Gladys Kalema (Munn and Kalema 2000). A map of Kasokwa Forest is shown in Fig. 11.1. Kasongoire Forest is some kilometres to the southwest of Kasokwa.

The Kasokwa chimpanzees live in a riverine forest strip which, although enjoying protected status, has been heavily encroached by the surrounding population. The population increase is due to the attractiveness to people of the Kinyara Sugar Works, which is contiguous with the Kasokwa Forest. Immigrants mainly from the north of Uganda have been settling here for some years, putting up houses and cultivating land that was

---

[67] David Pilbeam directed us to new work by Zihlmann *et al.* (2004) on age determination of wild chimpanzee from dentition, and I am grateful to all concerned for the help this provided.

hitherto empty. A survey of Kasokwa Forest Reserve by Lloyd and Mugume (2000) found that encroachers were cutting trees on a daily basis. Half of the chimpanzees in this forest had limb injuries, a shocking proportion.

On 6 June 2000 locals first observed Kigere, the alpha male of the Kasokwa community, dragging a large steel leg-hold trap which had closed over the fingers of his right hand. This unfortunate individual had been caught once before in a leg-hold trap and had lost his entire lower right leg below the knee. On 17 June 2000, Richard Kyamanywa found Kigere lying dead in thick bushes just outside Kasokwa Forest. He still had the trap on his fingers. A postmortem examination was carried out on 18 June 2000 by wildlife veterinarian Gladys Kalema. The trap (40 cm in length and weighing 10 kg) was attached across the distal metacarpals of the right hand. Based on the observations of local villagers it seems that the trap was attached for 10 days. As a result, the right arm had developed extensive gangrene which had led to overwhelming septicaemia and toxaemia. This was the worst case of chimpanzee suffering we at BFP had ever come across. Death was a welcome release for Kigere.

Another such case was reported in June 2000 near Nyakafunjo. A chimpanzee (probably not one of the Sonso community) was seen in a man-trap and this was reported to BFP. The director of BFP, Fred Babweteera, called the JGI at Entebbe and they dispatched a team of two veterinarians the same day. Although the chimpanzee could not be found, it was concluded that it was caught in a man-trap because it left behind a noticeable trail (indicating dragging of a heavy object) with some blood. It seems likely that this chimpanzee died a few days later. One local farmer confessed to having owned a man-trap in this area but said that he couldn't remember the exact site where he set this trap. It seems likely that this 'forgotten' trap killed the chimpanzee.

Deaths from spearing

Most injuries and deaths of chimpanzees caused by humans arise from snaring and trapping, but spearing happens too. We cannot tell how many deaths or injuries this has caused but we have one tragic example from Sonso and a second example from Kasokwa (see Chapter 11).

On 6 May 2003 I received an email from our Assistant Director, Sean O'Hara, with some exceedingly bad news: Jambo, our fine adult male, had been speared (Fig. 9.5). Jambo had been raiding a sugar cane field belonging to a man living in Nyakafunjo, who had planted cane right up to the forest edge, a real temptation to wildlife. The owner of the cane had employed a guard who had fatally speared Jambo. His cadaver was found two days after his death, in elephant grass at the forest edge to which he had run in his final moments. He had been speared from close quarters, the spear had penetrated his neck and gone down into his chest, probably reaching his heart as he died very quickly. Our staff collected his cadaver and buried him in the *Broussonetias* at camp. It was, for all, a very sad occasion; Jambo was a fine, uninjured, prime male, not power hungry but a very solid member of the community which was now weakened by his death. He had been born around 1975 and was thus 28 years old. For so long he had avoided snares and

Fig. 9.5:   Jambo — a fine Sonso male at the peak of his powers (photo: M. Emery Thompson).

traps, only to be killed by a sugar cane guard; for years we had been educating the people of Nyakafunjo and other villages about the importance of chimpanzees and the need to protect them; this man was new to the area, we were subsequently told. *The New Vision*, Uganda's daily paper, featured this killing and we hoped it would lead to a prosecution; at the time of writing this has not yet happened, although the two farmers on whose field Jambo was killed were ordered not to cultivate sugar cane on these fields and told that if they did so the sugar would not be bought by Kinyara Sugar Works; at the time of writing this stipulation is being adhered to.

The subsequent investigation brought another matter to light. In 1999 Vernon, the second ranking adult male and close ally of Duane, disappeared; he was last seen on 28 June of that year. It now transpires that a chimpanzee had been killed, also speared by a guard, in that very same field in 1999 at the time of Vernon's disappearance. The matter had been reported to the village authorities and the guard disappeared, never to be seen again. This time the same thing had happened again. That one sugar cane field may thus have claimed the lives of two of the finest chimpanzees in the Sonso community, though it was never proved that the earlier spearing was indeed of Vernon. As related more fully in Chapter 12, the Kinyara Sugar Works senior management were dismayed by these events and when Jambo was killed inaugurated a rule that no sugar was henceforth to be planted within 50 m of the forest edge. The offending field was harvested prematurely in the days following the spearing, the crop was cut and cleared, and the stubble was sprayed to prevent regrowth. More details are given by O'Hara (2003).

Effects of snare injuries on feeding and social life

A number of studies have been made on the effects of snare and trap injuries on the life of the Sonso chimpanzees.

First, feeding. Injuries reduce the ability of individuals to feed near the trunk of the tree in certain species such as *Ficus sur*, because it is harder for them to hold on to the trunk and large branches while feeding than it is for fully able-bodied chimpanzees (Burch 1994; Smith 1995). This is important in the case of *F. sur*, because not only is this a top-ranking food for chimpanzees but the majority of its fruits hang in pedicles from the trunk and large branches of the tree. High ranking, able-bodied chimpanzees tend to feed on the fruits hanging from the trunk and lower branches of *F. sur* trees, but injured chimpanzees cannot feed there. Smith (1995) recorded the feeding height of 15 uninjured and 10 injured Sonso chimpanzees at three heights in *F. sur* trees: low (on main trunk), intermediate (on large branches) or high (on small branches), over an 8-month period from February to October 1994. The mean height of each individual was plotted relative to the mean for all individuals in the tree (the group mean). This showed that uninjured chimpanzees fed significantly lower than injured ones (Fig. 9.6).

In some other species, such as *Celtis gomphophylla*, the injuries are less of a problem because the fruits occur only on the outer and smaller branches of the tree, where injured chimpanzees can support themselves more easily.

Smith (1995) found that the feeding rate (based on the number of fruits of *F. sur* eaten during 30-min sampling intervals) was the same for injured as for uninjured chimpanzees. Nor was there a difference in the amount of time spent in visits to feeding trees between injured and uninjured chimpanzees. As he points out, however, despite the disadvantage of not being able to reach certain parts of some tree species, the injured animals do not show a loss of condition such as poor coat quality or other signs of ill-health. He suggests that this may be because the fission–fusion system of chimpanzees allows an injured individual to choose to feed on less popular foods where competition is less intense.

A study of the relative disadvantage of injury on chimpanzees' feeding techniques and the adaptations made by individual chimpanzees to their injuries is that of Emma Stokes (Stokes 1999) who worked closely with Richard Byrne (Stokes and Byrne 2001). From a 13-month study (August 1997 to September 1998) based on 22 able-bodied and 8 injured individuals, these authors showed that chimpanzees with injuries modify the feeding techniques shown by able-bodied chimpanzees in their feeding behaviour, and this enables them to work around their impairments.

Stokes and Byrne (2001) focused on the techniques used by the Sonso chimpanzees when feeding on young leaves of *Broussonetia papyrifera*, a much preferred food. By breaking down the feeding process into its component parts, they were able to distinguish 14 different feeding techniques, two of which were preferred by all individuals. In this food species, leaf blades are eaten but the petioles are discarded. The two major differences in technique they noted were (a) stripping leaves towards and into the mouth, then discarding the petioles, and (b) detaching the petioles first, then stripping leaves

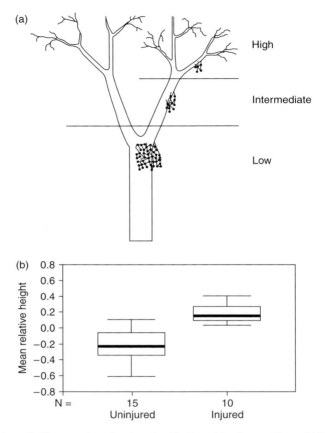

Fig. 9.6:   (a) Schematic *Ficus sur* tree showing classification of feeding positions. (b) Boxplot showing difference in relative feeding height of injured and uninjured chimpanzees on *F. sur* trees. The *y*-axis shows the mean feeding position of uninjured (left) and injured (right) individuals relative to zero. Zero represents the mean for all individuals (from Smith 1995).

away from the body and putting them into the mouth. Stokes and Byrne show in great detail how the process of feeding on leaves can be broken down hierarchically into several dozen component parts. The fact that all chimpanzees feed in the same two basic ways indicates a high degree of learned technique. With this degree of detail to explore, they were able to show that injured chimpanzees, while showing the same techniques as able-bodied individuals, modified them according to the details of their specific impairments.

For example, to pull *Broussonetia* leaves into range, different injured individuals used novel techniques such as knuckle hook, wrist wrap, back of wrist reach, lateral wrist reach, and wrist grasp reach. Because of loss of hand function, having reached and pulled in a branch for feeding, they supported the branch using the back of the wrist, hooked wrist, lateral wrist, wrist-wrap, knuckle hook, wrist grasp and a variety of

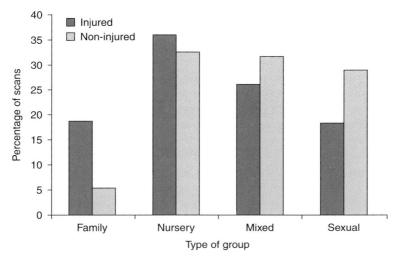

Fig. 9.7: Percentage of time spent by injured and non-injured chimpanzees in various group types (from Munn 2003).

two-handed or foot-and-hand grasps (Stokes and Byrne 2001: 23). In such ways do our Sonso chimpanzees that have had the misfortune to get snare injuries compensate for their disabilities and succeed, with great fortitude, in feeding themselves. The alternative is death from starvation, and we do not know how many chimpanzees, having extricated themselves from snares, either die of gangrene or other infections, or fail to adjust to the new situation and eventually die.

The effects of disablement on the social lives of Sonso chimpanzees was first studied by Quiatt *et al.* (1994). They noted that disabled individuals were nevertheless fully integrated socially and there was no evident discrimination against them by able-bodied individuals.

A wide-ranging and very detailed study of the effects of snare injuries on the Sonso chimpanzees is that of Munn (2003, in press). Munn found a number of hitherto unexpected differences between injured and uninjured individuals. Her study focused on the females of the Sonso community, and examined the effects of both snare- and trap-related injuries, looking at the ways injury affects social behaviour, mother–infant behaviour and locomotion.

Munn studied 12 females, all mothers with dependant infants, of which 5 were injured and 7 were uninjured, the latter being the control group. (The whole community consisted at the time of 49 individuals; all adult females were studied.)

In regard to social factors, Munn found that, compared with the uninjured females, injured females spent more time in small, family and nursery groups, and less time in the larger mixed and sexual groups (groups with oestrous females in them), as can be seen from Fig. 9.7. This was true whatever activities they were engaged in, i.e. resting, feeding or travelling. Family and nursery groups are in general quieter and more slow-moving than mixed and sexual groups, and this may be the reason they are preferred by injured females.

There was no avoidance of or discrimination against injured individuals by other members of the community; they were tolerated despite their injuries and were chosen as grooming partners with equal frequency as uninjured females. Dendrograms of association and proximity show that pairs consisting of an injured and a non-injured female were the commonest kind of dyads, indicating that injured females do not form an out-group but are fully integrated into the community.

Infants of injured mothers spent more time in social play than in the case of uninjured mothers, an intriguing and unexpected finding perhaps related to the relative lack of mobility of their mothers. By the juvenile stage (4 years and up) this was no longer the case.

Injured mothers spent more time moving arboreally compared with their uninjured counterparts. This may be due to difficulty in climbing up to feed, or to fear of predators, or (and perhaps especially) to fear of snares. Once caught in a snare and having escaped, it is small wonder that an injured animal lives in great fear of being caught again.

A flow chart (Fig. 9.8) summarizes Munn's findings and shows how they are related to each other.

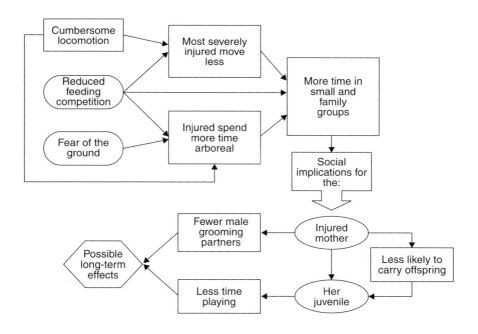

Square or rectangular boxes represent results found in the study. Rounded boxes show factors that are most likely to be affecting the results. Ovals represent the individuals affected by results. The arrowed box, and hexagon, describe the possible repercussions of these results.

Fig. 9.8:   How injuries affect the Sonso female chimpanzees (from Munn 2003).

Snare removal project

In 1962, the Busingiro chimpanzee community at Budongo studied by my wife and myself (Reynolds 1965; Reynolds and Reynolds 1965) had, we estimated, between 60 and 80 members. I may be wrong, but I don't recall seeing one with a missing hand or foot, or a ruined hand.

Sugiyama (1968: 230) reports on one chimpanzee, Mkono (the word means 'arm' in Swahili), of the Busingiro community who had no right hand. Suzuki (1971) mentioned the same individual, Mkono, and included a photograph of him showing that his right hand and wrist were completely missing. This unfortunate individual was later shot by an American researcher with an anaesthetic dart gun and died after falling to the ground (Suzuki 1972). The initial injury is very likely to have been caused by an encounter with a trap, probably while crop-raiding, although at that time, before the Amin wars, there were buffaloes in Budongo and the trap might have been set for one of them.

In other words, snare and trap injuries were rare 40 years ago. At that time, the traditional Bunyoro method of hunting was to go out communally from the village and hunt over the woodland savannah countryside using dogs and long nets, driving game into the nets. We did find some snares in the forest but not the large numbers that are found today.

Together with the students and staff of BFP, I have spent the years since we re-started the Project in 1990 trying to think of a way to solve the problem of snares and chimpanzees. On the one hand, as stated earlier, people do depend on meat for high-quality protein and children benefit from a protein meal when they can get it. Should we tell them that they must not set snares in the forest? Then there is the case of traps. Traps are mainly set to scare and kill baboons that raid farmers' fields. Baboons can do inestimable damage to growing crops (Hill 1997). Chimpanzees are less troublesome to farmers, being scarcer and taking less food (Johnson 1993; Hill 1997). But they too get caught by the traps. Should we therefore tell farmers they must not protect their crops with traps?

After years of discussion among ourselves and with the authorities, we at BFP decided the answer to both questions (concerning removal of snares and of traps) was 'Yes'. We should resolutely oppose the setting of snares and traps in and around Budongo Forest. In the case of leg-hold traps, these are actually illegal. We have been in discussion with the Masindi Local Government Environmental Protection Unit and they have passed a by-law to that effect.[68] Enforcement is the problem. Because these traps are set outside the forest, around farmers' private fields and in householders' private gardens where they have paw-paw trees and other fruits and vegetables to protect, there is little chance of finding and removing them.

---

[68] Section 26 (5) of Ordinance Supplement No. 2 to the *Uganda Gazette* No. 61, Volume XCV, dated 8 November 2002, which states that non-selective vermin control methods including snares, poison and metal traps shall not be used. Section 28 gives the penalty for infringement as a fine of one currency point: approximately 20 000 shillings (equivalent to less than £7 or $13) or imprisonment not exceeding one month. While applauding such legislation it has to be noted that the penalties are small and, more importantly, that at the present time enforcement is lacking.

In the case of snares, we at Sonso camp are surrounded by a state-owned forest which is full of them. Whether or not it is the responsibility of the authorities to remove them, during the 1990s no removal had ever taken place. We had discussed the idea of replacing snares with methods that would be able to catch duikers and pigs but would allow chimpanzees to release themselves and escape, but this was not viable. So we decided to go for snare removal. The decision was made easier by the fact that some funding was made available to us for this project by Debbie Cox, the JGI representative in Uganda. JGI was willing to let us have a grant to employ two men for snare removal. The work had been pioneered at Kibale Forest to the south,[69] and we were recommended to employ two local ex-hunters, who would know where snares are placed and would be able to find them. Two snare removers came over from Kibale Forest to train our men.

We were cautioned that we should not reward anyone monetarily for bringing in snares because that could lead to snare manufacture. We were also warned that reprisals might be a problem. Snare wire costs money and hunters would not appreciate having their snares removed. This warning proved correct.

We began the project in January 2000, employing two local hunters from Nyakafunjo, Ofen Anzima and Pascal Muhindo. In January they brought in 231 snares, in February they brought in 172, in March they brought in 111 : 514 snares confiscated from the forest in three months. The types of snares were sorted and counted by Julie Munn, to whom I am grateful for this information (Table 9.1). The snares themselves are shown in Fig. 9.9.

At that time, a student, Jeremy Lindsell, was doing his PhD at Sonso, a study of an understorey bird species, *Pluvel's Illadopsis*, in logged and unlogged forest. Lindsell was using some equipment to do the study. His logged-forest bird subjects were near to camp, but his unlogged-forest sample was in the Nature Reserve some 3 km away. In order to save carrying his equipment to and from camp he left some of it in the Nature Reserve where it had always been perfectly safe. Some of his equipment was left in the forest to collect long-term records, e.g. digital thermometers taking records of forest floor temperature. Now some of his equipment was stolen from the forest. In April when I visited camp I found he had lost expensive equipment on a number of occasions and this was beginning to have an impact on his budget and on his data collection. The village hunters were taking revenge on us.

Besides that, Ofen and Pascal had found small notes scrawled on paper near snares in the forest. One said (in Swahili): 'You are eating well now, you will not be able to eat soon.' This was a threat that the men might be attacked with pangas (machetes).

We discussed this at camp. Evidently things were getting out of hand. We decided to stop removing snares at least temporarily and do more intensive education of the village people to ensure they understood *why* we were removing snares — to protect chimpanzees and not to deprive them of meat. Fred Babweteera, our project director, and I arranged to go to some nearby villages and talk to the men about our work and why we were doing it. We found a lot of confusion about our project and what we were trying to do.

---

[69] Over a two and a half year period an average of 65 snares per month were removed from Kibale National Park by a team of two (Wrangham 2000*b*, quoted in Munn and Kalema 2000).

Table 9.1: Types of snares removed from Budongo Forest, January–March 2000.

Month, year	Thick wire	Thin wire	Cord or nylon	Fishing line or electric wire
January 2000	23	160	47	1
February 2000	7	147	14	4
March 2000	11	77	4	19
Totals	41	384	65	24

Fig. 9.9: Snare removers with three boxes of snares collected in the Sonso community range, January–March 2000 (photo: V.R.).

Ofen and Pascal were transferred from snare removal duties to education. We started a programme of weekly education meetings in villages and schools. This programme continues, in modified form, today. After three months, in July 2000, we re-started snare removal, recording the location of snares to obtain an idea of the areas of heaviest snaring. The education programme appeared to be having a big effect. During the 16 months from 22 August 2000 to 13 December 2001, just 399 snares were found, fewer than in the first three months of 2000. However, since that time numbers of snares brought in have begun to rise again. In May 2003, for example, over 300 snares were removed from compartment N4, adjoining the BFP's site and in the range of the Sonso community (Tumusiime 2002).

We now remove snares daily and continue the education programme. There has not been a repeat of the removal of scientific equipment from the forest, nor have our snare removers been threatened again. Initially there was an increase in snaring in the Nyakafunjo area at the southern end of the Sonso range.[70] More recently we have

[70] On a single day, 4 July 2003, our four snare collectors brought in a total of 64 snares all collected in compartment N4 near Nyakafunjo.

extended our snare removal activities further afield and at the time of writing we are finding a large number of snares in the Busingiro area to the west of the Sonso range.

There can be little doubt that the very high number of snares in Budongo Forest is a result of an increasing population resulting from the availability of jobs in the sugar and tobacco industries. Additionally, Budongo has seen a big increase in pitsawing in recent years. Pitsawyers combine this activity with vigorous hunting. They are hard to catch because some of them are legal pitsawyers, while others claim to be legal pitsawyers if questioned, while in fact they are hunters.[71]

Using a global positioning system (GPS), we have been recording the locations of snares removed since January 2000. Fig. 9.10 shows the location of 2333 snares collected between 14 February 2000 and 16 April 2004.

The locations where snares are found are influenced by a number of factors of which one can be clearly seen from this map: relatively few are found in the area to the north of camp where much BFP research activity takes place and our field assistants move around every day. The number of snares increases outside the area of the BFP trail system. Thus we in the BFP are having a negative influence on the setting of snares in the range of the Sonso chimpanzees.

This fact is borne out once again by recent data compiled by Fred Babweteera, which show the daily rate of snare removal in the areas where we now operate our snare removal programme. The data are shown in Table 9.2.

As can be seen from Table 9.2, the Busingiro area (compartments B1 and B2), which is some 5–10 km to the west of Sonso, had the highest daily rate of snares removed during this period. Compartment N6 also had a high rate; the reason here may be connected with the existence of a number of legal (and, probably, some illegal) pitsawyers in the area.

There is now a more insidious aspect of the increased snaring: a small but significant bushmeat trade has begun along the Masindi–Butiaba road which runs to the south of Budongo Forest. Until the mid-1990s there was no organized sale of bushmeat; wild animals were occasionally sold in the local markets by the hunters themselves or their wives. Today there is evidence of trade in forest meat. We have been informed that the traders set snares in the forest, collect the animals they have caught, and take them to particular places along the main road for sale. This is a new development and will inevitably lead to the setting of more snares and injuries to more chimpanzees. During the writing of this book, two of our young chimpanzees were caught in snares: in March 2003 Kana, the 3-year-old infant of Kutu,[72] was seen with a wire snare on her ankle, and in the same month Nora, the 7-year-old juvenile daughter of Nambi was seen with a wire snare on her wrist. The damage goes on.

An indication of the number of snares collected in the Sonso area over a recent two-year period can be seen in Fig. 9.11.[73] The number of snares recovered from

---

[71] F. Babweteera (pers. comm.)

[72] At 3 years of age this is the youngest of our chimpanzees to be snared. Indeed, in calculating snare-related losses, infants are normally excluded.

[73] The data on which this figure is based are simply the total numbers of snares collected in each month. This number should be corrected for the number of search days per month. However, there is no reason to think that such a correction would materially change the results. The same two snare removers were involved, and the number of search days each month was more or less constant during the whole period.

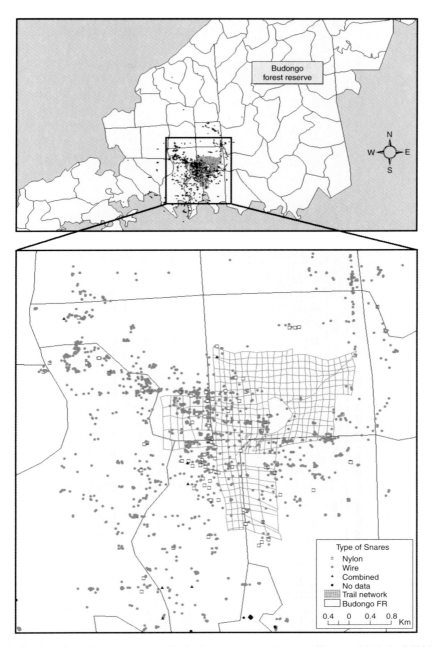

Fig. 9.10: Locations of snares removed in the Sonso community range, February 2000–April 2004, also showing snare materials (copyright Hazzah and Reuling).

Table 9.2: Number of days patrolled and rate of snares removed per day in compartments of Budongo Forest Reserve, October 2003–June 2004.

Compartment	Oct D	Oct S	Nov D	Nov S	Dec D	Dec S	Jan D	Jan S	Feb D	Feb S	Mar D	Mar S	Apr D	Apr S	May D	May S	Jun D	Jun S	Total D	Total S	Average Rate*
N1	7	15	4	37	1	2	4	25	3	21	3	6	6	83	4	5	3	11	35	205	6
N2	12	137	2	0	0	0	2	12	2	5	4	7	5	23	4	11	5	32	36	227	6
N3	1	0	2	23	1	7	3	0	3	5	2	9	4	9	6	10	3	12	25	75	3
N4	0	0	0	0	5	47	4	19	2	14	2	17	0	0	0	0	3	6	16	103	6
N5	0	0	0	0	4	12	3	23	0	0	2	27	0	0	0	0	0	0	9	62	7
N6	0	0	0	0	0	0	0	0	3	35	0	0	0	0	0	0	0	0	3	35	12
N15	0	0	8	36	10	13	2	7	0	0	8	52	5	43	2	8	4	27	39	186	5
B1	0	0	0	0	0	0	0	0	4	45	0	0	0	0	0	0	0	0	4	45	11
B2	2	38	1	18	0	0	0	0	0	0	0	0	0	0	0	0	0	0	3	56	19
W19	0	0	0	0	0	0	1	8	0	0	0	0	0	0	2	16	0	0	3	24	8
W21	0	0	3	21	0	0	0	0	2	13	0	0	0	0	2	24	1	0	8	58	7
W22	0	0	0	0	0	0	0	0	0	0	1	4	0	0	1	8	0	0	2	12	6
W23	0	0	0	0	0	0	0	0	0	0	0	0	0	0	0	0	1	0	1	0	0
Total	22	190	20	135	21	81	19	94	19	138	22	122	20	158	21	82	19	89	183	1089	6

Key: D = number of days patrolled; S = number of snares removed.
* Snare removal rate expressed as number of snares removed per day.

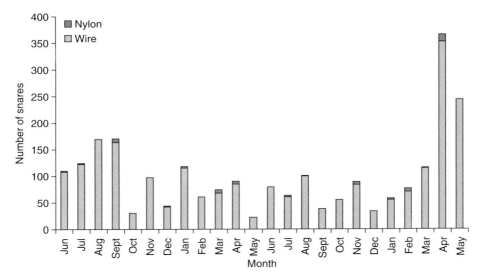

Fig. 9.11:   Snares recovered in the Sonso community range, June 2001–May 2003 (courtesy S. O'Hara).

compartment N3, in which the BFP is located, declined until December, 2002. This provides good evidence that the programme was having the desired effect. It was also contributing to an undesired effect by leading to higher levels of snaring elsewhere. We are counteracting this by having our snare removers work both close to camp and more distantly, if necessary camping overnight in the forest at more distant locations. The increase in 2003 remains to be explained.

Plumptre *et al.* (2003) in their survey of the chimpanzees in all of Uganda's forests found the greatest amount of hunting sign in Budongo and its neighbour to the south, Bugoma, of any of the western forests. Most of the sign was encountered around the Sonso area and to the east of Sonso, with an additional hot spot in the Busingiro area (we return to this later — see Fig. 12.4). These are areas where owing to the existence of trails made by illegal loggers extracting mahogany, hunters (often the illegal and legal loggers themselves) are able to move freely in the forest. Our own trail system is also, undoubtedly, used by hunters. If we compare Budongo and Kibale forests, both of which have research projects with trail systems, the situation is seen to be much worse at the present time in Budongo. One major difference is that Kibale Forest is now a National Park falling under the Uganda Wildlife Authority (UWA) and about 60 rangers patrol the forest looking for poachers. By contrast, Budongo Forest is a Forest Reserve falling under the Forest Department (FD, now renamed the National Forest Authority, NFA) and has only a modest number of rangers and forest guards whose role includes anti-poaching patrols, supervision of logging activities and stock-mapping. Forest guards can conduct patrols with armed policemen but it is rare for law-breakers in the forest to be caught. There is a request by NFA staff to be armed but this is still under consideration. More intensive co-operation between UWA and NFA is also now being discussed.[74]

The constant presence of researchers around the Sonso community has helped to reduce snare-setting. During the Christmas break when most of the staff and students are away, hunters tend to take advantage and set more snares around Sonso (they know that this area has many duikers). For this reason we recently decided to have some staff members stay on over the Christmas break. Of our team of four snare removers, two work throughout the Christmas period.

Live-trap project

While we have made no more than limited progress with the snaring problem, we have been somewhat more successful in regard to the problem posed to chimpanzees by leg-hold traps. In March 2001, BFP organized a Workshop for local farmers.[75] It was designed to facilitate interaction between BFP and the local community. In September the previous year, Fred Babweteera, the Director of BFP, and I had discussed how best to organize it. We felt that it should to some extent be driven by the community rather than us setting the agenda. In consultation with the local leaders, 14 villages bordering the Budongo Forest Reserve were selected to participate in this workshop. Each village identified and prioritized key issues they wished to be addressed and selected one representative to participate in the workshop. The same issues were raised by all 14 villages. The five topics for the Workshop were those rated most important in all the villages. For our part, we then looked for people with local expertise on each of these topics to introduce and guide the discussions.

The five topics selected by the representatives were as follows: micro-finance and self-help, seed collection and tree-planting, crop-raiding and vermin control, bee-keeping, and collaborative forest management. When March came, we had a highly productive week (Fairgrieve 2001). For present purposes the day on crop-raiding and vermin control was the important one. For crop-raiding local expertise was provided by Dr Kate Hill who had worked on this topic in the local community (Hill 1997, 1998, 2000). We shall discuss her work in Chapter 10. For vermin control we chose Christopher Byarugaba, Senior Game Assistant of Masindi District Local Government, and Head of the District Wildlife Management Unit.

Byarugaba turned up at Sonso before the day of his presentation to the Workshop with a truckload of poles and some local men, who under his direction constructed a sizeable wooden structure with a sliding wooden trap door at one end and a wire mesh roof. In his talk he discussed the problems faced by local farmers, the work of his unit, the designation of baboons, bush-pigs, vervet monkeys and porcupines as vermin, and the lack of sufficient resources (staff, transport, fuel) to do the job of vermin control

---

[74] I am grateful to Andy Plumptre and Fred Babweteera for the information in this paragraph. Responsibility for its accuracy remains, however, with me — V.R.
[75] We are grateful to DfID for funding this Workshop.

properly. He then introduced the 'Christopher' live-trap, his own design, and explained to us the type of trap this was and how it worked (see Fig. 9.12).

The Christopher live-trap catches alive whatever animal enters it. This animal can then be killed or released. The trap works as follows: the heavy door is held up by stout wire which runs down to a baited toggle inside the trap; when the animal touches the bait the door comes down, trapping it. It is simple and, if the trap is well constructed, effective. In fact, although new to us at Sonso, Chris had already made six such traps and placed them near farmers' fields in the vicinity of Masindi.

We from BFP listened with rapt attention. Could this be an answer to the problem of chimpanzees getting caught and injured in leg-hold traps?

After his talk, Chris gave us a demonstration. The trap was baited with green bananas (maize is also sometimes used) and we retired to the banda. Within 15 min we heard the trap door fall. Going over to the trap we found a baboon inside it, which we released by one man standing on top of the trap roof and lifting the door. This trap certainly seemed to work.

The farmers were universally enthusiastic and wanted such traps to put around their fields. The problem was that the poles were not freely available, they were quite expensive to buy and it would be illegal to go into the forest and cut them. Also there was the cost of the nails and the wire mesh which formed the trap roof. After the meeting broke up and the farmers went back to their homes, Chris stayed at camp and we had a discussion with him about whether we could put this trap to use to save chimpanzees

Fig 9.12:   Byarugaba with demonstration live-trap at Sonso (photo: V.R.).

from the injuries caused by the leg-hold traps. It seemed immoral not to try. Chris did not have a budget for this. We agreed that we would look for some seed money to start a pilot live-trap project in Budongo sub-county. The condition, we all agreed, was that any animal not constituting one of the species of officially declared vermin must be released, and, most importantly, there must be no setting of leg-hold traps in villages where live-traps were supplied, or the live-trap would be removed immediately.

Later in 2001 6 traps were constructed. Farmers were very enthusiastic indeed as they found the traps did in fact catch baboons, which they killed. There was an added advantage: if a baboon was caught in a trap, the individuals in that troop avoided the area for several weeks thus providing at least some temporary relief for the farmers. I should perhaps explain that not only is it legal to kill baboons in Uganda (they are classified as vermin) but people are extremely hostile to them and kill, maim and oppress them in every conceivable way. People actually hate baboons and would wish every baboon dead. The above points, often made to us at Budongo on visits to villages, were confirmed by a study of attitudes to live traps in the area surrounding Budongo Forest (Seabo *et al.* 2002) This study was made independently of the BFP by a small team of outsiders from Botswana and Eritrea as well as Uganda and it was thus of great interest to us at BFP to find that our live-traps were well received.

The attitude to baboons described above is in contrast to the attitude to chimpanzees which are traditionally tolerated even if they take food from people's fields and gardens. This positive attitude to chimpanzees is found today among older residents, some of whom link them with their ancestors. New immigrants, coming into the area from far and wide and from a great variety of different tribes and backgrounds, tend to be less tolerant and often have no traditional respect for chimpanzees. This especially applies to immigrants and locals who become sugar cane outgrowers; they are commercially motivated and show little respect for chimpanzees if they come crop-raiding.

Some individuals in Uganda use chimpanzee body parts for medicinal purposes, but they are in a minority. And even such people are less hostile to the small number of chimpanzees that occasionally visit their fields and gardens in comparison with the open hostility they show to the much larger numbers of baboons that inhabit the area and do serious damage to crops and fruits. These attitudes to chimpanzees and baboons have been studied by a number of our students (e.g. Kiwede 2001; Watkins 2001) and are explored further in Chapter 10.

Release of chimpanzees from live-traps

We do not know how many times chimpanzees have been caught in live-traps and released. In some cases, live-traps have caught mothers with infants; one of these releases occurred at Nyabigoma village just to the south of Budongo forest (Fig. 9.13). These photographs encapsulate the story of future chimpanzee survival and management in Africa. Unless and until we develop ways of combining the food security of rural people with the welfare and survival of chimpanzees, their numbers will slowly but inexorably be reduced to nothing. I hope that, faced with similar problems, people in other parts of

Fig. 9.13:   Release of an adult female chimpanzee and her infant, feet just visible as it clings below her belly, from a live-trap (photo: C. Byarugaba).

Uganda and perhaps other parts of Africa will adopt the live-trap approach to chimpanzee conservation.[76]

Snare injuries at other sites

Kano (1984) described a number of snare-related ape injuries in the bonobos of Wamba, DRC. Hashimoto (1999) has written about the injuries of the chimpanzees of the Kalinzu Forest: of 16 male chimpanzees identified, 2 had a wire snare embedded in their hand, while another 7 had injuries on their limbs probably caused by snares, including loss of hand or foot, claw hand or wrist, and loss of digits.

Wrangham and Goldberg (1997) quantified the amount of snaring of chimpanzees in Kibale Forest. Over a period of 300 'snareable chimpanzee years', 11 snares were recorded (snareable chimpanzees were defined as chimpanzees aged 6 years or over). Of 55 snareable chimpanzees in the Kanyawara community at Kibale, 18 showed snare damage as follows: lost hand, 4, lame hand, 5, lame finger, 2, crippled toes, 1. A sex difference was found, with 9/19 (47%) of males and 9/36 (25%) of females having snare damage. These proportions are horrific. We should note that Kibale Forest was not a National Park during most of the period when these snare injuries came about, it was a Forest Reserve until 1993, and that a bushmeat trade was already operational there.

---

[76] There is, it has to be said, a downside to live-traps. They require regular maintenance (replacing poles and nails that fall out). Responsibility for maintenance is not always clear when traps are placed between fields or on communal land. Also, baboons have an excellent record of learning how to avoid dangerous sites and objects, and we should not be surprised if in the future they avoid the baited live-traps completely.

Table 9.3: Numbers and types of injuries by age and sex resulting from snares and traps at 10 sites (combined data) (from Quiatt *et al*. 2002).

	AM	AF	SA	JUV	INF	Unknown	Total
Sample size	69	101	37	67	58	90	422
Loss of foot	1	3				1	5
Loss of hand	3	4		2			9
Loss of fingers	4	2	1	1		3	11
Damage to wrist	3	2	1				6
Died of injuries				1			1
Total injured	11	11	2	4		4	32
Per cent of age–sex group injured	15.9	10.89	5.4	6	0	4.4	7.58

Key: AM = adult male; AF = adult female; SA = subadult; JUV = juvenile; INF = infant.

In order to enquire what the current situation was at other sites, two colleagues and I sent out a questionnaire to nine chimpanzee sites: Gombe, Mahale, Taï, Kalinzu, Assirik, Ugalla, Chambura Gorge, Bossou, Kibale (Kanyawara), and we included our own data from Budongo. We requested information on community size, age/sex composition, injuries attributable to snares, nature of disabilities and local circumstances affecting type and frequency of snare and trap practices. The results are reported in Quiatt *et al.* (2002), and the data on snare injuries are shown in Table 9.3.

As can be seen, males had somewhat more injuries than females, probably because they cover larger distances and are bolder when it comes to crop-raiding. No infants were injured by snares (although, as stated earlier, we have had a snared 3-year-old infant at Budongo).

Chimpanzees were not the primary target of snares set in any of these places. A variety of prey species, from squirrels to hippopotami, were targeted at different sites. The preferred snare material was wire, from single strand to 14-strand cable, and there was some use of cord, from 2 mm to 14 mm diameter, and naturally occurring vines. Snares were of the non-sprung and the sprung type. Traps, where set, were of the leg-hold type. As is seen, the problem of snares is Africa-wide and needs to be addressed urgently.

10. *The human foreground*

Into forest at 7.45 with Kennedy who is going to show me where the illegal pit-sawing is taking place in the Nature Reserve. Follow 0B west to NR. At NR this becomes the zero trail, E-W. We continue for c. 1 km to line 11 then turn left and immediately find a pitsawyers' site. A large mahogany (*Entandrophragma utile*) has been cut and lies on the forest floor. One section 4 m long has been removed from middle of tree & rolled up on to a platform & partly sawn. I take photos. Diameter 82 cm where cut (20 March 2002).

Economic impoverishment can lead to over-use of natural resources (trees and wood, animals, water) which are seen as freely available. We know there is encroachment of Budongo Forest around its borders, that its valuable trees have been plundered, and its wildlife is being taken for food with increasing intensity. But who are the people doing this, and for what reasons?

Fortunately we have had some excellent studies on these matters by students working with the BFP.

Micro-demography of the local population

We have an insight into the nature of the local human population around Budongo thanks to a study by Heidi Marriott (Marriott 1999). This was a preliminary analysis of immigration, fertility, age of marriage, contraception, age of weaning and mortality. We are lucky to have these basic data on our peri-Budongo inhabitants; they are simply not available for most up-country areas like ours.

Marriott worked in Nyabyeya parish, just to the south of Budongo Forest. She studied five village communities: Nyakafunjo, Kyempunu, Nyabyeya 1, Nyabyeya 2 and Nyabyeya Centre (see Fig. 10.1). These villages are adjacent to the Budongo Forest, with some homesteads bordering the forest itself.

Most people are primarily subsistence farmers growing a variety of crops, including beans, millet maize, potato, groundnuts, banana and sorghum for household consumption. Surplus produce is sold at local markets. There are some families where heads of household have either temporary or permanent jobs which exclude them from farm labour, often taking them away from home. The pressures on women are clearly increased as a result of this arrangement as they may become solely responsible for agricultural production.

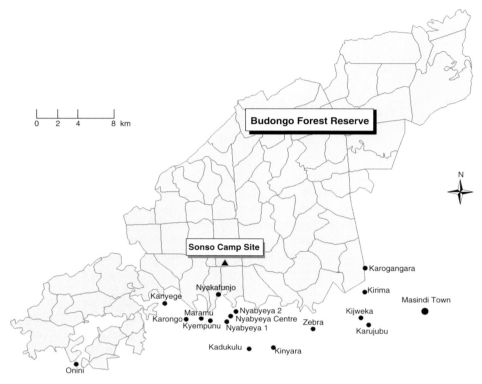

Fig. 10.1: Location of villages to the south of Budongo Forest (copyright Hazzah and Reuling).

Modern healthcare facilities are of a poor standard and dispensaries stock little other than painkillers and drugs for malaria (Sutton *et al.* 1996, see below). Traditional healers and herbalists are used more when witchcraft is believed to be involved, when modern treatment is unaffordable, and in villages where there are no clinics.

The population is characterized by a heterogeneous ethnic mix of people. Many of those living in the area were not born there but have migrated into the region as a result of civil unrest in Uganda, Sudan and the Democratic Republic of Congo (DRC), or in the search for better land and economic opportunity than are found in other more populous areas of Uganda.

Marriott interviewed 99 women over a six-week period in 1994 mainly in their homes, using a local, educated woman as an interpreter. They were interviewed alone. Their ages ranged from 14 to 60 years (mean age 31.3 years). None of the women asked refused to be interviewed, a remarkable 100% success rate: 56% had received no education, 38% had attended primary school, and only 6% had continued to secondary school. They came from the following tribes: Banyoro, Lugbara, Alur, Lendu, Okebu and Kakwa. Only 42% had been born in Masindi District; 20% came from the DRC or Sudan.

Table 10.1: Frequency (%) and mean numbers of children ever born to postmenopausal women (*N* = 69) showing surviving offspring and offspring who have died (from Marriott 1999).

Children ever born *N* (%)	Surviving children *N* (%)	Died *N* (%)
0 (13.6)	0 (13.6)	0 (27.3)
3 (18.2)	1 (9.1)	1 (22.7)
4 (4.5)	2 (13.6)	2 (18.2)
5 (13.6)	3 (4.5)	3 (18.2)
7 (9.1)	4 (13.6)	4 (4.5)
8 (9.1)	5 (9.1)	5 (9.1)
9 (9.1)	6 (13.6)	
10 (13.6)	7 (9.1)	
11 (4.5)	9 (9.1)	
12 (4.5)	10 (4.5)	
Mean 6.0	Mean 4.2	Mean 1.8

A total of 60% were currently married, with an additional 14% divorced or widowed, 20% co-habiting and 6% single. Mean age at first marriage was early at 15.7 years. Traditionally, at marriage bridewealth is paid by the groom's family to the bride's family, but in 38% of cases this had not happened: 58% of families lived in nuclear families, 23% in extended families and 12% in female headed families. At the time of the interviews, 42% of women were breastfeeding, 11% were pregnant and 22% were menopausal.

This latter (menopausal) group yielded an estimate of children ever born, including both surviving offspring and children who had died. Although the sample size is small (*N* = 69), the figures are of great interest (see Table 10.1).

Table 10.1 is of interest because it shows that this population has a high fertility rate and a low to moderate mortality rate. With a mean of 4.2 surviving offspring per woman, the doubling time for this population is equal to generation length, or approximately 25 years.

The population of the area around Budongo Forest, not counting migration, is doubling in size every 25 years. That, it seems to me, is a hugely significant finding.

Other findings in Marriott's survey were that the indigenous population had a lower birth rate than immigrants and that many women appeared not to have any knowledge of contraception. Even those who had heard of various methods of contraception were not using any. 'Current usage of contraception was zero', writes Marriott. Only a 'tiny minority' of women expressed a desire to stop or space their births. The interval between births was between two and three years. High value was attached to having offspring regardless of the sex of the child.

Marriott concludes:

High fertility rates and a pronatalist culture may be seen by a conservationist as a warning of a future which is ecologically unsustainable, and should, therefore, be challenged. However, unless the survival of children is ensured through access to reliable healthcare, and unless methods of birth spacing which are culturally acceptable are provided, there is no alternative strategy for any individual couple to take other than to maximize reproduction.

'No alternative strategy than to maximize reproduction' is a strong phrase with which to conclude the paper. Yet events since Marriott's study have been proving her right. There used to be spare land for housebuilding, now there is less and it is becoming costly. Building poles which used to be obtained free of charge (and legally so) from the forest are now sold for cash. Meat from the forest is now on sale at bushmeat centres along the road. Children are everywhere; the local Primary School at Nyabyeya is severely overcrowded.

Let us consider this school for a moment. A few years ago, the Ugandan Government introduced Universal Primary Education (UPE). Every Ugandan child is now supposed, by law, to attend primary school. In a letter (March 2003) I received from the school's Head she wrote me that numbers in the seven school years are:

P1   238
P2   150
P3   166
P4   168
P5   93
P6   60
P7   47

P1 is the youngest and P7 is the senior class. There are nine classrooms. The numbers show us both how overcrowded the school is and also the drop-off as children grow older. 'The school has no shutters [for the window spaces], no floor,[77] the walls have lots of cracks and the roof leaks terribly', she wrote. This is a common situation in rural African schools but no better for that.

Local uses of forest products

In 1991 a comprehensive study of the uses made by local people of the non-timber products of Budongo Forest was made by Kirstin Johnson (Johnson 1993). She interviewed 224 randomly selected community members of Nyabyeya parish. She found that one-quarter of her respondents were from DRC (Lendu, Okebu and Logo tribes), over one-quarter were from West Nile District in Uganda (Lugbara and Alur tribes), one-quarter were indigenous Banyoro, and the rest were from a variety of districts in Uganda.

She noted that many of these people had come to the area attracted by the prospect of work. In the 1950s and 1960s this work was provided by the Budongo Sawmills Ltd. (Our 1962 tracker, Manueri, a Bwamba from SW Uganda, was an ex-sawmill employee. He had left his wife in Bwamba in 1935 to seek money at Budongo and had never returned. We took him back there to die in 1995 — he looked for his wife but didn't find her which, in view of the passage of 60 years, is not surprising.) Today the lure of work at the Kinyara Sugar Works is bringing hundreds if not thousands more people to live around Budongo.

---

[77] Since then a church-based, fund-raising scheme in the English village of Alfriston, E. Sussex, has given the school a new high-quality cement floor.

Fig. 10.2:   Carpenter with furniture made from local materials (photo: K. Johnson).

Johnson studied the use of plant products obtained from the forest. In particular she drew attention to the income-generating use of the rattan climber (*Calamus deëratus*) found in swampy areas of Budongo. This, in combination with mahogany, was made into fine chairs and sofas which could be sold locally or in Kampala. Such furniture was still being made by local carpenters in the late 1990s (Fig. 10.2) although supplies were getting increasingly hard to find.

Another use of the forest was to make charcoal from slow-burning fallen trees, to collect herbs and other plant products for medicinal purposes, either for personal use or for treating others in the case of herbalists, and to collect particular barks and other forest items for either white (protective) or black (destructive) witchcraft. Among protective charms are some, still in use today, that will protect from 'night dancers', people with a grudge against you who dance outside your house at night and bewitch you. Destructive items are not talked about and it would indeed be dangerous to get involved with witchcraft in this part of Uganda because it remains a subject that inspires fear and trembling in many up-country folk, where sudden sickness or death are part of everyday experience and are often attributed to malice and witchcraft on the part of others.

Wild plants commonly used for food by local people and obtained from the Budongo Forest included fruits, green leafy vegetables, mushrooms and tubers. Of these, guavas and mushrooms were sold at the local Karongo market and thus contributed a little cash to some families. Forest plants were also used for medicinal purposes, e.g. the bark of the forest tree *Alstonia boonei* was reported by 37 respondents to be used for the treatment of various stomach complaints including stomach worms. *Balanites aegyptica* was used for the treatment of body pains and spirit possession. The bark of the forest tree *Canarium schweinfurthii* was used for the treatment of diseases caused by bad spirits or snakes. The most frequently used species was the forest herb, *Vernonia amygdalina*, used by 55 respondents for the treatment of malaria and eight other ailments. This is of interest because in some parts of E. Africa chimpanzees use this herb for medicinal purposes; curiously in Budongo they use a different one, *Aneilema aequinoctiale*, which was not listed by Johnson as being used by local people. Johnson (1993: 74–6) lists 65 forest plant species used medicinally. We shall return to consider health issues below.

Plant products are also used extensively in house construction. An average of 25–35 poles are needed for the construction of a traditional thatch roofed house: 71.6% of Johnson's respondents obtained their poles from the forest. Spear grass *Imperata cylindricum* was used for thatching and collected from the bush. *Pennisetum pupureum* grass was used to weave poles together. Sawn wood of a variety of species, e.g. *Maesopsis eminii*, was used for windows and doors and was obtained from pitsawyers, mostly illegally, involving some expense. Vines were used for binding parts of the house together.

Inside the house forest products are used to make pestles, spoons, stirrers, grinders, baskets, sleeping and other mats, mattresses, pillows, furniture, musical instruments, brooms and soap, and outside the house in the construction of pit latrines and bath houses. Many species are used, each with particular properties. Small baskets are made from the forest herb *Marantochloa leucantha*.

Perhaps the most important use of the forest in all households is for fuel. Nearly all cooking is done using wood. Old branches are preferred. They are brought to the house by adults (larger pieces) and children, and chopped up by both sexes using an axe. Even heavily pregnant women about to deliver will take up an axe and spend 15–20 min chopping firewood with a long-handled axe; this is considered perfectly normal. Many households try to build up a reserve of firewood. In recent years bundles of firewood have been offered for sale at the roadside, another indication of the increasing scarcity of essential materials.

Animals, as we have seen in Chapter 9, are obtained from the forest to supplement the mainly vegetarian diet, and a few domestic animals may be kept. These consist of chickens, goats and pigs. Cattle are not kept around Budongo, perhaps on account of the tsetse fly. Domestic meat is eaten only on special occasions. Hunting is particularly common among the families from West Nile and the DRC (Johnson 1996a, b). The following wild animals are eaten: cane rats (*Thryonomys swinderianus*), bush pig (*Potamochoerus porcus*), duiker (*Cephalophus* spp.), bushbuck (*Tragelaphus scriptus*) and squirrels of various species. In addition the groups from DRC hunted and ate primates, mostly baboons (*Papio anubis*) but also all local species of monkeys: blue monkeys (*Cercopithecus mitis*), redtail monkeys (*Cercopithecus ascanius*), black and white colobus monkeys (*Colobus guereza*), but not, I am glad to say, chimpanzees.[78]

Tumusiime (2002) conducted a survey of 240 households around Budongo Forest to collect socio-economic data on snaring and hunting. Factors positively associated with

---

[78] Since Johnson's study, cases have come to light in other parts of Uganda of Congolese traders hawking chimpanzee meat for sale as food. In one case a successful prosecution followed.

setting of snares were low education level, large family size and ethnic origin (with Congolese immigrants particularly involved in snaring and hunting).[79] The total proportion of the population setting snares was found to be 12%. Wire for snares was obtained illegally from Kinyara Sugar Works and was found around the old, defunct sawmills. Wild meat was considered by those involved to be of importance not just to their diet but also to their livelihood, indicating a clear economic aspect (sale of meat at markets and along the main road) to snaring.

All this hunting is being done illegally since all hunting and killing of animals except vermin is illegal in Uganda. But here is a case where the law has no teeth at all. Bannon (1997) interviewed hunters, non-hunters and local officials in and around Budongo and found that knowledge of hunting laws was limited or non-existent, laws were not enforced and were not policed, immigrants from Congo were skilled hunters who took hunting of wild animals for granted, and there were no facilities for education of the population on the law as related to hunting of wild animals. In addition there was a chronic shortage of resources and low morale among Forest Department staff so that wildlife issues remained a low priority for them. Since that time things have improved; there is now a move to bring local communities into forest management schemes, an Action Plan to protect chimpanzees is being hammered out, and there is a District Wildlife Management Unit based at Masindi. The Uganda Wildlife Authority (UWA) has the responsibility of protecting all wildlife in Uganda, including in the Forest Reserves, and whereas staff of UWA used to have difficulty working with staff of the Forest Department in field situations, things are now improving. Initiatives in wildlife legislation need to be well publicized and explained to up-country farmers; in the past confusion was rife at all levels (Bannon 1997).[80]

Local attitudes to the forest

It was clear from Johnson's study that around 90% of respondents valued the forest. Only 5.4% thought it should be cut down for agricultural purposes. It was highly valued for firewood (60.5%), for building purposes (52%), timber (51%), employment (pitsawing) (19.5%), animal resources (17.5%), attraction of rain (17%), water collection (12.5%) and medicines (11%). Many further uses were valued: wood for furniture, craft materials, charcoal production, fruit collection, soil fertility maintenance, mushroom collection, tuber collection, air cooling, scientific research, tourist attraction, prevention of soil erosion, seed collection, boat construction, wind protection, electricity poles and even fishing (there are fish in forest streams).

Johnson asked specifically about people's attitudes to chimpanzees. She notes that her link with the BFP may have inclined people to say nice things about chimpanzees. Chimpanzees were not thought to be destructive, were thought of as highly intelligent and admired by many: 65% of respondents supported chimpanzee protection; 43.8%

---

[79] It is to be hoped that Congolese immigrants will not introduce the eating of chimpanzee meat to this area.
[80] Some information on the history of Uganda's wildlife legislation can be found in Kamugisha (1993: 28–37), but this only covers the period to 1990.

thought they were not destructive; 33.5% stated they were valuable as a tourist attraction; and 8.2% of respondents thought they could be valuable as servants!

Health

Malaria is a major scourge of people living around Budongo Forest. While animal species have evidently developed genetic resistance to it, humans have only partly done so. The genetic condition known as sickle cell (Haemoglobin S), common in malarious areas, has the disadvantage that in the homozygote person (SS) it causes severe anaemia and early death. A proportion of infant mortality is due to this. The heterozygous person (AS) has protection from malaria. The normal non-sickle cell homozygote (AA) found in non-malarious areas has no resistance to malaria. Whereas travellers to Africa can take preventive drugs, local people cannot afford to buy them and so have to rely on native treatments or Western drugs for a cure after they have contracted the disease.

Malaria is known locally as 'fever'. In fact, when it comes to medical treatment this is unhelpful. Fever can be caused by sleeping sickness, typhoid, cholera, plague, yellow fever, encephalitis, heat stroke or snake bite, among other things. Nevertheless, in Uganda, 'fever' is most often interpreted to mean an attack of malaria. In Western-style clinics a blood test ensues to check for evidence of this disease, but elsewhere there is no test.

The most popular method of treatment, and the one most people automatically favour, is purchase of tablets at local dispensaries or drug shops situated in or near village centres. Johnson (1993) found that the vast majority of her respondents preferred tablets as against traditional herbal remedies or witchcraft (Table 10.2).

Whether they can afford a full course of treatment is another matter. A person with little money will buy just one tablet or two. The person in charge of a drug shop may not be qualified, and may sell drugs without a clear idea of what they are for. This is less likely to happen in a clinic or dispensary. In dispensaries there is more medical knowledge and malaria may be treated by direct injection of quinine. Injections bring their own problems. There is also the fundamental problem of diagnosis: if the diagnosis is wrong, no matter how effective the treatment for a particular disease it will be useless.

This aspect — the effectiveness of treatment with modern methods — was taken up by Sutton *et al.* (1996). They point out that Uganda suffered a rude shock to its health system when Idi Amin expelled the Asian population, and later alienated most other

Table 10.2: Frequency of types of disease treatments recorded by Johnson (1993).

Methods of treatment	Per cent of respondents (N = 224)
Tablets from dispensary	93.9
Self-applied plant medicines	36.2
Traditional doctor visited	8.0
Health centre visited	3.3
Witch doctor visited	1.4

foreigners who left the country, because most health professionals had in fact been foreigners. The civil wars of 1971–1986 disrupted medical supplies and the numbers of Western-trained doctors and pharmacists declined dramatically. Since 1986 much has been done to improve the situation but it is still very imperfect in up-country areas.

Sutton *et al.* interviewed 147 individuals in 13 villages. Their study showed that in many cases where a disease was labelled 'malaria' it might have been something else, and the same was true for respiratory and intestinal diseases. Accurate diagnosis is a serious problem in up-country districts outside the main clinics and hospitals.

Why does not everyone go to the nearest hospital? Because, again, of cost. If you go to hospital at Masindi, you have to take your food with you or pay someone to get it for you. You have to pay for your stay in hospital and for the drugs you are given. We in the BFP pay for the healthcare needed by our staff and so we know it is expensive. Sometimes outside Masindi Hospital there are people who have walked, sick as they are, to the hospital and then found they are unable to pay for treatment.

Medical personnel

A local doctor encountered by Sutton *et al.* (1996) was 'Dr X' (name withheld) who lived and worked not far from Budongo. Dr X was trained as a 'dresser' in a nearby hospital. Other than that he had no medical training. He was popular in the area, villagers trusted his treatments, considered them affordable and effective on the whole, and were prepared to travel long distances past local drug stores and other clinics to reach him. When interviewees were asked why they liked Dr X they replied that he always had a supply of medicines, he allowed credit, he always gave treatment of some sort, and he used a microscope to look at faecal and blood samples. In addition he was a local man. By contrast, they complained that the government clinics were often short of medical supplies, demanded payment at the time of treatment, and the health workers were outsiders, not from the local area.

As Johnson (1993) showed, traditional doctors using skills that include forest herbs can be expensive, charging a chicken or even a goat for their services. The charge depends partly on the perceived status of the sick person and their ability to pay. The advantage of traditional doctors is that they can treat complex diseases such as mental disorders, which may be brought on by 'charming', a form of sorcery or witchcraft. They can also concoct potions for people wishing to perform sorcery. I have myself witnessed twice the effects of a belief in sorcery, once in a man and once in a woman. The results can be devastating, including pains all over, shaking, and involuntary talking and shouting. One respondent explained 'charming' to Johnson as follows:

> Sorcery (when an enemy creates a concoction and places it along a path so the victim will walk over it) can result in swelling of the limbs, usually legs. For a cure of sorcery (charming) mix leaves of lenga (*Crinum kirkii*), kanja (*Alstonia boonei*) and munyama (*Khaya anthotheca*). Put half the mixture in tea and drink. Put other half in a sieve to obtain tiny particles and apply to small cuts made in the skin.

Why are we discussing these things in a book about chimpanzees? There are two reasons. First, looking at the details of the local population gives us an insight into the concerns and problems of the people to whom we must address ourselves if we want the forest to survive. We have to know why they value it, how they regard it and why *they* (not we) might want to keep it rather than see it cut down. Thus, if it offers cures for sorcery, that is something we need to know.

Second, there is a more practical reason. The health of people living around the forest is inextricably linked with that of the apes living inside it. Several chimpanzee sites have experienced disease epidemics and deaths from contacts with human beings living in the area, or visiting. Projects such as the BFP are currently rethinking methods of safe-guarding wild apes from the dangers posed by human beings who may bring with them viruses or bacteria to which the apes have no immunity. The Max Planck Institute of Evolutionary Anthropology in Leipzig, together with the Robert Koch Institute in Berlin and a number of other laboratories, is spearheading the use of new diagnostic methods to assess the health of great apes living in Africa, and advising projects such as BFP on how to deal with everything from mild disease outbreaks to severe epidemics in their study populations. In future we can expect to see collaborative, multi-site studies and the development of new ways of diagnosing diseases in the field and in the laboratory. It is already clear that the disruption of ape habitats caused by large-scale human immi-gration or other movements of people can lower the animals' resistance and render them more prone to human infectious diseases. We are starting with screening programmes for intestinal parasites, bacteria, viruses and blood parasites, working with non-invasive samples such as faeces and urine. As ape populations become more and more hemmed in (and this is happening everywhere), we need to do all we can to improve health and hygiene in the human population and to minimize the impact of human activities in forested and other areas where the apes live. This project has barely started at the present time.

Cash

If sorcery and 'charming' are unusual and puzzling to the Western reader, cash surely is not; we are all too familiar with the subject. A preoccupation with cash has in recent decades become the norm for the people around the Budongo Forest.

Just how cash-minded the newer and younger residents who live around Budongo Forest are was described by Mikala Lauridsen, who spent a year living in Nyabyeya Centre, in very close daily contact with her hosts, a Ugandan family, one of whom was her interpreter (Lauridsen 1999).

Lauridsen's background concern was with the question of collaborative forest management (Driciru 2001) and conservation around Budongo, but her foreground con-cern was with the people themselves, who they were and what their chief concerns were. She writes that much of the literature on local participation in forest management 'draws on the assumption that forest groups have a traditional relationship with the forest environment, and that there is important traditional knowledge available' (p. 1) but, as she points out, people around Budongo are living in a modern context and are participating in a global economy. In this economy 'work' is defined by wage-earning, while activities such as cultivating are no longer perceived as 'work'. High status goes with of having a 'job' and earning money (see also Parkin 1979).

The people of Nyabyeya are a diverse community of people from three countries and many tribes who have settled near the forest for a number of reasons all having something to do with the possibility (though for many not the reality) of earning cash. There is the remote chance that the sawmill will open again, with jobs. There is the (more realistic) chance of getting work in pitsawing, which employs men and women to fell and saw trees into 1-in. thick planks 4.2 m (14 ft) long, carry out the timber to waiting trucks, and drive the trucks to Kampala (on the mahogany industry, legal and illegal, see Singer 2002 — an exposé of the mismanagement of Budongo Forest in the 1990s). Some money has been made in the past from the sale of rattan (*Calamus deëratus*) collected in the forest, and we have already mentioned furniture-making by people with carpentry skills. There are two nascent eco-tourism projects that employ guides and campsite managers, but each is rather far away. The sugar cane works at Kinyara, 7 km away, is nowadays the best hope of men in Nyabyeya for employment.

However, there is not enough work for the large number of people living in Nyabyeya. Many of them arrived at the time the sawmill was employing hundreds of workers; they settled, married and had families. Today they regard Nyabyeya as home and do not plan to leave to go back to their native villages. As they have no work they live off the land as agriculturalists but should not be mistaken for traditional subsistence agriculturalists: they would leave very quickly if a real job paying cash became available somewhere else. They are migrant labourers. And these are the people laying snares in the forest, unless they are afraid to enter it.

Fear of the forest

Local people make use of forest materials as described by Johnson (1993), but not all of them are at home in the forest. Lauridsen (1999: 52) describes how women in particular tend to fear the forest. One informant told her 'God has given the forest strength, many different tribes get lost and die in the forest, people get killed when they enter, maybe by wild animals, maybe by something else.' There is fear of snakes and evil spirits, and women often enter the forest in groups to collect firewood. Even pitsawyers fear the forest. 'They will smear a special paste, *riji*, on their arms, which will help keep tree spirits from bothering them while cutting trees in the forest.' Some such beliefs were quite well established even before the migrants arrived, for example: 'It seems that since the early days of the sawmill, and perhaps even earlier for the Banyoro, there has been a strong belief that the *Mulimbi* tree (*Bombax reflexum*) walks and talks at night and brings trouble.' *Mulimbi* (the Lunyoro word) is described by Eggeling (1940*b*) as 'an uncommon forest tree' with bright red flowers. Such beliefs help us understand the complex relationship between people and the forest, but we should not over-emphasize these fears; the sheer number of snares laid daily in the forest shows that many people are willing to take any necessary risks in the search for wild animal meat.

Why stay?

There is a puzzle that is explored by Lauridsen in her study. That is the question: Why remain? A large number of her informants claimed to be looking for wage work but were not going far to find it. Some said they would like to go home but did not actually make the move. Some wanted photos of themselves and their families to send back home to show that they were OK. Lauridsen concludes that many families are living in hope, are not really settled, but are slowly becoming settled. I share this view. There are people living in the old sawmill houses behind our Sonso camp who are hoping the mill will reopen one day; they have remained there without more than a sprinkling of pay since the mid-1970s. They grow their own food, have no money, and wait. If an opportunity comes along to make a few shillings they seize it. For example, when the illegal pitsawing industry was at its peak and money was flowing, some of these local people made a lot of money from brewing illegal liquor which they sold to the pitsawyers, so that for about a year there was noise most nights around our camp as people drank, sang, argued and fought. Things are quiet again now. At that time too, some of the local women were able to trade sex for money. Right in the heart of the forest, in those days, youngish women wore bright town clothes and flashed eager smiles, a Western cigarette drooping from their lips — the surest sign of affluence.

So there is some work. Lauridsen gives some case studies. Edi, aged 67, worked at the Budongo Sawmill for 7 years as a carpenter, then moved to Masindi clearing bush for the Tsetse Control for the old Uganda Game Department for 10 years, then started fish-mongering, moving between Lake Albert and Nyabyeya. When this job grew too hard for him he set up a small shop in Nyabyeya Centre and also became a butcher. Lambera, aged 23, had been in the area for 3 years, had managed to get a number of jobs — one with timber, one as a tailor and one fishmongering. Batista was growing tobacco for BAT. And these days more and more people are turning their agricultural land over to sugar production as 'outgrowers' for the huge local sugarworks at Kinyara, or cycling each day to Kinyara to work at the sugar factory or in the fields. This has become one of the highest status jobs available.

In conclusion, Lauridsen was of the opinion that the critical factor in determining how people living around Budongo lived their lives was 'wanting to be seen as modern'. Having a monthly salary brought a measure of stability and the ability to plan for the future, to save money to build a house or save for a shop. We at the BFP have been paying salaries continuously since 1990 and have become one of the most prestigious places to work, even though our salary levels are not especially high. Understanding people's motivations has been one of our objectives on the Project, and has helped us maintain the loyalty of our staff, and to some extent the sympathy of the local population. We have raised some money from a charitable source to pay for housing materials for our staff; this has been a huge boost to morale. We have provided the local primary school with a new cement floor.[81]

---

[81] Thanks to money raised by the children of Alfriston Primary School in E. Sussex.

Beyond cash

It is not only cash that motivates local people. They are not, actually, money-minded. Like people everywhere, they enjoy socializing, especially having a good laugh at each others' expense, and they talk about anything and everything. They are interested in their children's education and we have been able to provide scholarships for a number of local children to continue at school into the sixth form.[82] People are very concerned about the security of their crops on which they depend for life and health; the live-traps (see Chapter 9) we have introduced have been a qualified success. Local people are much concerned with access to clean water: we have been able to help them by renovating local wells and boreholes.[83] Having cut down many of the local trees, people are now interested in supplies of firewood and wood for building; as a result, through our education programme,[84] they have taken up a longer perspective and embraced the idea of planting trees. Tree-planting was also one of the five topics selected for our Workshop in March 2001. Another was bee-keeping, and we now have a nucleus of bee-keepers in the surrounding villages. In such ways we hope to win friends and gain credibility when we tell people that they should not harm the forest and its chimpanzees.

On the question of forest management, it is clear that there is still a lot of groundwork to be done by the Forest Department[85] before participatory management schemes involving local people as managers can be expected to succeed. The locals are primarily interested in benefits for themselves, either immediate or in the future, but definite. They are used to an official economy and a 'black' economy where activities are illegal; they need to know that the official economy can function, that wages will be paid on time, that job security is a real prospect and that money, once received, can be kept safely until needed.

This last point was brought home to me not long ago when I discussed with a man who had received some money what he should do with it. I suggested he put it in the local bank in Masindi but he dismissed the idea. When I asked why, he said that all the people who had put money in the bank before Amin's time had lost it. He was absolutely convinced that if he put it in the bank he would lose it. The short-termism this attitude engenders is basic to many aspects of life for Ugandan wage earners: they rapidly turn money into goods and services rather than hold on to it for fear of losing it in any number of ways. One of the topics that we discussed at our farmers' Workshop was micro-finance: how to hold on to and manage money. This kind of basic training is not widely available in up-country Uganda and is much needed. In the meantime, people still deal in very small sums of money and most trade continues to take place either in small shops selling items such as matches, cigarettes, soap and candles, or in traditional markets (Fig. 10.3): enjoyable, colourful places that provide a social meeting place but have little if any connection with the wider urban economy in which people can really improve their standard of living and become what has sometimes been called Uganda's middle class.

---

[82] Grateful thanks to Cleveland Zoological Society which generously made these funds available.

[83] Funds for this were provided by Makerere University Faculty of Forestry through a grant from NORAD, part of which comes to BFP.

[84] Thanks to funding provided by JGI-Uganda.

[85] Reconstituted, in 2003, as the National Forest Authority.

Fig. 10.3:   Karongo marketplace (photo: V.R.).

Crop-raiding

From time to time I have mentioned the topic of crop-raiding. It is the reason farmers set leg-hold traps and the reason for our live-trap project. It is the reason farmers hate baboons more than any other species. Baboons are public enemy no. 1 in the area around Budongo Forest. These baboons (*Papio anubis*) live in the forest and emerge to raid crops. They cause a lot of damage, they take people's food which has to be bought instead, and because they may appear at any moment during the daylight hours it is necessary to be constantly vigilant. At night it's the turn of wild pigs to emerge and raid fields for root crops. So vigilance is needed by night too. Pigs are public enemy no. 2. Sometimes damage is so bad that crops have to be replanted. Kate Hill has made a special study of crop-raiding around Budongo Forest and what follows is largely taken from her work.

Her first study (Hill 1997) was made in seven villages in Nyabyeya Parish from July to September 1992. The villages were: Nyakafunjo, Nyabyeya 1, Nyabyeya 2, Nyabyeya Centre, Kyempunu, Maramu and Kanyege. Each village was visited for a total of 5–7 days. By interview, using two interpreters, questions were asked in Kiswahili, Lunyoro or English. A total of 245 people were interviewed. Only one person refused to take part.

Forty-three per cent of the sample were subsistence farmers, though some of these were growing cash crops such as tobacco; 22% were 'employees'. Others ran their own shops or pitsawing businesses.

Table 10.3: Top five species causing damage to farmers' crops (modified from Hill 1997).

Animal	Scientific name	Per cent farms raided	Farmers' rank as 'pest'
Baboon	*Papio anubis*	79.8	1
Wild pig	*Potamochoerus porcus*	78.2	2
Vervet monkey	*Cercopithecus aethiops*	33.3	3
Porcupine	*Hystrix cristata*	24.7	4
Birds: Weavers	*Ploceus* spp.	32.2	5
Doves	*Columba* sp. *Styreptoelia*		

Forty-nine different crops were cultivated, including grain and root staples, beans, vegetables, fruits, condiments and various cash crops. The four most commonly grown carbohydrate crops were maize, cassava (bitter and sweet), finger millet and sweet potatoes. Beans, taro, sorghum and bananas were also grown.

Ninety per cent of farmers reported some degree of crop losses due to raiding by animals. They were asked to rank the animals causing most damage. The top five 'pest' species are shown in Table 10.3. Chimpanzees ranked 15, with 8.6% of farmers reporting crop-raiding. Regarding vervet monkeys, these live outside the forest in open country and raid fields away from the forest.

Hill investigated the factors that make some farms more susceptible to crop-raiding than others. The location of the farm was very important. Farms closer to the forest edge were most at risk, particularly from baboons and wild pigs. Those farms within 100 m of the forest edge were at highest risk of raiding, while those farms 300 m or more from the forest edge mostly avoided being raided. These farms were buffered by the farms between them and the forest edge.

This raises a further important point. Traditionally farmers did not have their farms right up against the forest edge because of the known problem of crop-raiding. It is largely in the modern context of immigration that newcomers have been forced into this marginal land. Hill found that Lugbara people and people from DRC were most likely to have farms on the forest edge.

The fact that farmers plant their crops too close to the forest edge was also noted by Muhumuza (2002) who emphasized the need for community education programmes to inform people of the risk attached to this.

Hill found that the five crops most likely to be damaged, in order of rank, were as follows: maize, beans, cassava, sweet potatoes and finger millet. Maize, the favourite for wild animals, was eaten at all stages of development from the seeds to the mature cobs. According to local farmers, crops were differentially eaten by different species: maize was eaten by baboons, pigs, porcupines, guinea fowl, monkeys, civets and very occasionally chimpanzees, whereas sorghum was rarely damaged by animals other than birds. Bananas and taro were virtually ignored by all these vertebrate pests.

Methods of protecting crops included (not in rank order): snares, traps, shouting, ringing bells, clapping hands, lighting fires, throwing stones, chasing with pangas, chasing with dogs, scented soap laid around the perimeter of the field, goat dung spread around the field, clearing undergrowth around the field, weeding around crops, fencing using local rope or wire (sometimes with snares inserted in the fences), poisoned bait scattered among newly sown seeds to kill guinea fowl, bow and arrow and spear.

Why grow 'at-risk' crops?

As Hill puts it: 'If maize is really as vulnerable to animal damage as would appear from this study then why do people persist in growing it locally? Why not grow other, less vulnerable crops such as sorghum and taro?' (1997: 83). She suggests a number of reasons. In comparison with sorghum, the standing crop of maize cobs is resistant to bird damage and the stored crop to insect infestations. Sorghum requires more preparation time than maize. And there are cultural preferences which very often favour maize, despite it being fairly new in Uganda.

Baboons

Hill continued her work with a 12-month study (September 1993–August 1994) which followed up on the earlier survey in more detail, focusing on baboons (Hill 2000). During this year, 70 instances of crop-raiding were studied, 70% of these being by baboons. Maize and cassava were the most frequently raided crops. Of the 37 farms studied, only 15 experienced any crop damage by baboons. The average loss of crops raided per farm was 19% for maize and 25% for cassava. No damage was recorded for farms more than 450 m from the forest edge. Those farms lying near the forest edge were statistically more likely to suffer baboon raids than those further away ($p < 0.001$).

Attitudes to baboons were universally negative. They were perceived to be more destructive than other pests, to come in greater numbers and to be very persistent. 'Baboons come in large numbers and are not scared of people. When you chase them, they run and hide and then come back. When you chase other animals they run away for good', said one farmer. Baboons were considered to be intelligent: 'Baboons are a problem because of their skills, which are like those of humans — they check for the owner from tops of the trees, and when chased they just hide, and then return and take the crops.' 'After feeding, baboons destroy the rest of the crop. When there are no fruits, [forest] monkeys leave the crop; thus [forest] monkeys are better than baboons.' 'They just break and sit on any food they don't eat.'

Guarding against baboons is very time consuming, and it is easy to see why there has been recourse to traps and snares which have at least a temporary deterrent effect on the rest of the group. We, in the BFP, have been quite surprised at the hatred of baboons found in villages where we have put live-traps. People were universally pleased at our interest in their baboon problem. They have the vermin control people to whom they are supposed to refer these problems but because of shortage of resources they receive little or no help. This whole area of baboon control needs much more attention, and research needs to be focused on practical ways to help the farmers while considering the alternatives open to the baboons themselves if they are not to raid crops any more.

A further aspect relating to the intractability of crop-raiding by baboons is stressed by Hill (2002) where she refers to the issue of predictability. Contrasting baboons with birds, people told her: 'birds come to the fields early in the morning and again late in the afternoon/early evening — these are times when the children are free from school so can be sent to the fields to scare the birds'. She continues:

> when people talk of baboons they present a very different picture. Baboons are considered to be unpredictable — they can come at any time and they will eat whatever is in the field, and what they do not eat they destroy . . . . In addition to baboons causing more damage than other species they are also considered very difficult to deal with because 1. people cannot necessarily predict when or whether they will visit an individual farm, and 2. the protection methods available are not considered adequate (Hill 2002: 66).

Hill contributed to our Farmers' Workshop in 2001 and gave some useful ideas as to ways that farmers try to safeguard their crops in other countries (West Africa, Zimbabwe, Asia, Cameroon and Nigeria). These included scarecrows, trained dogs, planting non-food crops on the edge of land, fencing, planting thorn or sisal fences, digging ditches, control shooting, application of chemicals and using chilli peppers (either burned or chilli grease on strings).

In the ensuing discussion at the Workshop a number of points were made: the lack of resources to deal with the problem, the increase in baboons since sugar cane had been more widely planted around the forest, the issue of compensation for crop losses and the increase in the human population — indeed, humans could be seen as the root of the problem, for it is they who have increased in number on land previously foraged over by baboons.

We should remember that the 'problem of baboons' is as much a problem of human beings. The hatred people feel towards baboons is matched, I have no doubt at all, by a consuming hatred felt by baboons towards people.

Chimpanzees and humans

The preoccupation of people with baboons rather than chimpanzees was brought home to me when Fred Babweteera and I visited three local villages to address the communities about the need to protect chimpanzees and try to avoid snaring and trapping them. In one of the villages, we invited questions. I had noticed a decidedly cool atmosphere descending over the meeting and when question time arrived it became clear that during our talk the people had been thinking about baboons the whole time and had imagined we were trying to stop them from chasing them off their fields. We hastened to explain the difference between baboons and chimpanzees. Some people knew of chimpanzees but others did not. This village was less than a mile from the forest but presumably chimpanzees were rarely or never seen there.

Surveys of attitudes to chimpanzees must, therefore, be made inside chimpanzee habitat. Two studies have been made specifically focusing on chimpanzees, one in Nyakafunjo village (Watkins 2001) and one in Nyabyeya Parish (Kiwede 2001). A third study has been made comparing attitudes towards a number of crop raiding species including chimpanzees (Tweheyo *et al.* in press).

Cristy Watkins interviewed 53 adult individuals (23 women, 30 men) at Nyakafunjo village just outside the forest (Watkins 2001). Her interpreter was one of our field assistants from the BFP. Of the people interviewed, 45% saw chimpanzees on a daily basis. This happened in the forest, at the edge of the forest, or in mango trees or other crops in gardens and fields nearby.

> People tended to have either neutral interactions, where both the chimpanzee and the person ignored and passed each other, or fearful interactions where either the chimpanzee ran or the human ran. While some people might ignore the chimpanzee, let it eat or pass by, some mentioned that they sometimes stood to watch the chimpanzee's behavior. People seemed to identify with chimpanzee behaviors that were similar to human behavior, like food sharing, grooming and maternal behaviors. 86% of respondents felt that chimpanzees were not pests. Chimpanzees are most frequently seen in mango trees near human residences, or occasionally stealing a sugar cane stem, but people were adamant about the fact that 'according to the law of chimps' they take one helping for themselves and if they have a 'friend' or a 'wife', they will take two. Similarly if chimpanzees were raiding maize or sugarcane crops, people maintained that it was better to allow the chimpanzee to eat one stem, because when the chimpanzee was through it would leave the stem there, with the top part still intact. If people were to chase the chimpanzee away, it would flee with the stem, rendering the top useful part inaccessible to the farmer.
>
> Regardless of the manner in which humans and chimpanzees interacted, chimpanzees were still not considered pests. Similarly, occupation did not have an effect on the perception of chimpanzees as pests. The majority of all occupation categories felt that chimpanzees were not pests. The length of time respondents had lived in Nyakafunjo did not affect this perception either (Watkins 2001: 43–4).

Interestingly (especially in view of current debates in taxonomy!) 11.3% of Watkins' respondents thought chimpanzees were human rather than animal. Even those who thought they were animal believed they were related to people. Like baboons, chimpanzees were regarded as clever, but unlike baboons, 'chimpanzees have a good character, and are friendly, well behaved and considerate'. Whereas only 6% felt that a captured baboon should be spared, 100% felt that a captured chimpanzee should be spared. (Of those who wanted captured baboons spared, all qualified this by saying that they should be controlled and kept in the forest.)

Watkins asked whether chimpanzees were threatened: 45.3% thought that chimpanzees were not threatened at all, but 28.3% listed snares and other dangers from people as threats, while 13.2% listed drought, lack of food and illness as potential threats; 79.2% felt that simply leaving the chimpanzees alone would significantly help them. At the same time, 34% stated that tourism and research were the most important benefits from chimpanzees.

This last point is important. These people, living in close proximity to the forest, are aware of the benefits in terms of employment offered by the BFP and the nearby Busingiro eco-tourism site. The main general benefits of eco-tourism to local populations have been described by Grove (1995)[86] and Akhos (2000). Wathyso (2000) made a study of the benefits of eco-tourism around Budongo Forest and found that while there were no financial benefits for most people, they did support it because the Busingiro eco-tourism site had provided financial support out of its revenue to local primary schools.

There were other, less expected benefits arising from chimpanzees. Locals stated that wild pigs and baboons will keep their distance when chimpanzees are around, so that chimpanzees serve as 'guards' for crops; they accepted that this did not always happen, however.

Our second study to date of human attitudes to chimpanzees is that of Zephyr Kiwede, our senior field assistant at the BFP (Kiwede 2001). Kiwede conducted 120 interviews in the 10 villages comprising Nyabyeya Parish. He was especially interested in any possible changes in attitudes to chimpanzees following the widespread adoption of sugar cane growing by local villagers, as outgrowers for the Kinyara Sugar Works. Would the chimpanzees' known propensity for sugar cane have caused a hardening of attitudes among farmers? His sample contained 20 outgrowers of sugar as well as 100 non-sugar-cane growers. Being a Ugandan who knew the local languages as well as English and Swahili, Kiwede was able to conduct the interviews himself without an interpreter. The interviews included people from the Alur, Lendu, Kaligo and Okebu tribes of DRC, Kakwa from the Sudan, and among Ugandans Lugbara from Arua, Banyoro (the indigenous inhabitants of the Budongo area, here before the other tribes arrived), Bakonzo from Bundibugyo, Baganda from Buganda and Madi from Moyo.

Baboons and pigs were considered the most serious vermin. Chimpanzees were not regarded as a problem.

> One of the farmers reported that if two chimpanzees came to your garden, for example maize garden, each chimpanzee would take two cobs of maize. By greediness if one decides to take more than enough that causes a fight among themselves. So farmers considered chimpanzees as 'well behaved' animals.
>
> Farmers reported that chimpanzees were not a problem . . . only that they have always come for pawpaws and mangoes [which was] okay because there are many around. However, 21% of farmers complained of chimpanzees as serious vermin. Of these, 18% were sugar cane growers. 79% reported that chimpanzees assisted in chasing away the notorious baboons from their gardens (Kiwede 2001: 21 (also reported by Watkins 2001)).

The above point is important: 21% of farmers, and almost all of the sugar cane out-growers, considered that chimpanzees were 'serious vermin'. This is a new crop, and one that has only been grown for five years in the area since the Kinyara Sugar Works (KSW)

---

[86] Grove lists the following benefits: money for the local community, encouraging them to conserve the forest; job and training opportunities; some participation in forest management; and discouraging illegal activities such as pitsawing and hunting.

began its outgrowers scheme under which farmers are paid by KSW for the sugar they grow on their own fields. As suspected by Kiwede, the growing of sugar cane right by the forest has affected farmers' attitudes to chimpanzees in a negative way.

Kiwede asked farmers what their attitude would be if they found a chimpanzee in a trap: would they run away, rescue it, report to the local council, report to BFP or kill it? In reply, 55% of farmers said they would leave it in the trap and move away, the reason given was that chimpanzees could be very dangerous especially when trapped; 23% replied that they would report this to the BFP or to the local council so that action could be taken straight away; 15% replied that they would rescue it 'because chimpanzees are human. How can one leave his relative in the trap and move away?'.

However, 6% replied that they would just kill it and gave the following reasons:

- Bones of chimps when crushed and made into powder are good to be applied on fractures. Application: lacerate around the fractured area and apply the powder.

- Lacerate behind the palm and apply the powder. It's known to be good for boxers.

- The same powder can be used for relief of pain in vessels. Lacerate around where you are feeling the pain and apply the powder.

- The same powder is believed to chase away ghosts, if the powder is smoked near the patient.

At Nyakafunjo (one of the villages in Kiwede's study) a chimpanzee was killed a few years ago and when we made enquiries we were told it had been killed because its bones would be used for medicinal purposes. This was the first such case we had come across in our area. We were told that this was done by a visitor from Arua, to the north. The chimpanzee, probably a member of the Nyakafunjo community living to the south of our Sonso community, was never found.

Kiwede's conclusion: in general the attitude locally to chimpanzees is a positive one even among farmers. However, the spread of sugar cane outgrowing is changing attitudes from positive to negative. The negative attitude is one we encounter in extreme fashion when we consider, in the next chapter, the case of Kasokwa Forest, a southern offshoot of Budongo Forest, where hostility between humans and chimpanzees has reached severe, and lethal, proportions.

In the final study reported here, that of Mnason Tweheyo and his collaborators (Tweheyo *et al.* in press), 144 farmers around Budongo Forest were asked to rate crop-raiding animals according to how seriously they damaged crops. The results are shown in Table 10.4. This table is quite informative as it shows quantitatively the perceived difference of threat posed by chimpanzees as against baboons and bush pigs: the former are not considered to be either very destructive or even destructive, but enter the picture as being moderately destructive; by contrast both baboons and bush pigs occupy the two most destructive categories.

Table 10.4: Number of farmers (*N* = 144) categorizing wild animal species shown according to degree of destructiveness to crops (from Tweheyo *et al.* in press).

	Baboons	Bush pigs	Chimpanzees	Monkeys	Porcupines
Very destructive	84	20	0	2	0
Destructive	14	70	0	14	0
Moderately destructive	0	2	38	28	4
Mildly destructive	0	0	22	18	6
Least destructive	0	0	0	2	12

In this chapter we have looked at the human pressures on chimpanzees and seen how human perceptions of them differ in some respects from their perceptions of other crop-raiding species. In the next chapter we look at a situation in a village to the south of Budongo Forest where, because of excessive population pressure, perceptions of chimpanzees have changed for the worse, and actions towards them have become highly antagonistic.

11. *The Kasokwa Forest chimpanzees: a breakdown of trust*

I guess this has to be the moment when I'm most at peace with the world, and, yes, with myself. The fig tree is, for half an hour, a glorious orange colour as the western sun paints its branches each evening. The hornbills weave in and out, moving from the trees to the *Broussonetia* line. The monkeys jump from branch to branch. Behind me in the house Andy enters data — he's been doing it since 7 a.m., a 12-hour stint. Over at the staff houses logs are being split and dinner prepared by Priscilla and others for the field assistants, still not back from their full day in the forest. Sounds — the caw-caw of the hornbills, the crickets, the frogs, children, an axe on wood, a ball being bounced, the wings of a hornbill beating the air, voices of men and of women — nothing jars here, no telephone rings, no TV domineers, no cars flash by, what incredible luck to be in such a harmonious little world. Mind you there are realities here too. Like setting the rat-trap at bedtime: I'll try cheese tonight. Last night the rat ate the potato without setting it off (6 p.m., 31 March 1995).

In September 1999 I was introduced to Richard Kyamanywa, who lived in a village along the main road to Masindi. He told me he knew of a population of chimpanzees living in a forest fragment just outside the main Budongo Forest block. These chimpanzees were in some danger from local people. Soon afterwards he joined the staff of BFP as a field assistant.[87]

Kyamanywa told us that a small community of 13 chimpanzees was living in the Kasokwa Forest Reserve, a riverine forest 73 ha in extent, which was an outlier to the south of Budongo Forest not far from his home village, Karujubu (see Fig. 11.1). Since May 1999 he had been taking an interest in these chimpanzees since, he said, they were in danger of being killed by local people who were angry at their crop-raiding habits, they were also being displaced by the growing human population in the area. The Kasokwa Forest Reserve is located just east of the huge Kinyara Sugar Works. It is a riverine strip of forest following the line of the Kasokwa River. In fact it is part of a series of riverine forest strips lying to the south of Budongo, and was in the past joined to Budongo Forest main block along the line of the river. Today the human population has spread, creating whole new villages such as the village of Zebra, and has cut down the trees for 2 km between the main forest and the Kasokwa Forest Reserve.

---

[87] This was possible thanks to a grant from the National Geographic Society.

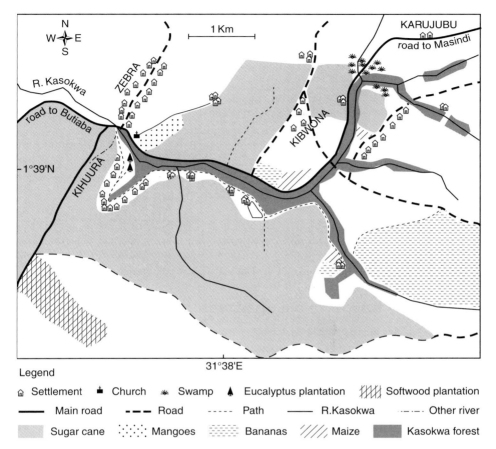

Fig. 11.1: Map of area to south of Budongo Forest, showing location of the Kasokwa Forest (from Chase 2002).

Because the boundary markers of the Kasokwa Forest Reserve could no longer be determined, because the forest was not being guarded by the Forest Department, and because there was some confusion about what areas were Forest Reserve and what were communal forests, immigrants to the area seeking work at the Sugar Works were encroaching on the Kasokwa riverine strip. From the road the forest looked intact but once you stepped inside it was being rapidly cut down and replaced by farmers' fields and homesteads. In this situation, the chimpanzees were coming under increasing threat from loss of habitat.

In this situation Kyamanywa was given the job of finding out more about the tiny community of Kasokwa chimpanzees: he would confirm there were just 13 of them, establish the group's demographic characteristics and its range, see whether they moved back to the Budongo Forest main block (which would entail travelling through village lands), find out what threats they faced from the local population and to what extent they engaged in crop-raiding on the sugar cane fields next to their forest, and try to work out how their survival could be ensured. He would observe them and make notes on them following the methods used by our field assistants at Sonso, and he would interview local people about them.

Since that time we have been collecting data on the Kasokwa chimpanzees. It appears they are confined to the Kasokwa riverine strip and the surrounding area which consists of gardens, farmers' fields and the Kinyara sugar plantation.

They subsist mainly on forest foods but are forced out of their little forest strip by food shortages there, and at such times they engage in crop-raiding. They have also been seen to hunt black and white colobus monkeys, and once, when they were successful, to eat the meat. Kyamanywa (2001) notes that females take part. He writes

> Both male and female (subadults and adults) participate in hunting. Some go and lay an ambush on the ground while some of the juveniles go up the tree where the prey are and try to provoke them, shaking the branches. In turn the prey, especially black and white colobus monkeys, chase the chimpanzees but are clever not to go down to the ground where the ambush is.

Tool-use by Kasokwa chimpanzees

The Kasokwa chimpanzees make leaf sponges to drink water, as the Sonso chimpanzees do. Also, like the Sonso chimpanzees, they eat honey, using sticks to obtain it.

> However, this attempt has proved dangerous as bees retaliate and bite or sting the chimpanzees forcing them to jump from a high tree to the ground. This is common with males (adult and subadult). They have also been observed, especially the males, carrying stones in their hands coming from crop raiding. But what they use the stones for is not clear. Perhaps it is a kind of defence in case they are attacked by humans while leaving the gardens after crop raiding (Kyamanywa 2001).

They feed in the Kasokwa forest when fruits are available, and have leaves, shoots and flowers as alternative foods. However, sufficient food is not always to be found, at which times they move out into the sugar cane to feed. They also visit local gardens to take paw-paws and mangoes. Kyamanywa (2001) writes:

> During mango fruiting seasons, they spend most of their day time feeding on mango fruits in villages around the forest. This happens mainly between 8.0 a.m. and 11.30 a.m. for the morning peaks and between 3.30 p.m. and 5.30 p.m. for the afternoon peaks...they cross through homesteads and feed on mangoes in gardens where people are farming. This exposes them to the danger of people throwing stones at them, hence in that way making them aggesssive.

Unlike the Sonso chimpanzees, the Kasokwa chimpanzees keep silent most of the time. They do not engage in the loud pant-hoot choruses characteristic of the Sonso chimpanzees. Kyamanywa speculates that this is because they live near people and fear them, and so they try to keep hidden as much as possible. He himself had some difficulty in observing them when he started his project, and they have remained elusive and hard to habituate.

Despite the difficulties, the constitution of the Kasokwa community was documented and each individual had been recognized and named by the end of 2000:

1 adult male: Kigere, alpha male, leader of the group

5 adult females: Kakono, Ruhara, Kemoso, Amooti, Kanyege

4 subadult males: Abooki, Sukari, Komuntu, Muzungu

2 juvenile males: Katoko (son of Kakono), Rukidi (son of Ruhara)

1 juvenile female: Amata (daughter of Amooti)

(An infant was born to Ruhara in August 2002).

Because of the hostility of local people to the chimpanzees, in 1999 Kyamanywa set up a conservation group called NACOPRA (Nature Conservation and Promotion Association), based on his home village of Karujubu. This organization has been successful in raising awareness of the need to conserve the chimpanzees and protect their habitat, but has encountered some intractable difficulties.

As a result of the continuing chimpanzee – human conflict in this area, a meeting was held in Masindi in April 2000 to discuss the survival of these chimpanzees. The major wildlife protection agencies attended: UWA, FD, JGI and local officials. Also among those attending was Dr Janette Wallis who subsequently secured further funding from the American Society of Primatologists for an assistant, George Otai, to help Kyamanywa with his work and took an increasing interest in this community and its problems (Wallis 2002*b*; Reynolds *et al.* 2003). The involvement of JGI led to the funding in June 2000 of two Wildlife Guards attached to the District Wildlife Management Unit to patrol the forest, remove snares and deter encroachers.

Kyamanywa also discovered other small clusters of chimpanzees in some of the forest fragments to the south and southwest of the Kasokwa Forest, including the Kasongoire Forest community. These remain unstudied at the present time. R. Kyamanywa (pers. comm.) estimates that the Kasongoire community may consist of 35 individuals. Whether there are interactions between the Kasokwa and Kasongoire communities is not known.

Chimpanzee–human conflict

In 1999, Kyamanywa had told us that many of the local people in the Kasokwa area were hostile to chimpanzees and would like to get rid of them. This was largely because of their crop-raiding activities, but also because in 1998 in Kibwona village neighbouring Kasokwa Forest a chimpanzee had attacked and injured a boy aged 6 years, injuring his genitals.

Soon afterwards there was further conflict. In June 2000 Kigere, the alpha male of the Kasokwa community, was found dead after his right hand was caught in a large leg-hold trap placed near a mango tree in a local garden; he dragged and carried the trap, which had snapped shut across his right hand, for seven days before dying of gangrene. His right leg had been lost previously in an earlier encounter with a leg-hold trap. His case and the post mortem that followed it were described in Chapter 9.

Within a short time of Kigere's death, a new adult male, together with an adult female and her infant, joined the community. Where this male came from is not known; he may have come from the main Budongo block to the north, or he may have come from one of the forest outliers further south. However, these individuals seem to have been attracted by the mango trees and not by the absence of an alpha male because they did not interact with the Kasokwa community and disappeared when the mango season was over, leaving the Kasokwa community without a mature adult male.

More conflict followed. In July 2000 one of the chimpanzees (thought to be either Kemoso or Sukari) seized an 8-month-old human baby boy near Kihuura village where immigrants had cleared a patch of forest and settled. The baby's mother had left it unattended while she dug her field by the forest edge. The baby was carried up into a tree in the forest, bitten, and subsequently dropped onto the forest floor where it was later found by one of our Kasokwa research assistants, Alfred. No part of it had been eaten. The baby was rushed to hospital but died on the way. Our project director attended the burial and made a contribution to the burial costs, explaining that this was on humanitarian grounds and not an admission of responsibility or any kind of official compensation. It was said by local people that the immigrants (or 'squatters') whose child it was had settled too close to the forest and had been in the habit of shouting and throwing stones at the Kasokwa chimpanzees when they came crop-raiding. The local people did not want squatters there, but were understandably aggrieved and hostile to the Kasokwa chimpanzees and vowed to kill any chimpanzee that came into their vicinity.

In part because of these conflicts, early in October 2000 the indistinct boundaries of the Kasokwa Forest Reserve were re-demarcated by officials from the Forest Department and encroachers were identified. There were 81 encroachers, in 25 families. They were given the alternatives of moving to new land, renting or leaving to return to their home areas. The encroachment that actually threatened the existence of Kasokwa Forest was thus temporarily halted.

In 2001, according to Kyamanywa (pers. comm.) the Kasokwa chimpanzees started to take chickens, 9 from one farmer and 3 from another. It began to look as if they were increasingly desperate for food. While their image in the eyes of local people is thus deteriorating, we should not blame them. Their actions are the direct result of human interventions in their habitat. The bad character now attributed to them in this area is wholly understandable; human beings equally threatened would react in similar ways.

Further cases of trapping

A report by Dr Richard Ssuna, veterinarian of the JGI and the Chimpanzee Sanctuary and Wildlife Conservation Trust (CSWCT), details a case of a chimpanzee trapped at the

Kasongoire Forest to the southwest of Kasokwa Forest in April 2002. The case had been brought to the attention of JGI by Kyamanywa.

> This chimpanzee was first sighted on Thursday 11th April by the Kinyara sugar cane guards. It was seen dragging a large man-trap on one of its limbs… They saw the chimpanzee up in a tree with the rest of the party staying around and extremely agitated. They were vocalizing and displaying all the time.
>
> On Monday 15th April Dr Richard Ssuna and Isaac Mujaasi set out for Masindi. We met both Richard Kyamanywa and the Agriculture Manager of KSW with whom we went out to find the chimpanzees.
>
> Before we entered the forest, the participants were addressed about all the possible approaches and the gravity of the task that lay ahead. A drill was carried out about the use of the net and our reaction in case the chimpanzee attacked. After that, we prepared our veterinary gear and set out tracking the chimpanzees.
>
> Just before we tracked them down, a faint foul smell was sensed. Chimpanzees were finally found feeding in a *Pseudospondias microcarpa* tree. We only identified two infants, a female in estrous swelling, a female with a missing hand and a normal adult male as the rest ran away after seeing us. In this same area we discovered the remains of a decomposed chimpanzee.
>
> There was a 40 cm long metallic trap, weighing approximately 12 kg with the teeth tightly locked along the entire row of the first phalangeal bones of digits one to four of the right hand (Ssuna 2002).

Ssuna recommends that 'use of man-traps should be completely outlawed. This follows the amnesty that was arrived at by the district administrators where locals were required to voluntarily hand in the traps for the manufacturing price'.[88] He recommends that personnel should be appointed to oversee the enforcement of the ban on these devices.

Another case of trapping was written up by Kyamanywa:

> On Sunday 26 May 2002, in Kasongoire Forest to the southwest of Kasokwa, a chimpanzee was reported caught in a mantrap [the name used locally for large leg-hold traps] by two Kinyara Sugarworks cane guards. The chimpanzee was seen carrying the trap with both hands, moving awkwardly on its feet through a buffer zone between the forest and sugar cane plantation. It was unclear which hand was trapped. The chimpanzee was moving in a group of three: 2 adults and 1 juvenile. On Monday 27 May the trap was found . . . and brought back to the offices of KSW. The metal trap weighed around 7 kg. The mechanism of the trap was weak, the teeth of the trap did not close tightly, this indicates that the chimpanzee probably pulled its hand free of the trap suffering only minor injuries to the hand. There was no evidence of blood or flesh on the trap or at the site where the trap was found. However between the teeth of the trap there were some hairs and a thin section of skin which looked to be from a chimpanzee (Kyamanywa 2002*b*).

On 25 July 2002 another chimpanzee was caught in a large leg-hold trap or man-trap in the vicinity of Kasokwa Forest but it escaped with injuries to its wrist and hand. Apparently this trap had been set with banana leaves in it, so that it would not cut off the

---

[88] In fact, a by-law was passed to this effect in 2001, but awareness of it, let alone enforcement, is minimal.

hand or foot of the victim and so allow it to escape. In fact the chimpanzee was able to extract its hand from the jaws of the trap because the leaves had become slippery.

More attacks by chimpanzees

Since that time there have been two more cases of chimpanzee attacks on humans. In each case the victim has been a girl and in each case the age was the same, 8 years old. In the first attack, in February 2002, near Kirima village, to the southeast of Budongo Forest (see Fig. 10.1), a girl was attacked by a chimpanzee in the forest while she was with her mother and other women and children collecting firewood in the forest. The girl was bitten on the upper leg, genitals and hands, and hospitalized for two months. In revenge for this attack, the villagers of Kirima hunted and killed a juvenile chimpanzee in March 2002.

In August 2002, in the same area, a second girl was attacked and bitten badly on the face and arms. Possibly the same chimpanzee, which appears to have been an adult male, was responsible for both attacks. The attacker was one of a group of 7–13 chimpanzees which had come out of the main Budongo Forest block for crop-raiding. Kirima village is on the forest edge — a new settlement of immigrants. The villagers have had to cut down their paw-paw trees, have removed passion fruit vines, and have abandoned planting pineapples because of the chimpanzees. People here, according to Kyamanywa, are very scared of the chimpanzees, more so than of baboons. Baboons raid and do great damage, but they rarely attack people.

Here is a more detailed account (names withheld) of the circumstances of, and what followed, the August attack:

> On 10th August 2002 I received a report from a resident of Kirima village that a chimpanzee had attacked and injured his brother's daughter. According to the report, he said it was on 9th August at around 3 p.m. when the attack occurred. He mentioned that the 8-year old obtained deep cuts on the face by the left eye, on the arm, and fingernails were pulled out. He said that the girl was obtaining medical treatment at the Doctor's clinic and a bill of 12,000/- (around £5) was outstanding relating to a vaccine for tetanus.
>
> While the story was being told, the father of the girl arrived with the girl on the back of his bicycle. Immediately he started quarreling bitterly and attracted the attention of Karujubu Trading Centre. In his quarrel he mentioned the following points:
>
> 'You are the people who are saying that we should protect chimpanzees and that we shouldn't kill them. But your chimpanzees are attacking and killing our children. Do you mean chimpanzees are more important than human beings? I am going to buy a kilo of nails for making arrows and tomorrow I am starting to hunt them and kill all of them!'
>
> I tried to direct him to the offices of UWA in Masindi but I was stimulating his temper. His demand was for compensation for the medical bills and feeding [in hospital] of his daughter. I witnessed the wounds and the medical form. The girl was really badly injured (Kyamanywa 2002*a*).

Kyamanywa adds the following significant points:

- Intervention such as holding village meetings for sensitization on the protection of chimpanzees may be futile.

- At the moment we are planning to ban the use of man-traps through an amnesty. If the above issues are not tackled, our mission is bound to fail.

- Government policies do not provide for compensation in such cases. As conservationists we should be flexible and think to the ground.

He made the following recommendations:

- A tentative provision should be put in place for compensation of such incidents directly or indirectly.

- A team from Masindi should visit the affected families for counseling and education and should remember to pay a small compensation. This is a temporary measure whilst comprehensive education and sensitization about chimpanzees is going on.

- An immediate programme should be broadcast on local radio focusing on the interactions between the local people and chimpanzees in order to reduce such accidents (Kyamanywa 2002*a*).

Can a solution be found?

To date there is still no compensation scheme for victims of chimpanzee attacks. BFP is reluctant to pay compensation because such payment indicates responsibility. An *ex gratia* payment was made to the father of the second girl attacked, but this is not the same as a proper government-backed or UWA-backed compensation scheme. I agree with Kyamanywa that if conservation organizations are going to protect chimpanzees, then even if some responsibility for the attacks rests with local people for growing crops close to chimpanzee habitat, there should be a properly sanctioned education programme and a compensation scheme funded by local or central government authorities or UWA.

There has been some movement in the direction of finding a solution. In April, July and August 2002 meetings were held in Masindi, attended by JGI, UWA, the local District Environment Office, Department of Agriculture, Vermin Control, Forest Department, BFP and NACOPRA. Plans have been made but much still needs to be done. Issues discussed have included:

- Where chimpanzees are located outside the main Budongo Forest block.

- Whether such chimpanzees are able to travel back to the main forest.

- Whether these out-forest chimpanzees are subgroups of a widely dispersed community and meet up with each other from time to time.

- Whether the out-groups have enough food inside their forest patches to survive or are forced to supplement their diet with human crops.

- How leg-hold traps can be banned.

- How people can be educated in chimpanzee conservation given the attacks by chimpanzees on humans.

- How to dissuade immigrants from settling close to the forest.

- How to reduce the level of encroachment on forest land.

- Whether a buffer zone could be established between the main forest and surrounding croplands, and what it would consist of.

- Whether tree corridors, perhaps along rivers where trees have been cut down, could be provided to enable chimpanzees to move safely from one area to another, and to the main forest block.

- Whether setting up more eco-tourism projects would improve the situation for people and chimpanzees.

- How best to demarcate areas of public or communal land and protected areas.

- How to enforce new regulations and ensure boundaries are known and respected.

These attacks are causing a hardening of attitudes towards chimpanzees around the Budongo Forest. It seems that when the density of the human population increases to the point where they start using marginal land adjoining the forest and planting this land with fruits, the gardens become irresistible to chimpanzees from the forest, the more so if those chimpanzees are short of food themselves. Then a real war begins, with fatalities on both sides. Chimpanzees use their teeth, humans use traps, stones and spears. To a zoologist this is a case of feeding competition between two sympatric, closely related, species. In the real world it presents a tremendous problem: the villagers are angry, those concerned with wildlife management are concerned and the chimpanzees are both afraid and defensive. Indeed, their very survival is at stake.

The worst incident to date occurred during the writing of this book. On Friday 4 July 2003, sugar cane harvesters set fire to a field of cane belonging to an outgrower in the village of Kijweka near Kasokwa Forest Reserve. When fields are burned in this way, the fire is set around the edges of the field and it is burned towards the middle. Any animals caught in the fire run out and are killed by the harvesters using spears. On this occasion three of Kyamanywa's study animals were in the field: a mother, Amooti, her juvenile daughter Amata, aged 5–7 years and her infant son Amanya, aged 2 years 1 month. The mother escaped with severe burns and is thought to have died later. The two offspring were burned to death. The juvenile female was seen to run out of the fire but on encountering the men with spears she fled back into the flames and died. Nothing more horrible can be imagined. Had any effort been made to establish whether there were chimpanzees in the field before the fire was set they could surely have been flushed out. Evidently the men concerned were not bothered. Pictures of the burnt individuals are too gruesome to publish. Sean O'Hara, Assistant Director of BFP, wrote to the General Manager and the Agricultural Manager of Kinyara Sugar Works on 11 July 2003 complaining about this incident, and it is to be hoped that some action has been taken to prevent a recurrence.

At the meeting in April 2000 already referred to, we discussed the question of attempting to relocate the Kasokwa chimpanzees (JGI has experience of relocating chimpanzees) but this policy was not agreed by some members present who felt the solution should be to find ways for people to co-exist harmoniously with the chimpanzees. In fact, operationalizing this, even with a small group of chimpanzees living in a 7 km, officially protected forest strip, has turned out to be a baffling problem. As more and more chimpanzee habitat becomes degraded and fragmented all over Africa, we can expect to see the Kasokwa picture repeated over and over again (Reynolds *et al.* 2003). Indeed, there have been attacks by chimpanzees on humans at a number of other sites (Gombe, Mahale and Kibale) and it is possible that in each case the cause has been excessive (to chimpanzees, that is) interference by people in their lives. By contrast, where, as at Sonso, Taï and some other sites, humans respect the chimpanzees, keep a goodly distance from them, and move away when they come close, violence by chimpanzees to humans has not (up to the time of writing) occurred, despite many years of daily observations.

The role of research

We return to the question of research. I have heard it said that research is a waste of time and money when species are endangered. This is almost never true.[89] Anything we can learn about endangered species is of value. In a case like that of the Kasokwa chimpanzees, where there is a real threat to their survival, research needs to be focused on protecting the animals. As can be seen from the whole of this chapter, there is still much we do not know about how to protect chimpanzees. But the focused research we have done so far has been all important. Our research has provided baseline data on the Kasokwa chimpanzees, incidents of trapping and attacks on humans that would otherwise have passed unrecorded. It has documented the deterioration of relations between humans and chimpanzees. Research will continue to be a vital part of efforts to protect these chimpanzees, and wide dissemination of the findings from Kasokwa (e.g. Wallis 2002*b*; Reynolds *et al.* 2003) will continue to be a primary aim.

I want to end, however, by mentioning a piece of research undertaken by an undergraduate team at Kasokwa, and written up by the leader of the team, Ella Chase (Chase 2002). From discussions with Ella and other members of her team (Alice Hawkins, Richard Gregory, Jonathan Shawyer) I am conscious of how acutely they were made aware of the human–chimpanzee situation while they were living in the village of Karujubu, not far from Richard Kyamanywa and George Otai.

This was an eight-week research project conducted in 2001 and results are based on 323 interviews conducted in five villages around Kasokwa. The sample constituted 6.3% of the population of those villages. Questions were asked about the forest in the

---

[89] See Chapman and Peres (2001) for a clear exposition of the role of scientists in primate conservation.

past and at present, attitudes to the forest, chimpanzees and attitudes to them, and trapping. Karujubu turned out to be a long-established village, whose population had, on average, been there for 32 years. By contrast in Zebra this was 13 years and in Kihuura it was 7 years. The median age group of men was 25–29 years, and 69% of the sample lived in grass-thatched houses while only 13% lived in the more desirable brick houses with iron sheet roofs.

This study drew the interesting conclusion that attitudes towards conservation did not necessarily correlate with attitudes to chimpanzees. For instance, whilst younger people tended to give positive environmental reasons for forest preservation, they were also more likely to give negative responses to chimpanzees. Gender too was important, with women having more negative feelings towards chimpanzees than men.

Chase also states that 'multiple attitudes' about chimpanzees can be held at the same time. She attributes this to the fact that chimpanzees cross a highly significant boundary: the boundary between their forest home and the homes (fields, gardens) of people. As an inhabitant of the forest, an environment where humans do not live, and one where danger lurks, chimpanzees are thought of as essentially alien. But just by being alien they are not hated: there was a respect for chimpanzees. There were people who valued them for their ancestry, their behaviour, their cleverness and their looks. In economic terms also, the chimpanzees were looked on positively, for their eco-tourism potential. And then there were the frankly negative attitudes to them as pests, thieves and even rapists and murderers (see also Naughton-Treves 1997), based in part on the competition between chimpanzees and humans over space and food, resulting in the chimpanzees being seen as a threat to the safety of women and children.

This, as Chase points out, marks a major difference between attitudes in Kasokwa and attitudes in the Western world. Our fascination for chimpanzees because of their close-ness to us is not extinguished by a negative attitude based on competition and distrust. We in the West do not have chimpanzees taking our food and attacking our children. Thus, when we come to Uganda and promote chimpanzee conservation we encounter a cultural difference in attitudes based on a different life experience, and unless we take account of this our conservation plans are bound to fail.

Chase ends her dissertation with a quotation, and I shall end this chapter with the same one:

> I like them! They are always around here and we try to see them, if we go to the garden they come nearby us ... they look like people, like living people, and they used to play there on the trees with the children ... sometimes they can do very bad things. Last year they killed somebody, a little baby nearby us here. That is the badness of them. People hate them when they have done bad things, for the moment they hate them. Then they love them after (Manase Oyamtao, 65-year-old retired driver from the Kinyara Sugar Works; words recorded by Chase 2002: 39).

12. *The future of Budongo's chimpanzees and of the chimpanzees of Uganda as a whole*

The party for Independence Day was a great success. We started with lunch at 2 — boiled leg of beef + matoke + sweet potatoes + greens, all plentiful and nicely cooked...After lunch the drinks appeared, 'tonto' (sour banana liquid, fermented with some sorghum put in as grains), and the local form of beer, made of millet and drunk hot from buckets through straws. The atmosphere was very nice, transect cutters, field assistants, Chris, Andy, Chris F, Duane, Kate, Peace, me, some sawmill people who'd come over, and Paitho. Lilian and Joy served the tonto in glasses, and also there now appeared waragi in glasses which circulated strictly among the Ugandans. People sat and talked, some went away and others arrived. George arrived and told us a long story about a tortoise and a monkey and a grinding stone. As the afternoon wore on the heavy rain came and we moved indoors and then it stopped and we moved out again. The women and children danced and sang for us muzungus. The men sat at one side, the women at the other. As it got later and dusk arrived Zephyr's tape recorder began to play and the dancing began in earnest....the usual jigging about. This went on into the evening and early night, children joining in and some falling asleep by the door of the house. Finally at 10 p.m. Kate, Andy and I made an omelette & toast and retired to our beds (10 October 1993).

The modern setting

In my book *The Apes* (Reynolds 1967) I foresaw the extinction of the great apes unless steps were taken to protect them. In Budongo, large-scale mechanized logging and poaching used to be the main problems. This situation has gradually changed, with illegal pitsawing, snaring and population pressure being the main threats in recent years (Reynolds, in press). Today, mechanized logging has all but ceased, and none of the four sawmills in Budongo is functioning. The illegal pitsawyers who had the run of the forest during the 1990s have removed most of the large mahoganies, even in the Strict Nature Reserve (Compartment N15, which was set up in 1944 and where many of our studies of unlogged forest have been done). This fine area of forest, where no logging of any kind had ever been allowed, remained sacrosanct until 2001–2002 when it came under pressure as the last place where fine big mahoganies could be found, and has been ravaged by illegal logging teams, working at night. Singer (2002) disclosed the extent of the illegal mahogany trade and the fact that members of the Forest Department had been involved in it; the matter was featured in the *New Vision* in November 2002;[90] at the time of writing matters have been taken in hand but the damage has largely been done. Perhaps, as suggested by Driciru (2001), it will only be when some form of Collaborative Forest Management is working on the ground that the mahoganies of Budongo will be safe; their safety depending in that case on the fact that local people will then have a stake in the profits derived therefrom. At present such benefits are denied to local people, as was shown by Obua *et al.* (1998) in a study of 200 households in eight villages around Budongo. Their conclusion was that current mistrust of the Forest Department would only be diminished when local communities 'could be empowered to co-manage and benefit from forest resources in their vicinity' (p. 113).

Shocking though the mahogany pillage is to us at the BFP, it remains a fact that this will not have directly affected the chimpanzees.[91] They did not use the Nature Reserve very much because the surrounding logged forest provided them with much more food. As explained in earlier chapters, selective logging has the effect of removing species such as mahoganies that are inedible to primates and they tend to be replaced with different tree species many of which provide edible fruits, flowers and leaves.

Poaching for chimpanzees themselves became a serious problem during the era of civil wars (1971–1986), with Budongo chimpanzee babies being smuggled to Entebbe airport in boxes in the back of vehicles and thence being exported world-wide. Today this trade is small to non-existent for Ugandan chimpanzees, but there remains a trade across Uganda of baby chimpanzees from DRC, some of which are found and confiscated by JGI at Entebbe and then cared for until they can be released on to Ngamba Island sanctuary in L. Victoria. For example, in May 2002 a chimpanzee smuggling sting netted three Congolese men and a woman in Kisenyi District in the southwest of Uganda who were trying to sell a 3-year-old chimpanzee for US $4000. Some of the very young chimpanzee babies die even if they are confiscated; all are in poor condition when they arrive at Entebbe and require (and receive) a lot of medical and personal care.

Today the threats to chimpanzees arise from two main sources: snaring (Chapter 9) and population increase coupled with habitat destruction. For example, along the Masindi–Butiaba road, where it passes through Budongo Forest, there appears to be encroachment of forest at various places (see Fig. 12.1). I made enquiries locally and was told that this was not Forest Reserve land but was private forest land, owned by an Asian whose name could not be remembered. In another place I was told it was owned by BAT, the British American Tobacco company, for whom many local people grow tobacco. The land being felled, burned and planted was certainly very close to the border

---

[90] By this time the *New Vision* had run a series of articles on the plunder of Budongo Forest — see, for example, the issues for 30 July 2002 ('Illegal loggers plunder Budongo'), 27 August 2002 ('Loggers bribe Budongo rangers'), and the Editorial Opinion column on 31 August 2002 ('Save Budongo Forest'). Also in August 2002 the *New Vision* had run a feature 'Budongo guards want guns', based on the fact that illegal loggers were resorting to violence in pursuit of their aims. I wrote to the Acting Commissioner for Forestry on 27 April 2002 and received a reply on 22 May 2002.

[91] They will have been affected indirectly. The value of the forest lay in its potential as a money-spinner for Uganda if the mahogany had been effectively managed. Loss of value of the forest decreases the effort that authorities will put into its careful management for long-term timber production and increases the chance that it will be cut down or sold off.

Fig. 12.1: Encroachment at the edge of the Budongo Forest (photo: J. Wallis).

of the Budongo Forest Reserve. This land is being chipped away every day as new immigrants arrive and search for a place to grow their crops. First they cut down the trees, then they burn them and the branches, and lastly they clear the undergrowth, hoe the land, plant their crops and build a house. The NFA has been making strenuous efforts to stop encroachment of Reserve forest and I hope they are able to do so.

Encroachment is not, of course, specific to Budongo. Forests containing primates are threatened world-wide by fragmentation (Marsh 2002). In Uganda, because of the constant, daily need for wood fuel and for building poles, not to mention sawn timber, forest land outside Protected Areas (and to some extent inside them) is being depleted of its trees. To the south of Budongo lies the Bugoma Forest where encroachment can be seen from the road. Many other forests — private, communal and national — have been encroached or even removed. From 10% of Uganda's land cover, natural forests were reduced to 2%–3% in the past (Hamilton 1984; Howard 1991) and this process is continuing. Often encroachment begins with forest outliers, as we have seen in the case of Kasokwa Forest Reserve (Chapter 11). Hamilton (1984) already showed such encroachment in his map of the area to the southeast of Budongo, which was based on a special study of this area (Fig. 12.2).

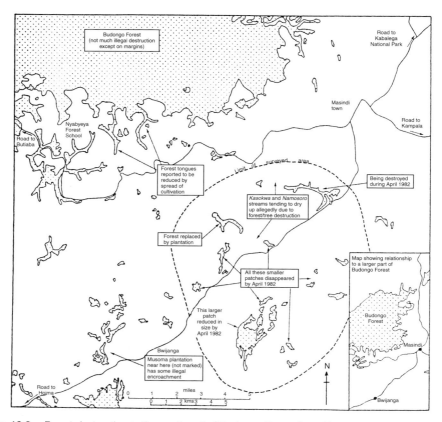

Fig. 12.2:   Forest destruction to the southeast of Budongo Forest (from Hamilton 1984).

The most recent data we have on encroachment in Uganda's forests comes from a national chimpanzee survey (Plumptre *et al.* 2003) which ranked the importance of the various threats to chimpanzees in each forest (see Table 12.1). In the case of Budongo Forest, threats were ranked from greatest (largest number) to least (smallest number) and, as can be seen, the top four threats were pitsawing, collection of forest products, encroachment and snares.

Management Plan for Budongo Forest Reserve

Since 1935 there have been four Management Plans (1935–45, 1945–55, 1955–65 and the current one 1997–2007) laying out the principles of management of this forest: how it is to be demarcated, where the borders are, licensing arrangements for tree-felling, enrichment by removal of weed species and planting of desirable species, and so on. Here I want to focus on the aspect of the current Plan that is central to our interests: the management of Budongo's chimpanzees.

Table 12.1: Threats to Budongo Forest Reserve (from Plumptre *et al*. 2003).

Threat	Area	Impact	Urgency	Total	Overall Rank
Hunting					
Dogs/nets/spears	2	2	3	7	1
Snares	13	5	10	28	10
Charcoal burning	4	9	7	20	6
Crop-raiding	9	1	8	18	4
Encroachment	3	13	13	29	11
Fire	5	10	6	21	7
Firewood collection					
Household	10	7	9	26	9
Grazing	1	4	4	9	2
Medicinal plant collection	8	3	5	16	3
NTFP					
General	11	8	11	30	13
Rattan	12	6	12	30	13
Pitsawing					
Illegal	14	12	14	40	14
Legal	6	12	2	20	6
Research	1	4	1	6	1
Sawmilling	7	14	1	22	8
Tourism	4	3	1	8	1

Numbers in columns are ranks; highest number = greatest threat.

The fourth Management Plan (Kamugisha and Nsita 1997) is current at the present time, covering the period July 1997–June 2007. It goes beyond its predecessors in its emphasis on wildlife protection. For a start, the amount of Strict Nature Reserve (SNR) has been increased considerably by the addition of several new compartments. In addition, surrounding the SNR areas we now have buffer zones (BZs), where only light harvesting can take place. The 'Conservation Working Circle' is described in Chapter 13 of the current Management Plan. It comprises 31 compartments, 17 for SNR and 14 for BZs. The total area now allocated to SNR and BZs is 30 853 ha. Of this, 11 722 ha are SNR and 5293 ha are BZs. This is a huge increase in the size of the protected area in Budongo (N15 itself, the main SNR until the present Management Plan, is a mere 747 ha) but we should remember the underlying problem of enforcement so that what appears to be a great step forward has yet to be translated into a reality. Illegal activities still go on undiscovered and not prevented. However, in terms of future development of the forest, the increase in protection is surely a good thing.

The bulk of the new SNR is in Waibira Block in the northern part of the forest. As the Plan states: 'From the ecological view point, large single blocks of forest provide better habitats for conservation of forest taxa than several isolated small blocks' (p. 75). This is surely true (Chapman and Peres 2001; Mittermeier and Cheney 1987). However, being in the north of the forest and hence far from the authorities, Waibira Block will be very hard to police.

The Management Plan gives a picture of what is intended for biodiversity conservation in Budongo generally, without mentioning chimpanzee protection in particular, except in relation to eco-tourism.[92] Biodiversity conservation in Budongo is further explored in the excellent report by Howard *et al.* (1996) but with reference to trees and shrubs, birds, small mammals, butterflies and moths, i.e. without specific reference to chimpanzees.

PHVA

In January 1997, a four-day Population and Habitat Viability Assessment (PHVA) for Uganda's chimpanzees was held in Entebbe (Edroma *et al.* 1997). PHVAs have been conducted for a number of endangered species. The organizers are the Conservation Breeding Specialist Group (CBSG), a subgroup of IUCN based in the USA. Fifty-seven participants, from Uganda and eight other countries, attended. Besides presentations, much time was devoted to establishing what was known and not known about the distribution and numbers of chimpanzees in Uganda.

The objectives of the PHVA included:

- to formulate priorities for a practical management programme for survival and recovery of the chimpanzee in wild habitat;

- to develop a risk analysis and population simulation model for the chimpanzee which can be used to guide and evaluate management and research activities;

- to identify specific habitat areas that should be afforded strict levels of protection and management.

Recommendations included the following:

- Budongo Forest should be one of four key forests given priority for chimpanzee conservation (the others being Bugoma, Kibale and Kasyoha-Kitomi).

- Some Ugandan Forest Reserves, about whose chimpanzees little or nothing was known, were designated for censusing.

- The importance of some *unprotected* areas for chimpanzees was stressed.

- A policy was needed for chimpanzees living in forest outliers in agricultural areas.

- Conservation education programmes focusing on chimpanzees should be developed.

One of the great achievements of this PHVA was to put together a series of tables outlining the status and distribution of chimpanzees, and the threats facing them in the different forests. Tables 12.2 and 12.3 provided an invaluable guide to further work on censusing the chimpanzee population that followed.

---

[92] Chimpanzee tourism has yet to reach its full potential in Uganda. The two tourism sites in Budongo Forest are seriously under-utilized; at one of them, Kaniyo–Pabidi, chimpanzee viewing is very successful indeed. At other sites, notably Kibale Forest, chimpanzee viewing is more popular than at Budongo at the present time.

Table 12.2: Ugandan chimpanzees: status and distribution (from Edroma *et al. 1997*).

Site	Status	Area (km²)	Altitude	Habitat	Continuity	Chimpanzee populations					Human populations		
						Presence	Quality	Density	N	Patrols	Density	Immig.	Cult.
Mt. Kei	CFR	200?	Medium	W	C	?		L?	0?	0	L	–	0
Otzi	CFR	50?	Medium	W	C	+		L?	0?	0	L	–	0
Rabongo	NP	2.5 (200 tot.)	Medium	G	R/C	+	S	L	20	F	L	–	0
Budongo	CFR	430 (825 tot.)	Medium	THF/G	C/R	+	C	H	650	S	L	+	0
South of Budongo	CFR & no protect	10	Medium	W	F/R	?	L	0?		0	L	+	S
Bujaawe–Wanibubaya	CFR & no protect	15	Medium	G	F/R	?	L	25		0	L	–	S
Bugoma	CFR	365	Medium	THF/G	C	+		H	450	S	M	+	S
Kasato	CFR & no protect	250?	Medium	Mosaic	F/R	?	L?	100?		0	M	+	S
Kagombe–Kitechura–Matiri–Ibambaro	CFR & no protect	300?	Medium	Mosaic	C/F/R	+	S	L?	100?	S	M	+	S
Itwara and surroundings	CFR & no protect	>8?	Medium–High	THF	C	+	S	H	100	S	H	–	S
Buyaga area (N)	No protect	400?	Medium	W	F	?		L?	50?	0	M	–	H
Mwenge	No protect & private	700?	Medium	G/W	F	+	L?	50?		S/0	H	–	H
Kibale	NP	400 (760 tot.)	Medium–High	THF/G	C	+	C	H	550	F	H	–	S
Semliki Valley	WCA	15 (200 tot.)	Low	G	R	+	S	L	60	F/0?	L	–	0
Semliki	NP	219	Low	THF	C/F	+	S	L	150	F	L	–	0
Ruwenzori North	CFR	2.5	High	THF/G	C	+	S	L	0	0	H	–	0
Ruwenzori	NP	120 (966 tot.)	High	THF/G	C	+	S	L	50	F	H	–	0
Dura R., E. Kasese	NP	200?	Medium	G	R	+		L	30	S	M	–	0
Kasyoha–Kitomi, Kyambura	CFR & NP	399	Medium	THF/G	C/R	+	S	H	500	S	H	–	0
Kalinzu–Maramagambo	CFR & NP	580	Medium–Low	THF	C	+	C/S	H/L	350	S	H	–	0
Ishasha	NP	0.2	Low	G	R	+	S	L?	25	F	H	–	0
Bwindi	NP	321	High	THF	C	+	S	L	100	F	H	–	0

Key:

Status: CFR, Central Forest Reserve; NP, National Park; WCA, Wildlife Conservation Area

Area: Chimpanzee habitat, numbers in parentheses show total protected area of site if larger than suitable chimpanzee habitat area

Altitude: Low, <1200 m; Medium, 1200–1800 m; High, >1800 m

Habitat: W, Woodland; G, Galley forest; THF, Tropical high forest

Continuity: C, Contiguous habitat block; R, Riverine strips; F, Fragmented forest habitats

Data Quality: C, Census data; S, Survey only

Density: L, Low; H, High

Patrols: 0, None; S, Some; F, Frequent

Density: L, Low; M, Medium; H, High

Cult. (Cultivation): 0, None; S, Some; H, High levels

Table 12.3: Ugandan chimpanzees: threats and conservation priorities (from Edroma *et al.* 1997).

Site	Threats				
	Habitat loss*	Poaching[†]	Disease[‡]	Political instability[§]	Tourist potential**
Mt. Kei	L	0	0	H	0
Otzi	L	0	0	H	0
Rabongo	L	0	L	L	L
Budongo	L	1.3	H	L	H
South of Budongo	L	0	L	L	0
Bujaawe–Wanibubaya	L	0.5	M	L	0
Bugoma	L	2.5	M	H	L
Kasato	M	?	H	L	0
Kagombe–Kitechura–Matiri–Ibambaro	M	2.5	H	H	?
Itwara and surroundings	H	2.5	H	L	L
Buyaga area (N)	H	?	H	L	0
Mwenge	H	0.5	H	L	0
Kibale	M	1.3	H	L	H
Semliki Valley	L	0.5	L	H	L
Semuliki	L	5	M	H	L
Ruwenzori North	L	2.5	L	H	0
Ruwenzori	L	2.5	M	H	L
Dura R., E. Kasese	L	0.5	L	L	0
Kasyoha–Kitomi, Kyambura	L	0.5	L	L	L
Kalinzu–Maramagambo	L	1.3	H	L	H
Ishasha	L	0	L	H	0
Bwindi	L	1.3	H	L	L

* (Unprotected areas): L, Low; M, Medium; H, High
[†] Percent of total population annually
[‡] Probability of human-induced disease epidemic: L, Low; M, Medium; H, High
[§] H, High; L, Low
** 0, None; L, Low; H, High
From these data the following areas are designated high priority for chimpanzee conservation in Uganda:
1. Budongo; 2. Kibale; 3. Kasyoha–Kitomi; 4. Kalinzu–Maramagambo; 5. Bugoma.
The following areas have a high priority for further surveys and represent potential chimp conservation areas:
1. Kagombe–Matiri; 2. Kasato and neighbouring unprotected forest areas.

A map of chimpanzee distribution in Uganda was published in the PHVA volume. It showed the forests and other areas where chimpanzees were known or believed to live. This map has been superceded by a census-based map (Plumptre *et al.* 2003), shown in Fig. 12.3.

The PHVA additionally produced a number of scenarios showing population projections and extinction probabilities for the chimpanzees of Uganda, based on varying assumptions of population size, habitat type and number and degree of threats. It explored the extent to which chimpanzees from one forested area were able to travel to, and hence mix genes with, other populations, again influencing their survival probabilities. These initiatives have not so far been followed up; they deserve to be.

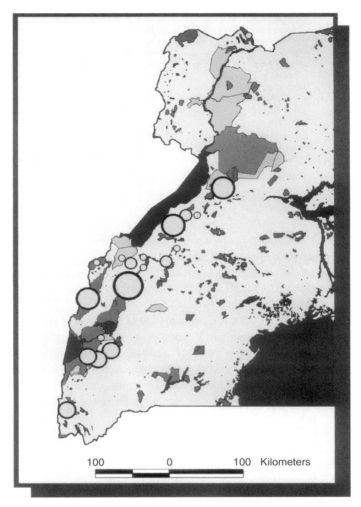

Fig. 12.3: Map of chimpanzee distribution (larger circles represent larger chimpanzee populations). (courtesy of A. Plumptre, WCS from Plumptre *et al.* 2003).

Conservation education

As we have seen, one of the recommendations of the PHVA was for improved conservation education. This idea was taken up by the BFP and we have been running a conservation education programme in the villages and schools of Budongo sub-county for the past five years. One of our existing staff, senior field assistant Zephyr Kiwede, together with a senior teacher at the local Primary School at Nyabyeya, have conducted this programme, which visits schools and villages once a week, on Wednesdays and

Table 12.4: Conservation topics mentioned by respondents (from Muhumuza 2002).

Topic	Respondents ($N = 118$)
Protection of wildlife	6
Forest conservation	5
Tree-planting	51
Bee-keeping	5
Tree nursery	2

Sundays respectively.[93] Issues that have been discussed at village meetings (often at the request of villagers) have included methods, locations and species for tree-planting, problems of crop-raiding by chimpanzees and other species, requests for live-traps, requests for seeds to plant and polythene pots to grow seedlings in, and help with vermin control. Several objectives have been achieved, e.g. a group of farmers in Bwinamira III village has established a woodlot of Eucalyptus with 100 stems; farmers in Owini village raised and sold over 700 Eucalyptus seedlings and now plan to grow Musizi (*Maesopsis*). At schools, issues discussed have included requests to visit the BFP and the forest itself, Murchison Falls National Park and the Uganda Wildlife Education Centre at Entebbe.

Geresomu Muhumuza, one of our most senior field assistants at BFP, wrote a dissertation on conservation education while studying at the Nyabyeya Forestry College (Muhumuza 2002); what follows is taken from his work. He interviewed 118 individuals living in six villages in Budongo sub-county. The majority of his informants were aged between 31 and 50 years: 59% had received primary education only, 17% had gone on to secondary or even in a few cases tertiary education; 81% were subsistence farmers; 5% were professionals; 9% were jobless.

Of these people, 59% had received conservation education from the BFP's programme. Of the 41% remaining, Muhumuza writes that they were mostly 'women who never turn up for community meetings'. Indeed, he continues 'The great and most depressing constraint in conservation is not habitat loss or over exploitation but the human indifference to socio-cultural problems' (p. 12).

The topics mentioned by respondents as ones from which they had learned new information from or benefited from the BFP's Conservation Education Programme are given in Table 12.4.

Clearly there is a long way to go, but a good start has been made with our emphasis on the need for tree-planting (we have also distributed seeds and seedlings, and referred them to Nyabyeya College nursery for tree seedlings and to BUCODO,[94] who have established tree nurseries in some villages). Of the 51 who expressed an interest, 12 had actually planted trees.[95] The interest in bee-keeping relates back to our Farmers'

---

[93] We are grateful to Oakland Zoo and JGI-Uganda for making funds available for this programme.

[94] Budongo Community Development Organization.

[95] Tree-planting is encouraged by Masindi District under Ordinance 2, Part VI, Section 30 (1), 2002, which states: 'Every land owner or land user shall plant trees on at least 10% of his or her land.'

Workshop in March 2001 when we held a one-day course on bee-keeping and distributed hives to 18 local farmers.

In general, despite some indifference, the attitudes towards conservation encountered by Muhumuza were 'highly positive'. The main constraints given for lack of progress were lack of resources, poor communication and transport facilities, and lack of funding for projects. Among the recommendations for wildlife conservation suggested by Muhumuza were that farmers should not be allowed to buy or use land around the edges of Budongo Forest or if they did, they should grow unpalatable (to wildlife) crops such as millet or sorghum. His study was conducted in the villages but he also stressed the need to continue the work BFP is doing in schools so the idea of conservation is present in the minds of the next generation.

Census of chimpanzees in Uganda

Arising from the PHVA, a nationwide survey was made of the chimpanzees of Uganda (Plumptre *et al.* 2003). The great advantage of this survey is that it was done using one set of methods, by one set of people, with the same set of concepts about extrapolating from transect-based counts to overall calculations of population size. This is better than trying to get an idea of the population size of chimpanzees in various forests in Uganda from a number of independent surveys using different methods and of varying reliability.

As with earlier censuses of chimpanzees, the basis of the national census was nest counts. The method used was the one pioneered by Plumptre in his work in Budongo Forest (Plumptre and Reynolds 1996, 1997), namely the 'Marked Nest Count' method. Walks were made along a series of five 4-km carefully sited transects at 2-weekly intervals over a total of 3–4 months, marking each chimpanzee nest seen using a ribbon and a stake beneath the nest. The results were ratcheted up to the forest as a whole using the computer package DISTANCE which can calculate density using perpendicular distance data, and correcting for two factors:

1. that infants up to 4 years of age do not make nests;

2. that a certain percentage of nests are day-beds and these must be discounted.

A figure of 1.1 nests per day per nest-building chimpanzee was derived from a study based on dawn-to-dusk follows of individual Sonso chimpanzees (Plumptre and Reynolds 1997).

The corrected results of this national survey are shown in Table 12.5. In this table, all the major forests are included, as well as all the intervening areas where chimpanzees are known to occur. The result: Kibale Forest (a National Park) has the highest population of any forest in Uganda with 1430 chimpanzees, compared with 639 for Budongo Forest Reserve. Uganda as a whole has 4962 chimpanzees, or, with 95% confidence limits applied to the distribution, between 4000 and 5700 chimpanzees. This is a higher estimate than the 3300 total estimated at the PHVA in 1997.

Table 12.5: Estimates of chimpanzee population sizes in major forests with 95% confidence limits (courtesy of A. Plumptre, WCS, from Plumptre *et al.* 2003).

Forest	Distance walked (km)	Density (km²)	Population in forest	95% confidence limits
Budongo FR	513.7	1.36	584	356–723
Bugoma FR	511.2	1.90	570	424–769
Kagombe FR	165.8	0.71	80	29–218
Itwara FR	126.6	1.35	120	67–215
Kibale NP	564.1	2.32	1,298	817–1615
Kasyoha–Kitomi FR	477.7	0.92	370	250–530
Kalinzu FR	311.0	1.55	220	120–380
Bwindi NP	285.6	0.43	140	49–566

Hunting and pitsawing

We have discussed threats to Budongo's chimpanzees above, and also in Chapter 10 and Chapter 11. Here we look at threats across the whole country.

Plumptre *et al.* (2003) recorded signs of human activity per kilometre walked during their forest surveys: 'Encounter rates associated with pitsawing (pitsaw pits, stacked timber, cut trees for props, porters carrying timber, campsites for pitsawyers) and hunting (snares, pitfall traps, skinned animals, hunters encountered, nets, dogs) were summed to provide a measure of the relative abundance of these two threats' (p. 31).

Fig. 12.4 shows the extent of hunting sign encountered in Budongo Forest. Alas, as remarked in Chapter 10, Budongo Forest, together with its neighbour, Bugoma Forest, have much more hunting sign than any of the other forests surveyed. This, together with the survey results on population sizes, says much about the future of Budongo's chimpanzees. These forests are being ravaged by the human population. The level of hunting is damaging the Budongo chimpanzee population. Fortunately efforts are now underway to try to improve the situation, with discussions between UWA and NFA about employment of anti-poaching rangers. Let us hope that such co-operation will become a reality in the near future.

Pitsawing activities also cause chimpanzee deaths. In the above survey, Budongo came out with the most signs of pitsawing activity. Quite apart from their main activity, pitsawyers need food; their snares put out for duikers and pigs right in the heart of the forest end up catching chimpanzees. We know the story only too well.

Solutions

We are indeed fortunate in having good data for Uganda. Since the Amin and post-Amin wars there has been excellent research in Uganda on the problems faced by chimpanzees. Besides the work described above, high-quality research has also been carried out by the BFP, by JGI-Uganda, by the Wildlife Conservation Society and by other people and organizations. But it has to be said that solutions to the problems faced by chimpanzees remain elusive.

Fig. 12.4: Hunting sign (encounter rate per kilometre walked) in Budongo Forest Reserve. The larger the circle, the larger the encounter rate. Signs included snares, pitfall traps, hunters encountered, dogs and nets (courtesy of A. Plumptre, WCS, from Plumptre *et al.* 2003).

The BFP's live-trap project is one, partial solution to the problem of chimpanzees being caught in leg-hold traps and snares near farmers' fields. Other ideas have not progressed beyond the drawing board stage. One such is the idea of buffer zones between sugar cane fields and the forest. Such zones are not without their problems, especially in the context of immigration, and raise complex issues of land use and management (see, for example, Peterson 2001; Curran and Tshombe 2001; Richard and Dewar 2001). However, each situation demands its own solution and lessons from one area may not be wholly relevant to another.

Buffer zones

In September 2000, members of the BFP, together with relevant local people, worked on two ideas that would have benefits for chimpanzees. The first was a buffer zone between sugar fields and the forest where at present the cane comes right up to the forest edge. We prepared a document on the subject, for consultation with the Kinyara Sugar Works (KSW) management, the relevant local and central authorities and funding agencies.

This document explains that, with the expansion of production by KSW, sugar has been grown by private landowners (outgrowers) as well as by the main KSW itself. Many of the fields of the outgrowers are located to the northern side of the Masindi–Butiaba road, and they extend in many cases as far as the southern borders of the main Budongo Forest block.

A survey of the locations of 40 outgrowers' plots close to the boundary of Budongo Forest Reserve was conducted in 2003 by a BFP field assistant, Gideon Monday, and a representative of KSW. The results are shown in Table 12.6. As can be seen, of these 40 sugar cane fields, 9 were either partly inside the boundary or on the boundary of the Reserve. As a result, chimpanzees (as well as other species such as baboons) are able to move from the forest where they normally live into the sugar cane fields in order to raid the crops. While this is something KSW is familiar with in the case of baboons, and for

Table 12.6: Distance of sugar cane outgrowers' fields from the boundary of Budongo Forest Reserve (courtesy of Gideon Monday, unpublished data).

Distance from boundary (m)	No. of plots
<0 (inside the boundary)	2
0	9
1–5	22
5–10	5
>10	2

some years KSW has been killing baboons to try to control the problem, it is new for any appreciable number of chimpanzees to be raiding the sugar. In past times, chimpanzees would occasionally take a small amount of sugar from the main KSW plantations during the course of their movements along the riverine forest strips to the south of Budongo. They were seen by staff of KSW and their presence was tolerated. When they occasionally took mangoes and paw-paws from village shambas they were tolerated because they did not take much food and their presence was appreciated by the indigenous population.

With the expansion of sugar production outgrowers have been co-opted by KSW to increase sugar production on privately owned fields right up to the edge of the main block of the Budongo Forest. The owners of these small fields receive payment from KSW according to the amount of sugar they produce. They are therefore eager to maximize their production and are unwilling to allow animals to crop-raid. The result is a proliferation of snares and traps (including large 'man-traps') in the sugar fields on the border of the forest. Any prevailing attitude that the chimpanzees can be tolerated seems to have disappeared in the new conditions. In one village it was even said that the chimpanzees were a greater threat than baboons (F. Babweteera, pers. comm.). Evidently because the outgrowers are not local people but immigrant workers, and because they are dependent on the income from sugar to live on (having no alternative income and perhaps also no land on which to grow food crops), they have a negative attitude to the chimpanzees as well as baboons and other crop-raiding species and are ready to kill them or maim them in order to dissuade them from approaching their fields.

The proposal suggested in the BFP's document was to set up a buffer zone between the main Budongo Forest Reserve and the sugar fields. This had already been discussed a number of times between BFP personnel such as A. Plumptre, C. Bakuneeta, V. Reynolds and F. Babweteera, and KSW personnel such as G. Pollok, G. Macintyre and P. Wyatt. At the Second Budongo Conference held at Nyabyeya Forestry College on 7–8 September 2000, attended by the KSW Sugar Outgrowers Manager and many other local officials, the idea was discussed again. All agreed on the need for a buffer zone of some kind. From the point of view of BFP, a buffer zone would to some extent at least protect the chimpanzees. From the point of view of KSW, a buffer zone would assist in reducing the costs incurred in loss of sugar to animals and payments to patrol guards.

A meeting was held at KSW on 20 September 2000 between BFP and KSW. We agreed on the principle of a buffer zone and discussed how long and how wide it should be, and what it should consist of. Serious damage to cane from animals extends some 25 m into fields, so the width should be 25 m. An initial buffer zone could be 1 km long. It was suggested that the buffer zone might be planted with sugar cane. A thorn hedge could be planted between the buffer zone and the adjacent production zone. Animals proceeding beyond the thorn hedge would be subject to patrol guards. But in the buffer zone they would be free to feed on the sugar cane.

At present the area of the buffer zone is privately owned. Compensation to farmers whose land would be confiscated was discussed. The question arose at the meeting at KSW of who would manage the buffer zone? It would need to be carefully managed because sugar would have to be planted and weeded, and replaced after each three-year period because of disease. In addition, patrol guards would be needed to deter local hunters from placing snares and traps in the buffer zone to provide meat. KSW would probably not be interested in managing the buffer zone because its remit is to produce sugar. Some cost savings to KSW would arise from the buffer zone and these might form the basis for a financial contribution by KSW to the management of the zone. BFP is not in a position to manage the buffer zone. It therefore needs to be explored who would manage it. It was suggested that local people should be made responsible for this. An alternative might be NFA.

Money would be needed for a number of purposes, e.g. purchase of buffer zone land, compensation to landowners for loss of income, planting of sugar or other crops on buffer zone land, managing the crop, guarding the buffer zone against human activities, and planting and maintaining the thorn hedge. For these purposes a conservation fund is needed. An initial suggestion was ECOTRUST, the Environmental Conservation Trust of Uganda. This organization may be able to acquire land and offer environmental easements in areas requiring conservation, and hold such land in trust for suitable management purposes.

On 23 September 2000 F. Babweteera and I met with officials from ECOTRUST in Kampala and had a discussion of the possibilities. To date nothing further has been heard of this initiative. We remain hopeful that some progress will be made with a view to reducing the incidence of chimpanzee deaths and injuries resulting from the fact that sugar cane is planted too close to the forest edge at the present time. At least this idea is worthy of further exploration.

A small step forward was made in 2003 after the death of Jambo, described in Chapter 9. After this tragedy, senior members of BFP met with senior representatives of KSW. The staff of the Sugar Works were very concerned that one of our fine males had been speared to death by one of the field guards and noted that this was directly related to the fact that outgrowers were planting fields of cane right up to the forest edge. They agreed to harvest the sugar cane crop early (it would have normally been harvested three months later), to plough the field and the one next to it, and later to burn it over.

These two fields would not be used for growing sugar cane in future. They also agreed that all new contracts for outgrowers will have the regulation that plots must be at least 50 m from the forest and the land between must be kept clear so as not to provide chimpanzees with cover to cross (O'Hara 2003). This 50 m rule is certainly a move we rejoice in, even though it remains to be seen whether 50 m is sufficient to deter chimpanzees from crop-raiding. However, a message has now been sent to sugar outgrowers that the forest border must be avoided and that crop-raiding species must be identified before action is taken. Indiscriminate killing of anything that moves in the cane is not permissible, and our education programme continues to stress the value of chimpanzees and the need to treat them differently from other wildlife.

Tree corridors

A second idea that arose at around the same time was for a tree corridor. This came up in relation to the problems faced by the Kasokwa chimpanzees, but I include it here rather than in the chapter on Kasokwa (Chapter 11) because it has wide applicability to anywhere that chimpanzees are living in forest outliers or fragments or riverine or gallery forests. This idea has not so far met with any take-up in the Budongo area but it is important to pursue the possibilities in future.

A tree corridor for chimpanzees was set up by Japanese primatologists in the area of Bossou, Guinea, in 1998, with a view to enabling chimpanzees to move between two areas of their habitat that have become separated by agricultural activity: the Nimba Mountains and the village of Bossou (Hirata *et al.* 1998). This project is called the 'Green Passage Plan'. The distance between the two areas to be joined is 4 km, and the corridor is 300 m wide. Locally occurring species have been used for the corridor, all of them found in the chimpanzees' habitat. Local farmers as well as field assistants with the Bossou chimpanzee research project have worked to make the scheme take off, and financial rewards are being paid to farmers to ensure the survival of their young trees. Time will tell whether this scheme is successful; at present the trees, which were grown by the project in a special nursery, are still young. This is surely an excellent initiative[96] and is being watched by others with similar interests in other parts of Africa.

Fred Babweteera and I had some preliminary discussions with the office of the Environmental Protection and Economic Development (EPED) in Masindi about this corridor, and later presented them with a document about it, but to date the matter has not been taken further. As with the buffer zone, there is something here that is worth exploring. And, importantly, if a breakthrough could be made here for one small corridor, it might have implications for the other places in W. Uganda where corridors are needed to ensure gene flow, and indeed survival, of the animals, for example between forested areas along the Albertine Rift.

---

[96] Funds were received from the Japanese Government for this project.

Who is responsible for chimpanzee protection?

One of the things that will have occurred to the reader by now is the question: To whom should these ideas for chimpanzee protection be addressed? That is a very good question indeed. The fact is: everyone and no one is responsible for chimpanzee protection and conservation. The Uganda Wildlife Authority is responsible for protection of all endangered species in Uganda. The National Forest Authority takes some responsibility for endangered species living in its Forest Reserves. Some of the big non-government organizations (NGOs) that exist to study, protect and conserve wildlife have been active in chimpanzee work in Uganda, such as WCS, CI, NGS and JGI. Many zoos have helped, directly or indirectly; directly in cases such as the support we receive at Budongo from Cleveland and Oakland Zoos, indirectly such as the support received from zoos all over the world that enabled the Entebbe chimpanzee PHVA to take place in 1997. Non-Ugandan government agencies can be extremely helpful in supporting chimpanzee conservation indirectly: the BFP was helped by USAID in renovating and building its camp in 1991, from 1991 to 1997 we received core funding for the BFP from the British DfID (then ODA) through its forestry research programme, and since that time we have received core funding for the Project in our role as Field Station for Makerere University Faculty of Forestry and Nature Conservation from NORAD. Direct non-Ugandan government funding for chimpanzees also exists: the US Great Ape Conservation Fund channels US government funds into chimpanzee conservation projects, including for example a Ugandan national chimpanzee census (Plumptre *et al.* 2003).

I seem to have re-interpreted the question of responsibility in terms of funding. There is, evidently, a strong link. People do not fund chimpanzee protection unless they feel some sense of responsibility for it. However, funding and responsibility are not the same thing. Funding is a bit further removed from action than responsibility. A responsible person or institution has to have the human and practical resources to exert responsibility. External funding agencies have the financial means but not the human and practical resources on the ground. Who has?

In the forests of Western Uganda where the apes live we find two groups of people who are responsible for the protection of chimpanzees: employees of UWA and NFA, and members of research and conservation projects such as BFP. Direct responsibility lies with UWA (a parastatal[97]) and NFA (a government department until 2003 when it became a parastatal).

At the present time the relationship between government agencies, parastatals, NGOs and projects has not been properly worked out. There are often rivalries and suspicions, both between government agencies and between NGOs. Small organizations such as BFP may have very clear ideas of what needs to be done to protect chimpanzees but lack the funds to do it. Larger NGOs can and do help projects in their capacity as funding agencies giving grants for which the smaller projects apply. Co-operation between

---

[97] Meaning that it has responsibility for funding its own operations through income it raises itself rather than being dependent on government funding.

NGOs and small projects can indeed be very good, as for example in the case of the work done by BFP thanks to funding by NGS or JGI and other NGOs.

The weak link at present is between these NGOs and the projects they support on the one hand, and the Ugandan governmental and parastatal agencies on the other. In my opinion properly organized chimpanzee protection will not happen in Uganda until projects and NGOs work hand in hand with these agencies. At present there seems to be an atmosphere of mutual distrust and suspicion between the two, yet they ought to be working together as partners. This distrust is wholly unnecessary. FD regarded BFP with suspicion all through the 1990s; indeed, some members of FD wished we were not there.[98] We tried our hardest to work with FD but largely failed. Fortunately those days are now over and we are working with NFA at all levels, and we now have a research agreement with them. Such agreements can provide the basis for collaboration, shared goals and shared discussion of the means of achieving them.

Action Plan

At the time of writing, the most recent initiative for chimpanzee conservation is the national action plan for chimpanzee conservation: 'Conservation for Uganda's Chimpanzees 2003–2008' (Uganda Wildlife Authority 2003). This gives the number of chimpanzees in Budongo Forest Reserve as 639 (95% confidence limits 392–796) out of a total population for Uganda of 4950 (95% confidence limits 4000–5700). Analysis of satellite imagery between the mid-1980s and 2000/2001 shows that approximately 800 km^2 of forest has been lost in chimpanzee habitat areas, almost twice the area of Budongo Forest. Most of this loss has occurred outside of protected areas, i.e. on communal or private land, but the loss to chimpanzees is huge and must make their movements across country, on which (outside the forests) they depend for survival, much more hazardous than ever before.

This Action Plan embodies a bold Vision, a five-year set of objectives, as follows:

- Reduced fragmentation of habitat

- Reduced conflict between local communities and chimpanzees

- Promotion of awareness of chimpanzees

- Reduction of human-caused deaths and injuries

- Improved corporate responsibility

- Minimized potential spread of disease

Each of these objectives is described and the actions needed to achieve it are listed, together with indicators of success. The key stakeholders, including BFP, and the potential funding agencies for achieving these objectives are listed. This is indeed a thoroughly

---

[98] I was told by a senior member of FD in 1993 that we should not have built our camp in the forest.

worked out and deeply insightful document and if its ideas and objectives can be achieved, even in part, it will have played a large part in ensuring the survival of Uganda's chimpanzees. As can be seen, things are moving along, but the pace of modernization and change is fast and the pace of conservation, hitherto, has been too slow. If chimpanzees are to survive in Uganda, we need to speed up our efforts to protect them. That is the challenge facing us.

Appendices

A.   The Sonso chimpanzee community

The members of the Sonso chimpanzee community, from 1990 to 2004, are as given in Table A.1. The names of members who have died or disappeared are shown in italics. The table shows names, codes, sex, dates when named, estimated dates of birth, accuracy of the estimates and notes. Offspring are indented and follow the name of the mother; a double indent implies a second-generation offspring. Paternities, where known, are shown under Notes.

Table A.1: The Sonso chimpanzee community in October 2003 (including dead/disappeared individuals).[*]

Name	Code	Sex	Date named	Estimated D.o.B.	Accuracy of D.o.B.	Notes
Andy	AY	M	21-Jun-94	1982	6 months	Found dead in forest 28 July 00
Banura	BN	F	20-Oct-93	1968	3 yrs	Swollen r. club foot
Zefa	ZF	M	25-Nov-93	1982	6 months	Snare caused damage to 3rd digit of r. hand. Son of Nkojo.
Shida	SH	F	21-Oct-93	1990	1 yr	Father: Duane. Snared June 98, l. hand missing. R. leg snared, nylon snare present
Beti	BT	F	2-Oct-96	01-Oct-96	3 days	Scars on r. hand from snare injury
Ben	BE	M	1 Oct-01	13-Sep-01	3 days	Last seen 31 Oct 01
Black	BK	M	9-Oct-93	1974	2 yrs	Father of BH, NR, KN & CT
Bwera	BW	F	28-Apr-92	1975	3 yrs	Last seen 9 Aug 94
Biso	BI	M	10-Jun-92	1990	1 yr	Disappeared with mother
Bwoba	BB	M	2-Sep-92	1987	1 yr	Father: Magosi
Bwoya	BY	M	5-Jun-92	1965	5 yrs	Found dead in forest 24 Dec 01
Chris	CH	M	19-Jul-94	1967	3 yrs	Disappeared 9 Aug 97
Clea	CL	F	27-Aug-97	1986	2 yrs	
Clint	CT	M	1-Apr-02	27-Apr-02	2 days	Father: Black
Duane	DN	M	1-Oct-93	1966	2 yrs	Father of RS, KE, SH, JT & KM
Emma	EM	F	14-Oct-98	1991	2 yrs	Immig. Sept 97, disappeared Jan 03
Flora	FL	F	8-May-03	1979	5 yrs	Missing r. hand, slit in r. ear
Fred	FD	M	8-May-03	1994	1 yr	
Frank	FK	M	8-May-03	1999	1 yr	
Gashom	GS	M	20-Jul-94	1987	1 yr	Father: Nkojo
Harriet	HT	F	11-Sep-96	1978	5 yrs	Slit in l. ear
Hawa	HW	M	11-Sep-96	1993	1 yr	Father: Kikunku
Helen	HL	F	23-Feb-01	Feb-01	3 wks	

Table A.1: (continued)

Name	Code	Sex	Date named	Estimated D.o.B.	Accuracy of D.o.B.	Notes
Jake	JK	M	13-Sep-94	1988	1 yr	Last seen 4 Feb 95
Jambo	JM	M	20-Oct-93	1978	2 yrs	Died from spear 4 May 03
Janie	JN	F	28-Sep-95	1984	1 yr	Cut in r. ear
Janet	JT	F	25-Oct-99	01-Oct-99	3 days	Father: Duane
Jogo	JG	M	14-Nov-92	1986	1 yr	Last seen 23 Apr 94
Juliet	JL	F	1-Jan-02	1990	1 yr	L. hand snare injury
Kalema	KL	F	28-Apr-92	1979	1 yr	R. hand snare injury
Bahati	BH	F	16-Dec-94	Dec-94	1 wk	Father: Black
Kumi	KM	F	15-Sep-00	17-Sep-00	2 days	Father: Duane
Kigere	KG	F	11-Mar-92	1976	3 yrs	Missing r. foot. Slit on r. ear.
Kadogo	KD	M	11-Jul-92	1990	6 months	Last seen 5 Nov 99
Unnamed	n/a	?	11-Sep-97	10-Sep-97	2 days	Premature, born dead
Keti	KE	F	22-Sep-98	01-Sep-98	1 wk	Father: Duane
Kuki	KI	F	19-Sep-03	16-Sep-03	1 wk	
Kikunku	KK	M	30-Sep-92	1976	1 yr	Disappeared 6 July 98. Father of HW & KA
Kutu	KU	F	8-Aug-92	1979	2 yrs	Slit on top of left ear.
Kato	KT	M	13-Oct-93	Sep-93	3 wks	
Kana	KN	F	29-Oct-98	29-Oct-98	1 day	Father: Black. Snared right foot Feb 03
Kasigwa	KS	M	14-Nov-03	15-Aug-03	2 wks	
Kwera	KW	F	5-Aug-92	1981	2 yrs	
Kwezi	KZ	M	28-Jan-95	07-Jan-95	3 wks	Father: Muga
Karo	KR	F	1-Nov-01	01-Nov-01	1 day	Father: Maani
Maani	MA	M	5-Aug-92	1958	5 yrs	Father of KR
Magosi	MG	M	5-Aug-92	1960	5 yrs	Found dead in forest 4 July 99. Father of NK, BB, GZ, MS & ZG
Mama	MM	F	17-Jun-95	1970	5 yrs	Mother of Muhara. Disappeared 13 Aug 98
Muhara	MH	F	28-Aug-95	1985	1 yr	Last seen 3 Apr 01
Matoke	MT	F	28-Apr-92	1962	5 yrs	Last seen 5 Feb 93
Toto	TT	F	28-Apr-92	1988	1 yr	Disappeared with mother
Melissa	ML	F	22-Oct-01	1982	5 yrs	
Mark	MR	M	10-Jan-01	1997	1 yr	
Unnamed	N/a	?	n.a.	8 July 02	1 wk	Born dead
Monika	MN	F	18-Jul-03	22-Jun-03	2 wks	
Mukono	MO	M	22-Apr-92	1970	5 yrs	Deformed right wrist. Last seen 12 Feb 94
Mukwano	MK	F	28-Apr-92	1980	3 yrs	R. foot lame from snare
Monday	MD	M	1-Nov-02	Nov-02	1 month	Disappeared after Apr 03 when MK temporarily left Sonso
Nambi	NB	F	21-Jun-94	1962	5 yrs	Slit on upper l. ear
Muga	MU	M	8-Aug-92	1977	1 yr	Last seen 23 Mar 00. Father of KZ
Musa	MS	M	21-Jun-94	1991	6 months	Father: Magosi
Nora	NR	F	9-Apr-96	Feb-96	3 wks	Father: Black
Night	NT	F	6-Feb-03	06-Feb-03	exact	
Nkojo	NJ	M	4-Sep-92	1968	5 yrs	Slit on lower left ear. Father of ZF, GS & RE. Disappeared 30 Nov 03.
Polly	PL	F	1-Jul-03	1984	5 yrs	First seen Jan 03, possibly Dec 02
Pascal	PS	M	1-Jul-03	1998	1 yr	First seen with mother
Ruda	RD	F	14-Nov-92	1976	5 yrs	Died of peritonitis, 9 Nov 01
Bob	BO	M	14-Dec-92	1990	6 months	
Rachel	RE	F	9-Jul-97	Jun-97	1 month	Father: Nkojo
Ruhara	RH	F	10-Apr-93	1968	5 yrs	
Nick	NK	M	3-Jan-95	1982	1 yr	Father: Magosi
Grinta	GT	F	10-Apr-93	1990	1 yr	Disappeared 9 June 99
Rose	RS	F	15-Nov-97	15-Nov-97	1 day	Father: Duane
Ramula	RM	F	13-Sep-02	06-Sep-02	1 wk	

Table A.1: (continued)

Name	Code	Sex	Date named	Estimated D.o.B.	Accuracy of D.o.B.	Notes
Sabrina	SB	F	23-May-01	1981	4 yrs	
Sally	SA	F	1-Apr-01	1996	1 yr	
Sean	SN	M	Sep-02	Oct-02	1 month	Disappeared after Aug 2003
Salama	SL	F	20-Oct-93	1981	2 yrs	Disappeared in 1995, reappeared in 2001, disappeared
Sara	SR	F	27-Jan-95	1980	5 yrs	Last seen 29 Feb 00
Tinka	TK	M	30-Sep-92	1960	5 yrs	Both hands injured by snares. Eye and skin problems.
Vernon	VN	M	26-Apr-94	1961	2 yrs	Disappeared, possibly speared, 28 June 99
Vita	VT	F	9-Apr-93	1990	1 yr	Disappeared 26 May 98
Wilma	WL	F	27-Nov-01	1981	5 yrs	R. hand missing. Wart under r. eye.
Willis	WS	M	27-Nov-01	1996	1 yr	Disappeared Dec 01
Zana	ZN	F	2-Sep-92	1981	5 yrs	Snared both hands: thumb only on r. hand, l. hand severely damaged
Zalu	ZL	M	12-Jul-95	29-Jun-95	3 days	
Zed	ZD	M	1-May-01	May-01	1 month	
Zesta	ZT	M	17-Feb-93	1980	1 yr	Killed by senior Sonso males, 4 Nov 98
Zimba	ZM	F	11-Aug-93	1968	5 yrs	Back of r. hand swollen.
Kewaya	KY	F	28-Apr-92	1983	1 yr	Father: Kikunku. R. hand severe snare injury
Katia	KA	F	30-Dec-98	30-Dec-98	exact	
Gonza	GZ	F	11-Aug-93	1990	1 yr	Father: Magosi. Disappeared Jan 02
Zip	ZP	?	28-Jun-96	15-Jun-96	2 wks	Last seen 17 Aug 96
Zig	ZG	M	7-Jul-97	24-Jun-97	2 wks	Father: Magosi. Snared on r. hand
Zak	ZK	M	1-Dec-02	21-Nov-02	1 day	
Unnamed	n/a	?	1-Aug-97	Jul-97	1 wk	Being carried by KW, already dead
Unnamed	n/a	?	12-Aug-97	Aug-97	1 wk	Being carried by SR before disappeared

* I am grateful to several people for help with this table, notably N. Newton-Fisher, M. Emery Thompson, Sean and Catherine O'Hara, Lucy Bates and Crystal Davis. Where actual or putative kinship has been used to estimate ages of older individuals, consideration has been based on the age categories specified in Table 2.4. For adult females, a minimum age is given, assuming the oldest infant is the first-born and was born when the female was 15 years old, except where other information indicates to the contrary. The accuracy of the D.o.B. in these cases is not ± but −. In other cases, accuracy of D.o.B. is ±.

B. Sonso chimpanzees: plant food species

The main source of our data on the foods eaten by the Sonso chimpanzees are the records of foods seen eaten by our field assistants and students in the course of their daily observations. From the start of the BFP we have collected samples of food species and many of these have been photographed fresh for our Plant Identification Album, and subsequently dried and mounted by Andy Plumptre for the BFP's herbarium at Sonso camp. In cases where the identification was uncertain they were identified at Makerere University's herbarium, mostly by A.B. (Tony) Katende and then returned to the Sonso herbarium.

In 1997 Rhiannon Meredith and Susie Whiten made a list of 76 species of plants providing food for the Sonso chimpanzees. These records are based on observational data and do not imply anything about seasonality of the plant species, nor that chimpanzees do not eat certain species at times left blank on the table; they have simply not so far been observed eating them in these months.

Table B.1: Chimpanzee food species recorded (Sonso community) 1990–2004.

Food species	Plant	Part eaten	J	F	M	A	M	J	J	A	S	O	N	D	No. months
*Acalypha neptunica**	Shrub	L	1	1	1	1	1	1	1	1	1	1	1	1	12
Adenia tricostata	Climber	L	1	1					1						3
Allophilus dumeri	Tree	F									1				1
Aframomum sp.	Herb	F, P				1	1	1	1	1	1	1	1	1	9
Alafia laudolphoides	Climber	F	1	1							1	1	1		5
Albizzia ferrungunea	Tree	F													
Alstonia boonei	Tree	B, L						1	1						2
Aneilema aequinoctiale†	Herb	L													
Aningeria altissima	Tree	F													
Antiaris toxicaria	Tree	F, L	1	1	1								1		4
Antrocaryon microcaster	Tree	Fl, F, B	1				1		1	1			1	1	6
Aphania senegalensis	Tree	F				1									1
Balsamocitrus dawei	Tree	F					1					1	1		3
Bosqueia phoberos	Tree	F											1	1	2
Broussonetia papyrifera	Tree	F, L, Fl	1	1	1	1	1	1	1	1	1	1	1	1	12
Calamus deerratus	Climber	F, P						1	1						2
Caloncoba schweinfurthii	Tree	F, S					1	1	1	1	1		1		6
Celtis gomphophylla	Tree	Fl, F, L, B	1	1	1	1		1	1	1	1	1	1	1	11
Celtis milbraedii	Tree	Fl, L, F	1	1	1	1	1	1	1	1	1	1	1	1	12
Celtis wightii	Tree	L, Fl	1		1	1	1		1			1		1	7
Celtis zenkeri	Tree	F, L		1	1	1	1	1	1	1	1	1	1	1	11
Chrysophyllum albidum	Tree	F		1	1				1				1	1	5
Chrysophyllum gorungosanum	Tree	F		1	1	1						1	1		5
Chrysophyllum muerense	Tree	F													
Chrysophyllum perpulchrum	Tree	F			1	1									2
Cleistanthus polystachyus	Shrub	F, W										1			1
Cleistopholis patens	Tree	F, W	1	1	1				1			1	1	1	7
Coffea sp.	Tree	F							1						1
Cola gigantea	Tree	F, S, W						1	1	1					3
Cordia millenii	Tree	F, Fl		1	1	1	1	1	1	1	1	1	1		10
Crossonopholis africana	Tree	S			1										1
Croton macrostachys	Tree	F, S						1			1	1	1	1	5
Croton sylvaticus	Tree	F						1	1			1			3
Cynometra alexandri	Tree	L, Fl, S, B	1	1	1		1	1	1			1	1	1	9
Desmodium sp.	Herb	L	1	1										1	3
Desplatsia dewevrei	Tree	F, L	1		1	1	1	1	1	1	1		1		9
Dialium excelsum	Tree	F													
Ekebergia capensis	Tree	F	1	1	1			1	1						5
Entandophragma cylindricum	Tree	S, L										1			1
Entandrophragama utile	Tree	F													
Erythrophleum suaveolens	Tree	S										1	1	1	3
Ficus asperifolia (urceolaris)	Shrub	F							1						1
Ficus barteri	Tree	F		1		1	1				1	1	1	1	7
Ficus exasperata	Tree	F, L, G	1	1	1	1	1	1	1	1	1	1	1	1	12
Ficus lingua	Tree	F											1	1	2
Ficus mucuso	Tree	F, L, B	1	1	1	1	1	1	1	1	1	1	1	1	12
Ficus natalensis	Tree	F		1		1	1				1	1	1	1	7
Ficus ottonifolia	Strangler	F											1	1	2
Ficus polita	Tree	F		1								1	1	1	4
Ficus sansibarica (brachylepis)	Tree	F	1	1	1		1	1			1	1	1	1	9
Ficus saussureana (dawei/lutea)	Tree	F	1	1	1	1	1	1					1	1	8
Ficus sur (capensis/vogelana)	Tree	F, L	1	1	1	1	1	1	1	1	1	1	1	1	12
Ficus thonningii	Epiphyte	F										1	1	1	3
Ficus trichopoda (congensis)	Tree	F			1		1							1	3
Ficus vallis-choudae	Tree	F	1	1	1	1		1	1						6

Table B.1: (continued)

Food species	Plant	Part eaten	J	F	M	A	M	J	J	A	S	O	N	D	No. months
Ficus variifolia	Tree	L, F		1		1			1		1	1		1	6
Gynura scandens	Climber	L			1		1	1							3
Holopteria grandis	Tree	L							1						1
Iodes africana	Climber	F													
Khaya anthotheca	Tree	B, G	1	1	1	1	1	1	1	1	1		1	1	11
Klainedoxa gabonensis	Tree	F								1	1				2
Lantana camara‡	Shrub	F									1	1			2
Lasiodiscus mildbraedii	Tree	L								1	1				2
Leptaspis sp.	Grass	P											11		1
Macaranga schweinfurthii	Tree	L, F					1								1
Maesopsis eminii	Tree	F	1	1	1	1	1	1	1	1	1	1			10
Mammea africana	Tree	F							1	1	1				3
Marantochloa leucantha	Herb	F, P	1	1	1								1		4
Marantochloa purpurea	Herb	F, P	1	1	1								1		4
Megaphrynium sp.	Herb	P													
Mildbraediodendron excelsum	Tree	F				1	1	1	1	1	1	1			7
Milicia excelsa	Tree	F, L	1	1	1	1	1								5
Mimusops bagshawei	Tree	F									1	1			2
Monodora angolensis	Tree	F										1			1
Monodora myristica	Tree	F, Fl	1	1				1	1						4
Morus lactea	Tree	F, L, Fl	1	1	1	1	1					1			6
Myrianthus holstii	Tree	F				1	1	1	1	1	1	1	1		8
Olea welwitschii	Tree	F													
Pennisetum purpureum	Grass	St						1	1						2
Physalis minima	Herb	F													
Piper guineense	Climber	P													
Platycerium elephantotis	Epiphyte	L	1												1
Pseudospondias microcarpa	Tree	F				1	1	1	1	1					5
Psidium guajava	Tree	F		1		1	1			1					4
Pterygota mildbreadii	Tree	F													
Raphia farinifera	Tree	W, P				1	1	1	1	1	1				6
Ricinodendron heudelotii	Tree	F							1				1	1	3
Sarcophrynium schweinfurthianum	Herb	P	1							1				1	3
Setaria megaphylla	Herb	P	1	1											2
Solanecio angulatus	Climber	L													
Sterculia dawei	Tree	S, L	1	1											2
Strychnos mitis	Tree	F		1		1									2
Syzygium guinense	Tree	F				1	1	1	1						4
Tectaria gemmifera	Herb	P			1	1									2
Tetrapleura tetraptera	Tree	S			1										1
Trachyphrynium braunianum	Shrub	P													
Treculia africana	Tree	F						1							1
Trema orientalis	Tree	F													
Trichilia martineaui	Tree	F						1	1						2
Trichllia rubescens	Tree	L	1	1					1						3
Urera cameroonensis	Climber	Fl, F, G	1	1						1			1		4
Uvariopsis congensis	Tree	F				1	1		1						3
Whitfeldia elongata	Shrub	Fl, L											1	1	2

* Not a food as such but its leaves are chewed occasionally throughout the year, sometimes in the context of leaf-sponging.

† Not eaten but used for medicinal purposes.

‡ Fruits of this species were not recorded being fed on between 1992 and 2003 but in 2004 were eaten vigorously (V. Kosheleff, pers. comm.).

Key: B = Bark; F = Fruit; Fl = Flowers; L = Leaves; G = Gum; S = Seeds; St = Stem; W = Wood; P = Pith.

Chimpanzees were seen to eat parts of the following kinds of plants: trees, shrubs, herbs, climbers and epiphytes. They fed on the following plant parts: fruits, leaves, flowers, bark, seeds, stem/pith, gum and wood. Of these, ripe fruits and young leaves made up the bulk of the Sonso diet.

Several studies have been made under the auspices of the BFP of the foods eaten and the feeding behaviour of the Sonso chimpanzees (see Chapter 4). Here, for completeness and for comparisons with other sites, I include an alphabetical list of the plant food species we have seen the Sonso chimpanzees eat (Table B.1). The key to the table is at the end. Besides species name, the table includes type of plant, part of plant eaten and the months in which we have recorded the species being eaten.[99] The final column indicates how many months each species is eaten and therefore to some extent how important it is in the diet. This table does not include non-plant foods such as meat, insects and soil, and it does not include cultivars obtained wheni crop-raiding.

I have used the work of those mentioned above, revised by Zephyr Kiwede, Geresomu Muhumuza, and also Newton-Fisher (1997), Fawcett (2000) and Tweheyo *et al.* (2004). I am most grateful to these people and to others who have contributed to this project.

C. Genetics of the Sonso community

With the advent of non-invasive methods of obtaining DNA, we can now genotype individual chimpanzees without needing to take blood from them. DNA can be obtained from hair samples (the hair follicle contains DNA), wadges (a wadge is a bolus of rejected food, e.g. fig seeds or rattan fibres, that is formed in the mouth and spat out; DNA is present in the oral mucus in and around the wadge) and faeces (which contain DNA from the anal mucus they receive during defecation).

Our genetic studies on the Sonso chimpanzees are still in progress. Studies of relatedness and of paternity are being done by Linda Vigilant, Dieter Lukas and her team at the Max Planck Institute for Evolutionary Anthropology in Leipzig, Germany.[100]

Since 2000 we have been collecting samples (hairs and wadges initially, latterly faeces) for analysis of microsatellite sequences of nuclear DNA which give information on maternal and paternal related-ness. The analysis, which is painstaking, slow, and can lead to false conclusions unless the greatest care is taken at both the stage of laboratory analysis and interpretation,[101] has enabled us to make advances in two areas: the extent of relatedness of our males compared with that of females, and relationships of kinship in the community.

It is generally assumed that the degree of relatedness of the males of any chimpanzee community will be greater than that of its females. The reason is that whereas the males stay put in their home range and do not emigrate, females very often do emigrate at adolescence. Thus what we see in any community is a succession of females entering the community and leaving it, whereas the males do not leave, they remain in the community until they die. Thus it seems obvious that the males in a community will be more closely related to each other than the females are to each other. The males will be related as fathers

---

[99] Where no month is entered, this means the item has been seen eaten but the date was not recorded or has been lost.

[100] We are greatly indebted to Linda Vigilant and the Leipzig team, and to Christophe Boesch who originally offered to genotype our samples there.

[101] For details of the method, which has since been refined, see Morin *et al.* (2001). Figure 4 in that paper is especially useful in outlining the method used. For recent refinements, see Vigilant (2002), Bradley and Vigilant (2002), Thalmann *et al.* (2004) and Nsubuga *et al.* (2004).

and sons, brothers, male cousins, uncles and nephews, etc. In the case of females after puberty, there is no reason to think they will be mothers and daughters, sisters, or aunts and nieces, because they disperse far and wide at puberty.

Vigilant *et al.* (2001) initially tested this prediction in three communities of the Taï chimpanzees of Ivory Coast and in the Gombe chimpanzees. In contrast to earlier studies, they found that the males were not more closely related to each other than were females. This was attributed to two factors: gene flow between communities mediated by females, and the inability of individual males to monopolize females so that offspring are fathered by a variety of individuals.

We are now finding the same at Sonso. The original model of chimpanzee population dynamics, with males more closely related than females, may be wrong. Our Sonso males appear to be no more closely related to each other than females are (Lukas *et al.*, in press).

This question of the relatedness of males is of especial interest because it has been believed hitherto that the reason that community males often associate with each other and move around together more than females do is that the males are close kin, whereas the females are not. The same argument has been used to explain the communal hunting of monkeys and other prey by males, the communal patrolling and defence of the community's range by males, and their concerted attacks on members of other communities. It now seems not to be close genetic relatedness as such that underlies these actions. In addition, Mitani *et al.* (2000, 2002) have shown that males who are close maternal kin are no more likely to co-operate for grooming, or keep close to each other, than less closely related males. They conclude that rank and age are more important than kinship in determining male–male affiliations. It seems therefore that we shall have to re-think our theories of the basis of the co-operative behaviour we see between community males.

What have we been able to discover about the kinship relationships that do exist between Sonso individuals? This can be broken down into questions of (a) maternal kinship and (b) paternal kinship.

Maternal kinship

In the case of offspring born since the Project began, we know the mothers. For individuals born before the Project started, in several cases we allocated them putative mothers on the basis of their behaviour; in the following cases genetic evidence has confirmed that they are indeed mother–offspring pairs: Nick is the son of Ruhara; Gonza is the daughter of Zimba; Musa is the son of Nambi; Bob is the son of Ruda; Kewaya is possibly the daughter of Zimba.

Paternal kinship

We have no evidence that the fathers recognize their offspring and so we conclude that father–child relationships are not known to chimpanzees. As we saw in Chapter 6, males compete for access to sexually receptive females and in consequence there is differential reproductive success, which is to some extent linked to status. We therefore expect that alpha males will have a disproportionate number of offspring. The two alpha males since we have been observing the Sonso community have been Magosi who was alpha male until 1995, and Duane who took over in that year.

Table C.1 shows the paternities that can be attributed with confidence on the basis of the genetic results we have obtained so far. Not all offspring are included; in some cases samples have yet to be obtained, in others the DNA has yet to be sequenced. Mothers are also included in the table.

Table C.1: Sonso paternities in order of birth date (mothers also shown).

Offspring	Birth date	Mother	Father
Zefa	1982	Banura*	Nkojo
Nick	1982	Ruhara	Magosi
Kewaya	1983	Zimba	Not assigned[†]
Bwoba	1987	?	Magosi
Gashom	1987	?	Nkojo
Gonza	1990	Zimba	Magosi
Shida	1990	Banura	Duane
Bob	1990	Ruda	Not assigned[†]
Musa	1991	Nambi	Magosi
Kato	Sept 1993	Kutu	Not assigned[†]
Hawa	1993	Harriet	Kikunku
Bahati	Dec 1994	Kalema	Black
Kwezi	Jan 1995	Kwera	Muga
Zalu	June 1995	Zana	Not assigned[†]
Nora	Feb 1996	Nambi	Black
Rachel	June 1997	Ruda	Nkojo
Rachel	June 1997	Ruda	Nkojo
Mark	1997	Melissa	Not assigned[†]
Zig	June 1997	Zimba	Magosi
Rose	Nov 1997	Ruhara	Duane
Keti	Sept 1998	Kigere	Duane
Kana	Oct 1998	Kutu	Black
Katia	Dec 1998	Kewaya	Kikunku
Janet	Oct 1999	Janie	Duane
Kumi	Sept 2000	Kalema	Duane
Karo	Nov 2001	Kwera	Maani
Clint	Apr 2002	Clea	Black

* Probably Banura is the mother of Zefa but the genetic evidence is incomplete.
[†] Not assigned: the offspring's genotype is not compatible with any Sonso male available for testing.

From the table we can see that, at the time of writing,[102] paternity has been assigned to 22 individuals in the Sonso community. Magosi sired 4 of the 10 infants for which we have data during the period when he was the alpha male (until mid-1995), while Duane has sired 4 of 16 offspring since he became the alpha male in mid-1995. Reproductive success for these two periods is shown in Table C.2.

As Table C.2 shows, the alpha male is not able to monopolize females when they are ovulating and most likely to conceive. He may be mating with them around that time but other males are also mating with them. They may have better timing than him, or more luck, or we may be seeing the results of sperm competition rather than mating competition in these findings. Black sired 3 offspring in the second period, compared with Duane's 4. Black was third ranking male until Vernon disappeared in June 1999, after which he became second ranking male. He was an aggressive individual, more so than the other adult males and, as seen in Chapter 6, was aggressive in some of his sexual encounters and may have succeeded in sequestering some females on safaris. Maani, often an alliance partner of Duane

---

[102]   October 2004.

Table C.2: Reproductive success for the periods before and after mid-1995: (a) while Magosi was alpha male, (b) while Duane was alpha male.

Name of father	No of offspring
(a) *While Magosi was alpha male (prior to mid-1995)*	
Magosi	4
Nkojo	2
Black	1
Muga	1
Duane	1
Kikunku	1
(b) *While Duane was alpha male (after mid-1995)*	
Duane	4
Black	3
Nkojo	2
Kikunku	1
Maani	1
Magosi	1

in this period, was a rather calm individual. He, and the other middle ranking males during this period, Nkojo, Kikunku and Magosi, were able to sire offspring. These facts show that while rank is important it is not essential to be a top-ranking male to achieve reproductive success.

Indeed, some males fail. Vernon, Duane's close ally and the beta male until he disappeared in June 1999, sired no offspring. Nor did the high-middle ranking male Jambo who allied with Black until his death in May 2003. Bwoya, who lived from 1965 to 2001, died without offspring. Tinka, the omega male, born in 1965 and still alive at the time of writing in 2004, has had no offspring. Zesta, the male who was killed by his fellows as a young adult male, likewise did not produce an offspring with Janie (see Chapter 8); her daughter Janet, born 11 months after his death, was sired by Duane.

Can we link births with specific copulations?

In five recent Sonso births where conception cycles were identified with endocrine data and pregnancy test strips, gestation length averaged 229 days (range 210–241; Emery Thompson, unpublished data). This is the first such estimate for wild chimpanzees and is remarkably consistent with previous reports for captive chimpanzees (229.4 days, $N = 8$, Martin *et al.* 1978; 233.3 days, $N = 3$, Shimizu *et al.* 2003).[103] The actual fathers of these five infants have not been identified at the time of writing.

Offspring not sired by Sonso males

As seen in Table C.1, five of the individuals whose genetic sequences we have studied in the Sonso population have fathers who are not among the Sonso males. That is to say, their nuclear genetic sequences cannot be assigned to (or derived from) any of the Sonso males sampled. This includes

---

[103] I am grateful to Melissa Emery Thompson for permission to use these findings.

males who are now dead or disappeared, from all of whom we took samples. We now look at these five in turn:

1. Kewaya was a subadult female when we first identified and named her in 1992. She is possibly the daughter of Nambi, currently alpha female at Sonso; if so it would seem that Nambi moved into the Sonso community having already conceived outside it.

2. Mark arrived with Melissa in October 2001 at which time he was already a juvenile. Melissa was one of three parous females who arrived together at this time and we concluded they came from a community that was disintegrating (see Chapter 5). His father would thus be expected to be outside Sonso.

3. Bob was first seen in 1992 at which time he was already an Infant 2 (aged 3–4 years) with his mother Ruda. Possibly his father disappeared very early in our study, before being identified (identification began in 1992), after siring Bob.

4. Kato was born in late September 1993, his mother is Kutu. His paternity outside the Sonso community may be explicable in the same way as Bob's, i.e. his father disappeared before we had identified him.

5. Zalu is harder to explain. He was born to Zana at the end of July 1995. Zana has injuries to both hands and is relatively immobile compared with other females (Munn 2003). She is rather solitary and ranges on the outskirts of the community's range. This might have given her an opportunity to mate with a non-Sonso male.

MHC genetics

A second, more recent, genetic study ongoing at Sonso focuses on the genetics of the major histocompatibility complex (MHC) in relation to health.[104] Products of the MHC class II genes are crucial components of the immune recognition system, presenting peptides of extracellular origin to helper T cells. The extensive polymorphism and allelic diversity of these genes are believed to increase the number of different peptides bound and recognized by the immune system, thus conferring a wider repertoire of resistance to disease. For this study, we are collaborating with Dr Leslie Knapp of the Department of Biological Anthropology at the University of Cambridge, with fieldwork by Catherine O'Hara who has developed new techniques for the study of non-invasively collected samples and who is currently analysing the data. The objective of this research is to build up a database of individual MHC genotypes that may be used to investigate how the MHC influences resistance to infection, social status, mating behaviour and parity in wild chimpanzees. At this stage of the research, individual haplotypes are being generated and ultimately, using the MHC data, the following hypotheses will be tested:

*   MHC diversity is favoured by overdominant selection with heterozygotes being more resistant to parasitic infection;

*   MHC-based disassortative mating results in more MHC heterozygote progeny than expected.

Like the Leipzig work, the Cambridge study is based on totally non-invasive methods, the DNA being extracted from faecal samples. While difficult to achieve, these results will improve our understanding of health and reproduction in the Sonso chimpanzees.

[104] This study is funded principally by the Leakey Foundation.

D. Report of (a) necropsy on Ruda and (b) outbreak of respiratory disease

(a) Report on the necropsy of a female chimpanzee, Ruda, at Sonso, Budongo Forest on 30 October 2001

Background

Melissa Emery who was undertaking a PhD at Sonso, Budongo Forest contacted me (Wayne Boardman — WB) at 11 am on 29th October 2001 to report that a chimpanzee in the research group was lying face down on the ground and was unable to move. WB immediately contacted Dr J. Okori of UWA who thought that it would be good if WB and Drs Ssuna and Apell could attend to undertake a necropsy. We contacted Mr Apophia of UWA by phone later that afternoon to seek permission to go which was granted. There was not enough time in the circumstances to obtain written permission from UWA.

WB contacted Melissa Emery later that evening and she related that the chimpanzee was worse and was being eaten by safari ants. The research assistants would stay with her overnight.

At 2 am on 30/10/01 the field assistants left the animal to return to camp. She was barely breathing and unable to move. At 7 am that morning, they again visited her. She was found dead and several other chimpanzees were calling nearby.

Professor Bwangamoi and Dr Ssebide Bernard Joseph from UWA were asked if they would like to accompany us.

Necropsy (see Fig. D.1)

CSWCT necropsy worksheet

Animal Information.

Species:	Chimpanzee	Euthanased:	NO
Animal ID:	Ruda	State:	Fresh
Weight and Sex:	est 25–28 kg: F	Dissector:	Drs Bwangamoi, Ssuna, Apell,
Age:	c.30 years		Ssebede and Boardman
		Date & Time of	from 2.30 pm to 5.30 pm
		Necropsy:	on 30/10/01

History/circumstances of death

The chimpanzee, Ruda, was found dead in Block 30 approximately 400 m from Sonso Camp. Her history showed she had a 4 year old female offspring called Recho and a 10 year old male offspring called Bob. Her appetite appeared to be good but she had begun to lose condition from August. She had occasionally had diarrhoea but this could not be attributed to anything specific at the time. It was noted that in the past, she had always been shy but on losing condition over the last two months she had come much closer to the researchers.

At 10 am on 29/10/01 she was seen in ventral recumbency, twitching and unable to move significantly when approached. She deteriorated further throughout the day. Safari ants were attacking her. She died between 2 am and 7 am on 30/10/01. When she was found dead she was sprayed with insecticide spray to discourage insects and covered in cloth which was kept wet to keep the cadaver as cool as possible.

She was hung from a tree in the forest, which was determined to be the coolest place in order to reduce the effects of decomposition.

External findings

The chimpanzee is a sexually inactive female in an emaciated condition. There are many bruises due to safari ant attack on the ventral aspects of the head, neck, chest and groin. Rigor mortis is not yet apparent. There is external autolysis, which is not extensive. The teeth are present and in good condition except for normal wear of the tabular surfaces.

Body condition

Hydration: very dehydrated.
Fat Stores: no fat stores.
Muscle Mass: emaciated.

Internal findings

The ventral skin is strongly adherent to the adnexa because of lack of subcutaneous fat.

There are multiple fibrous adhesions from the mesentery and omentum to the ventral cadaver wall. Viscera are entrapped in the adhesions. There is a 2 cm diameter abscess in the left lateral cadaver wall. Abdominal muscle surrounding the navel contains pockets of yellow-orange pus and caseous material. There is an irregular consolidated swelling in the dorsal aspect of the spleen.

Both the right and left diaphragmatic lung lobes have bullous emphysema extending to the cardiac lobes. There is bilateral atelectasis in the ventral lung lobes. The parenchyma of the lung contains small dark spots. The trachea contains small quantities of inhaled soil.

There is a focal pericardial adhesion to the left ventricle. The right side of the heart appears flabby and the chordae tendinae appear weakened. There are multiple hepatic lymph node abscesses and multiple focal calcified abscesses in the parenchyma of the ventral liver lobe, which contain caseated material. In addition, there are several 3–4 cm diameter abscesses with thickly fibrous capsules containing creamy pus in the dorsal parenchyma. Focal telangectasia can be seen in the diaphragmatic lobe. There is a considerable amount of ingesta throughout the entire alimentary tract. She had a good appetite until the last 24 hours.

A large 40 cm long taenia-like tapeworm is seen in the duodenum. There are focal petechial haemorrhages in the stomach mucosa. Many 2 cm long nematodes with dark anterior portions were seen in the jejunum. Many 2 cm long white nematodes were seen in the caecum associated with haemorrhagic ulcers. Nematodes are attached to the centre of the ulcer in places. There are multiple haemorrhagic spots and ulcers on the mucosa of the colon and caecum. Diffuse congestion is seen throughout the colon. There are multiple 5 mm diameter nodules in the caecal wall which are associated with nematodes which extend from the nodular surface. There is a 2 cm diameter abscess on the serosal surface of the mid colon.

All other tissues appear normal. The brain and spinal cord were removed and examined.

Samples

Cytology: 2 × umbilical abscess smears and 1 × hepatic abscess smears (to Veterinary Pathology Dept at Makerere University)

Bacteriology: 2 × umbilical abscess (1 to Veterinary Bacteriology Dept at Makerere University and 1 to Ebenezer Lab in Kampala) and 1 × hepatic lymph node (to Veterinary Bacteriology Dept at Makerere University)

Faeces: Fixed in 10% formalin (to Veterinary Pathology Dept at Makerere University)

Blood: 15 ml serum (to Uganda Viral Research Institute, Entebbe)

Parasites: Fixed in 70% alcohol — jejunal, caecal and colon nematodes and tapeworm (to Veterinary Pathology Dept at Makerere University)

Frozen tissues: Spleen, lung, liver and kidney in aluminium foil (to Uganda Viral Research Institute, Entebbe)

Fixed tissues: 3 × brain whole, spinal cord, skin, heart, pericardium, lung, liver, lymph nodes, abscesses, umbilical abscess, kidney, thyroid, adrenal, bladder, uterine body, stomach, duodenum, jejunum, caecum, colon.
1 × to Veterinary Pathology Dept at Makerere University (whole brain and spinal cord only)
1 × to WARM Dept, Makerere University
1 × to CSWCT

Carcase: Fresh to WARM Dept, Makerere University.

Interim diagnosis

Diffuse chronic active peritonitis with extensive chronic adhesions.
Chronic navel and cadaver wall abscessation.
Multiple focal chronic hepatic abscessation.
Hepatic lymph node abscessation.
Left-sided ventricular–pericardial adhesion.
Duodenal taeniasis
Mild jejunal nematodiasis.
Severe caecal and colonic nematodiasis.

Cause of death

The cause of death can be attributed to right-sided heart failure secondary to severe chronic active peritonitis, severe hepatic abscessation, severe caecal and colonic nematodiasis and a focal pericardial adhesion.

It is considered that the primary problem may have occurred from a penetration of the abdominal wall adjacent to the navel by a foreign body, which caused extensive peritonitis. It is likely the hepatic abscesses arose due to the haematogenous spread of bacteria from the primary site.

Acknowledgements

We thank Melissa Emery, Robin May and Wilma van Riel from Sonso for their information and advice, Zephyr Kiwede for detailed information on the group and the Budongo Forest Project for hospitality. We also thank CSWCT for hiring a vehicle for us all to travel to Budongo Forest and UWA for giving permission for us to perform the necropsy.

Compiled by: Wayne Boardman(WB)[1], Professor Bwangamoi[2], Dr Ssuna[1], Dr Apell[1] and Dr Ssebide[3]

[1] CSWCT, PO Box 369, Entebbe
[2] Dept of Veterinary Pathology, Makerere University
[3] UWA, PO Box 3530, Kampala

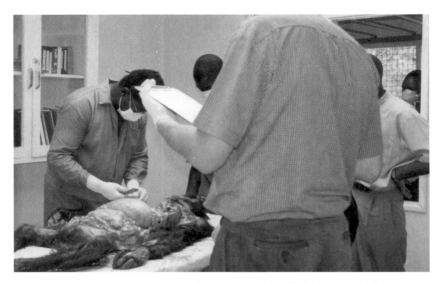

Fig. D.1:   Necropsy in progress on Ruda in the Budongo Forest Project's camp at Sonso (photo: M. Emery Thompson).

(b)   Report on the investigation into an outbreak of respiratory disease in chimpanzees of Budongo Forest, Uganda

by Gladys Kalema, Veterinary Officer, Uganda Wildlife Authority and Wayne Boardman, Wildlife Veterinary, Uganda Wildlife Education Centre, Entebbe.

Introduction

A respiratory disease was observed in the chimpanzees at the research site at Sonso, Budongo Forest, from the 11th of November 1999. The chimpanzee community at this site consisted of 53 chimpanzees. The frequency of coughing and the numbers affected increased over the next few days. A mother and infant did not leave the night nest for a long time on one day, which, together with continued coughing, stimulated the decision to visit Budongo Forest research camp to undertake a veterinary investigation on 23rd and 24th November 1999.

This is the first recorded incidence of an epidemic of chimpanzee respiratory disease in Budongo Forest since habituation started in 1990.

Methods

1.   Discussion with staff at Budongo forest project.

2.   Observation of chimpanzees for 2 consecutive days.

3.   An intervention was not performed because the existing policy is not to intervene unless the problem is life threatening. The respiratory disease appeared at this stage not to be life threatening. There was indirect evidence that the respiratory disease could have been human caused.

Discussion with the staff at Budongo Forest Project

The staff of BFP were questioned about the epidemic. This indicated that the outbreak of coughing started on 11 November with NK coughing on that day. It spread rapidly and by Nov 21, 23 individuals had been observed coughing. The following chimpanzees were affected: NK, BH, AY, KW, KZ, TK, CL, JB, MH, GS, MU, ZG, KY, GZ, BK, HT, MA, NB, NR, MS, KL, KA, SR.

Symptoms

Symptoms varied between a dry cough to a wet, moist productive cough. GS has a bilateral mucoid nasal discharge; BK, NB and MS had a unilateral mucoid nasal discharge. Some chimpanzees had paroxysms of coughing, consisting of prolonged coughing of up to 40 seconds duration. No dyspnoea or tachypnoea (difficulty in breathing) was observed. No increase in lethargy, depression or anorexia was observed. These observations were strengthened by discussions with field assistants and visiting researchers. An unidentified adult female and infant on the 17th November were reluctant to get out of the nest until later in the day, and were coughing.

The chimpanzees did not lose vigour and were alert and responsive, including the infants, which were seen to be active and continued to suckle.

Frequency of coughing in infants was greater that in adults. During the observation period, infant NR coughed 3 times an hour with prolonged bouts and the mother, NB, coughed 4 times an hour. Infant KA coughed 3 times an hour, and the mother, KY, coughed 2 times an hour. Infant ZG coughed 4 times an hour, and his older sister, GZ, coughed 5 times an hour, however their mother, ZM did not cough.

Proportion of individuals observed coughing during the epidemic, 11–24th November 1999:

Date	Observed coughing rates (% of all individuals observed on that day)
11th	33
16th	25
17th	60
18th	75
19th	36
20th	66
22nd	80
23rd	70
24th	53

From 11th November to 24th November, 23 animals out of 34 observed (67.6%) were seen to be coughing regularly. 11 animals (32.4%) were never observed coughing.

Only one animal, NK, had a determined period of disease. He was observed to start coughing on the 11th November, and stopped coughing 9 days later. In this case the course of the disease lasted for 9 days. For the rest of the individuals it was difficult to determine when the coughing started and ended, as they were not observed every day prior to the start of coughing. From 24th November the incidence of coughing declined, after which the epidemic slowly disappeared with no loss of life.

Observations

From the 15th to 18th of October, a new temporary member of staff came to Sonso, R.C., who was sick for 3 days while at Sonso during which time he went into the forest to observe the chimpanzees with a field assistant. He had a cough. On the 4th day, he went to hospital for treatment. A field assistant (not the one he entered the forest with) began coughing 3 days later, and did not go to the forest for one day, as he was feeling sick.

No changes in weather conditions were noted during the time of the epidemic.

There was no evidence of respiratory disease in any of the other species of primates in the Sonso area: blue monkey, black and white colobus monkey, red tailed monkey and olive baboon.

An unidentified immigrant female in oestrus came into the group on 10th November and mated with several males including NK, who was observed to have a cough on 11th November. However this female did not cough.

Assessment

Based on morbidity data and a lack of serious symptoms, it is likely that the outbreak can be attributed to one of the following diseases:

Respiratory Syncytial Virus (RSV)—Chimpanzee Coryza Virus
Paramyxovirus 1, 2 or 3.
Influenza A and B

Symptoms were mainly upper respiratory signs, cough and discharge. The disease did not progress to more serious symptoms. RSV, Paramyxovirus and influenza viruses are often self-limiting if there is no secondary bacterial infection. Other diseases would have had more serious symptoms, and may have needed antibiotics or other supportive therapy to clear the symptoms. These 3 viruses have a zoonotic potential: there is a risk of human to non-human primate infection and minimal risk to humans.

Conclusion

It is likely that the outbreak of respiratory disease based on available information at the time, was due to an anthropozoonotic virus infection. A respiratory infection in humans, locals, researchers, field assistants, may have been transmitted to the chimpanzees by unknown means.

Recommendations

Efforts to prevent disease spread from humans to chimpanzees should be reviewed. Any sick people, including field staff, should not go to study or habituate the chimps, until they have fully recovered, as they could very easily spread the respiratory or any other disease to the chimpanzees which are often unexposed to the human diseases and therefore have little to no immunity to fight the disease. The 7 m minimum distance between humans and chimpanzees should also continue to be observed.

E.   Other primate species of the Budongo Forest

This book is about the Sonso chimpanzees. It makes no attempt to describe the other primate species living in the Budongo Forest, even in the range of the Sonso community. We have mentioned other primate species in their role as prey of the Sonso chimpanzees. We have mentioned baboons as crop-raiders, but mainly to compare them with chimpanzees and show how, in relation to traps, the fate of chimpanzees is intimately bound up with that of baboons. In terms of dominance, chimpanzees appear to be dominant over all the monkey species which generally move away when chimpanzees appear in a tree where they have been feeding. Between the forest monkeys I do not know what the dominance order is, but baboons with their larger size and aggressiveness are probably dominant over the three forest monkey species.

Vervet monkeys (*Cercopithecus aethiops*) are not included here because they are not forest monkeys, staying outside the forest all the time. They are fairly common around Budongo Forest and sometimes raid farmers' fields and household gardens.

In this appendix I shall do little more than outline the studies of prosimians and monkeys made so far, almost all under the auspices of the BFP.

Prosimians: pottos and bushbabies

The commonest species of prosimian living in Budongo Forest is the potto (*Periodicticus potto*). This species can be seen at night moving slowly in the canopy by shining a torch at it, when its large eyes reflect the light with an orange glow. Nothing much appears to be known about the behaviour of this nocturnal species in Budongo, though Ambrose (pers. comm.) observed a potto occupying a tree hole.

Surprisingly, efforts to find other species of prosimians e.g. *galago* spp. (bushbabies) which are common in some other Ugandan forests such as Kibale and Kalinzu, have met with little success in the Sonso area, and it seems that they are not at all common there. In 1970–71, Judy Harris did a survey of prosimians in Uganda's forests. In 1970 she observed four individuals in the Busingiro area of Budongo Forest, and in 1971 she observed three individuals in the same area, each time using a method whereby a colleague shone a 12 V spotlight out of a vehicle and looked for eye-shine. She concluded that there were 'very few [bushbabies] compared to other forests' (pers. comm.).

In 2001, Lesley Ambrose recorded two unidentified galagos, one was heard along the Royal Mile in the range of the Sonso chimpanzees and the other was identified by rapidly moving eye-shine on the logging road that runs north from the Sonso sawmill (pers. comm.).

Monkeys

Four species of monkey live in the Budongo Forest: baboons, blue monkeys, redtail monkeys and black and white colobus monkeys. All are relatively common; all four species would very likely be seen in a one- or two-hour walk in the forest. The only comparative study to date of the three forest monkey species, blues, redtails and colobus, together with chimpanzees, is that of Plumptre (in press), focusing on dietary preferences and overlap, in logged and unlogged forest. He found that dietary overlap was significantly higher than expected from random neutral models, indicating that for most comparisons, competition for food is unlikely to be important. All the monkey species occurred at higher densities in logged than in unlogged forest. The two *Cercopithecine* species, *C. mitis* and *C. ascanius*, preferred fleshy, ripe fruits over unripe ones. Interestingly, *Colobus guereza*, often assumed to be primarily a leaf-eating species, ate more of the fruit as of the leaves of *Celtis gomphophylla*, a common tree species in Budongo (see below).

Baboons

The Budongo Forest contains a large population of baboons of the species *Papio anubis*, the olive baboon. Their density was found to be 11.7/km² (Plumptre and Reynolds 1994). They have been studied primarily by Jim Paterson from the University of Calgary, and also by Adam Alberts Okecha from Makerere University. They range deep inside the forest and subsist on forest foods. Those living in the Sonso area are frequent visitors to the Sonso camp and the houses around the sawmill, where they eat food waste thrown out by people and constitute a considerable nuisance by stealing human food.

Studies of baboons at the BFP include the following (for full details see References at end): Paterson (1996, 1997*a*, *b*, *c*, 2001), Paterson and Teichroeb (2001), Paterson *et al.* (2001), Okecha (2000, in press).

Blue monkeys

The first study of blue monkeys (*Cercopithecus mitis stuhlmanni*) in Budongo Forest was that of Aldrich-Blake (1970), made in the Busingiro region. Aldrich-Blake described the ecology and behaviour of the monkeys in that area, together with details of their social organization: he found that they lived in one-male groups.

Today, blue monkeys are common in Budongo Forest when compared, for example, with Kibale Forest. Their density in Budongo was found to be 43.9/km² (Plumptre and Reynolds 1994) and is higher in selectively logged forest (58.2/km²) than in unlogged forest (15.6/km²). In a study of primate densities before and after pitsawing, blue monkey densities were unchanged after the pitsawing ended (Beresford-Stooke 1999).

Fairgrieve (1995*a*, 2003) studied the behaviour, ecology and social organization of blue monkeys in great detail, and also made comparisons between blue monkey groups living in logged forest in the Sonso area and unlogged forest in the Nature Reserve. He showed that they live at higher densities in logged forest, their ranges are smaller in logged forest (see also Plumptre *et al.* 1997: 43; Plumptre, in press), and there is a higher density of the kinds of foods they prefer in logged forest. Fairgrieve (1995*b*) observed the takeover of one blue monkey group by an outsider adult male; the male fought and drove out the resident male, after which he killed and ate an infant from the group he took over. There is fierce antagonism between rival adult male group leaders. One male killed as a result of a fight between two rival males (or possibly killed by a predator — the fight occurred after dark and could not be seen) had the following characteristics when examined within 12 h of death (unpublished data): Length (tip of tail to front of head) 62.5 in.; tail length 40.5 in.; body length 22 in.; front inside leg length 9 in.; rear inside leg length 11 in.; hand length 4.5 in.; foot length 6.5 in.; weight 7.2 kg. This animal had cuts on the face, both rear legs, one hand and a deep and severe wound in the belly which had punctured its stomach.

Fairgrieve (1997) describes how on 1 February 1994 he came across a group of excited blue monkeys centred on an adult male who was holding and eating the carcass of a Lord Derby's flying anomalure (*Anomalurus derbianus jacksonii*) — a flying squirrel. The remains were examined and were still fresh, indicating that the male had caught and killed it. Blue monkeys were also observed to eat guinea fowl eggs during this study.

The fondness of blue monkeys for ripe fig fruits in particular but also for young fig leaves is described in Tweheyo and Obua (2001).

Blue monkeys are occasionally hunted and eaten by chimpanzees — see Chapter 3 of this book and Newton-Fisher *et al.* (2002).

Studies of blue monkeys at the BFP include: Fairgrieve (1993, 1995*a, b, c*, 1997).

Redtail monkeys

Redtail monkeys (*Cercopithecus ascanius schmidti*) are the smallest of the forest monkeys in Budongo. Their density in the forest as a whole is 33.3/km², and as with blue monkeys they are found at higher densities in selectively logged forest (46.4/km²) than in unlogged forest (8.3/km²) (Plumptre and Reynolds 1994; Plumptre, in press). They are found in smaller, overlapping ranges in logged forest, whereas in unlogged forest their ranges do not overlap (Plumptre *et al.* 1997: 44). Beresford-Stooke (1999) in a 'before and after pitsawing' study found an increase in density of redtail monkeys after pitsawing.

The ecology, behaviour and social organization of redtails were studied by Sheppard (2000). She found that redtail densities were three times higher in selectively logged than in unlogged forest, and that the selectively logged forest contained higher densities of preferred redtail food trees. As in the case of blue monkeys, redtails live in one-male groups, with antagonism between male group leaders and their rivals who live outside the group. Redtails and blues are often found together in mixed associations, and frequently alarm calls by one species are taken up by the other. Such mixed associations and mutual alarm calling can also include the third forest monkey species, the black and white colobus.

As in the case of blue monkeys, redtails are fond of ripe fig fruits and also young fig leaves (Tweheyo and Obua 2001).

An unusual form of interaction involving a redtail monkey was recorded by Tinka and Reynolds (1997). On 4 September 1996, Gonza, then a subadult female chimpanzee, initiated play with an adult redtail monkey. At 8.47 a.m. 'she played with the tail folding it around her neck', shook it, and groomed the body of the redtail. 'The monkey stretched its legs to allow easy reach for Gonza. She groomed it under the abdomen, chest and back. The grooming was mixed with playing e.g. beating the sides of the monkey, pulling its legs, and Gonza rubbed her vagina against the anus of the monkey. The whole process lasted 20 minutes...Gonza was the actor only...At 9.07 the monkey terminated this activity by moving away' (p. 6).

Studies of redtail monkeys at the BFP include: Sheppard (2000), Sheppard and Paterson (2001*a, b*).

Black and white colobus monkeys

Black and white colobus monkeys (*Colobus guereza*) with their flowing black and white robes are spectacular primates as they move in the canopy. As with the redtails and blues, they are plentiful in Budongo. Their density is 39.3/km² for the whole forest, and as with the blues and redtails they are more numerous in selectively logged forest (44.2/km²) than in unlogged forest (27.0/km²) (Plumptre and Reynolds 1994). Beresford-Stooke (1999) in her 'before and after pitsawing' study referred to above found an increase in density of black and white colobus monkeys after pitsawing.

Whereas colobus monkeys are often characterized as leaf-eaters, in contrast to the cercopithecine fruit-eaters, in Budongo Forest this opposition breaks down. A close comparison of diets of the three forest monkey species showed that more fruit than leaves was eaten by colobus. This is in contrast with the situation in Kibale Forest and at other sites. One of the major items in the diet of black and white colobus monkeys at Budongo was the fruit of *Celtis gomphophylla* (previously *durandii*)

(Plumptre *et al.* 1994: 71; Plumptre, in press). The five commonest items in the diet of *C. guereza* at Budongo were:

1. *Celtis gomphophylla* — ripe fruit — 16.65% of diet

2. *Celtis gomphophylla* — unripe fruit — 14.61%

3. *Celtis gomphophylla* — mature leaves — 6.90%

4. *Cynometra alexandrii* — young leaves — 6.52%

5. *Cynometra alexandrii* — seeds — 5.51%

We therefore have fruit-eating colobus monkeys at Budongo.

Preece (2001, in press) studied the density of black and white colobus monkeys in Budongo Forest in relation to the abundance of lianas. Past studies, e.g. at Kibale Forest, have shown a higher density of black and white colobus monkeys in logged areas and a partial explanation for this has been sought in the abundance of lianas with mature leaves in disturbed forest. Preece examined whether this source of leaves was an important food source for the colobus monkeys of Budongo Forest. Lianas, with a diameter of 1 cm or greater, were counted in a number of sample plots, and correlated with the density of *Colobus guereza* in the areas concerned. He found that whereas abundance of lianas did not correlate with density of colobus across compartments (large areas of forest), within compartments there was a good correlation. He concluded that the effect of liana abundance on colobus density is masked across large areas by confounding variables such as habitat type, but that within smaller areas, forest lianas, especially the larger ones that provide large quantities of leaves in the canopy, are an important food source of *C. guereza* in Budongo and do in part determine density. He found a greater abundance of lianas in logged forest than in unlogged forest, due to the creation of gaps by logging activities and agreed that this could help to explain the higher densities of colobus found in logged forest than in unlogged. However, we need to recall that Budongo colobus are mainly fruit-eaters and so liana abundance can provide only a supplementary explanation, the main one being (as for the other forest monkeys) the increased abundance of edible fruiting trees in the logged forest.

The detailed ecology, behaviour and social organization of black and white colobus monkeys in Budongo Forest remains to be studied. This is an interesting species and its characteristic loud, low-pitched gurgling sound is one of the delights of Budongo. Every morning, at around 7 a.m., all of the colobus groups around camp join in a 5-min morning chorus which is one of the most enjoyable sounds of the forest and a very pleasant way to start the day. I suspect it is the dominant males that call. I do not think these colobus monkeys live in one-male groups; more likely they live in multi-male groups. This may in part be because of the protection this affords from predators, notably chimpanzees.

Chimpanzees in Budongo Forest are not great hunters but the majority of monkey prey are colobus (Newton-Fisher *et al.* 2002). Young individuals are preferred. Colobus-eating has been described in Chapter 3. I end with a small tribute to the colobus: on one occasion, the day after we had found the hairs and some bones of a young colobus monkey in the faeces of one of our adult male chimpanzees, we encountered this individual, together with three other adult male chimpanzees, in a food tree with two adult male colobus monkeys. They appeared to be sparring — the colobus were rushing towards the chimpanzees and the chimpanzees were responding in a similar way. After a while the two colobus males made a joint concerted rush at one of the four chimpanzees which backed down and fled, with the colobus in hot pursuit. This led to the other three chimpanzees leaving also, in as dignified way as they could. It was undoubtedly a triumph for these two brave colobus males over four much larger, but cowardly chimpanzees. Possibly the colobus males were enraged by the loss of one of their young and were protecting the rest of their group which was nearby. In any case, it seems that the Sonso chimpanzees do not attack colobus unless they are sure of an easy kill, when a young one is on its own and offers little resistance.

Non-primate species

It is not possible in this book to discuss the studies made by the BFP on non-primate species. The majority of these have been summarized very succinctly by Zoe Wales in BFP Report No. 81, titled *What We Know About Budongo Forest*. In this excellent set of summaries are to be found details of BFP studies of invertebrates, amphibians, birds and small mammals up to 1990. Studies of non-primate species under BFP are listed on the BFP's website, www.budongo.org, which includes reports, publications, dissertations and theses.

Let us end this Appendix with a eucnemid beetle. In 1995 Thomas Wagner identified a new genus of the subfamily Melasinae which he recovered from the crown of a tree not far from our camp at Sonso. This species is defined by three cornered hypomera and partly uncovered metepimera. The specimen, a male, was described by Lucht (1998) and given the name *Nebulatorpidus wagneri*.

F. The Budongo Forest Project

History of BFP

In 1998 Shirley McGreal of IPPL sent me a cutting from the *New Vision*, the Ugandan newspaper, about two infant chimpanzees that had been smuggled from Uganda to Dubai *en route* to a dealer; the whistle had been blown and they had been returned to Uganda where they were being cared for by a young couple in Entebbe. The article stated that the source of the chimpanzees was thought to be the Budongo Forest.

In 1999 I had a sabbatical term and spent much time writing to all the big wildlife agencies suggesting a research project to conserve the chimpanzees of Budongo. No luck: they had other priorities. A few organizations helped, however, and so, with many setbacks, we slowly got off the ground. The Boise Fund paid for my first return visit to Budongo in 1990, we got a grant from the Jane Goodall Institute which enabled Chris Bakuneeta to join the project, we obtained much needed support from the National Geographic Society's Committee for Research and Exploration for our chimpanzee studies, and a grant from the Rainforest Action Fund, a small organization that gave us some support for our first members of staff at the outset. In 1991 the Project's forestry research received a grant from the British Government's Overseas Development Agency under their Forestry Research Programme which enabled us to begin a study of how past logging of Budongo Forest had affected wildlife, including endangered species such as chimpanzees, and to what extent the wildlife in turn were essential dispersers of the seeds of forest trees. Andy Plumptre joined the project (BFP) as research director in 1991 and set up a number of long-term studies, some of which are still running today. It was at that time that the name Budongo Forest Project (BFP) was coined; and it was then that the project's logo (shown above) was designed.

We have been able to attract many excellent students from Uganda and from overseas (see Table F.1). In 1997 BFP made a link with the Faculty of Forestry and Nature Conservation at Makerere University in Kampala, and since then we have been receiving support from the Government of Norway under its overseas aid programme, which supports forestry and forestry education in Uganda. At that time we established a Steering Committee to oversee the work of BFP and to take final responsibility for decisions. The current Chairman of the Steering Committee is John Kaboggoza, Dean of the Faculty of Forestry and Nature Conservation at Makerere University, and recently elected Chairman of the National Forest Authority. Also on the Steering Committee are the Science Secretary of the Uganda National Council for Science and Technology, the current Director and Assistant Director of BFP, and myself as Founder and Adviser. The Project's Director is currently Fred Babweteera, and the Assistant Director, who is normally an expatriate researcher. Recent Assistant Directors have been Sean O'Hara and Nick Newton-Fisher.

Table F.1: Research students, senior scientists and volunteers who have worked at the BFP (1990–2004).

Name	Status	From	Area of interest	Time at Budongo
Chris Bakuneeta	Research officer	Makerere Univ.	Chimps/Co-director	April 90–June 97
Kirstin Johnson	M.Sc. student	Oxford Univ.	Use of local forest products	July–Sept 91
Dr Andy Plumptre	Research officer	Oxford Univ.	Forest ecology/Co-director	Sept 91–Sept 96
Dr Jake Reynolds	Postdoc volunteer	Oxford Univ.	Construction of research site	Mar–Oct 92
David Bowes-Lyon	Undergrad Biol	Oxford Univ.	Mapping	Mar–June 92
Gladys Kalema	Vet. student	Royal Vet. Coll.	Chimpanzee parasitology	Sept–Oct 92
Charles Walaga	M.Sc. student	Makerere Univ.	Forest soils	Oct 92–Jan 94
Chris Fairgrieve	Ph.D. student	Edinburgh Univ.	Blue monkey ecology	Jan 93–Sept 94
Isaiah Owiunji	M.Sc. student	Makerere Univ.	Birds	Feb 93–April 95
Paul Musamali	M.Sc. student	Makerere Univ.	Small mammals	Feb 93–April 95
Dr Peter Miller	Senior scientist	Oxford Univ.	Dragonflies	Mar–Apr 93
Dr Kate Hill	Senior scientist	Oxford Univ.	Crop-raiding around Budongo	July–Dec 93
Prof. Duane Quiatt	Senior scientist	Colorado Univ.	Chimpanzee behaviour	Sept–Dec 93
Dr Liz Rogers	Senior scientist	Edinburgh Univ.	Herbaceous diets	Nov–93
Bob Smith	Research student	Oxford Univ.	Chimpanzee ecology	Jan 94–Dec 94
Jo Thompson	Ph.D. student	Oxford Univ.	Field methodology	Jan–94
N. Newton Fisher	Ph.D. student	Cambridge Univ.	Chimp male behaviour	June 94–Dec 95
Angela Stanford	Ph.D. student	Bristol Univ.	Squirrel ecology	May 94–Mar 96
Dr Heidi Marriott	Senior scientist	Oxford Univ.	Human demography	July–Aug 94
Druin Burch	M.Sc. student	Oxford Univ.	Chimp injuries	July–Aug 94
Theresa Price	Undergrad Biol	Oxford Univ.	Small mammals	July–Sept 94
Josephine Wong	M.Sc. student	Oxford Univ.	Chimp vocalizations	July–95
John Waller	M.Sc. student	Oxford Univ.	Chimp injuries	July–Aug 95
Judith Knight	Soc Anth student	Oxford Univ.	Trapping methods	July–Aug 95
Patrick Boston	Research visitor	Oxford	Butterflies	Jan 95–Feb 96
Thomas Wagner	Senior scientist	Museum Koenig	Insects	June–July 95
Rebecca Sutton	Undergrad Hum Sci	Oxford Univ.	Medical aspects of villagers	July–Aug 95
Juliet Ferguson	Undergrad Hum Sci	Oxford Univ.	Medical aspects of villagers	July–Aug 95
David Bowes-Lyon	Postgrad Zool	Oxford Univ.	Medical aspects of villagers	July–Aug 95
Ruth Prince	Undergrad Hum Sci	Oxford Univ.	Medical aspects of villagers	July–Aug 95
Michelle Barrows	Vet. student	Glasgow Univ.	Intestinal parasites	July–Aug 96
Malcolm Starkey	Undergrad Zool	Oxford Univ.	Amphibians	July–Sept 96
Louise Aukland	Undergrad Zool	Oxford Univ.	Amphibians	July–Sept 96
Rob Ingle	Undergrad Zool	Oxford Univ.	Amphibians	July–Sept 96
Sharon Jones	Undergrad Zool	Oxford Univ.	Amphibians	July–Sept 96
Hugh Notman	M.Sc. student	Oxford Univ.	Chimpanzee vocalizations	July–Aug 96
Harriet Bennett	Anthrop. student	Edinburgh Univ.	Kinship in chimps	April–Sept 96

Table F.1: (continued)

Name	Status	From	Area of interest	Time at Budongo
Andrew Brownlow	Hum Sci graduate	Oxford Univ.	DNA from hair in chimps	Sept–Dec 96
Jeremy Lindsell	Ph.D. student	Oxford Univ.	Understorey bird ecology	Sept 96–Apr 00
Jim Paterson	Senior scientist	Univ. of Alberta	Forest baboon behaviour	Sept–Dec 96
Lucy Bannon	Law student	College of Law	Understanding of hunting laws	Sept–Dec 96
Rebecca Wingate-Saul	Assistant to L.B.	London	Understanding of hunting laws	Sept–Dec 96
Duane Quiatt	Senior scientist	Colorado Univ.	Chimpanzee behaviour	Oct–96
Katie Fawcett	Ph.D. student	Edinburgh Univ.	Female chimp ranging	Jan 96–Aug 98
Clea Assersohn	Ph.D. student	St Andrews Univ.	Infant chimp learning	July 96–Sept 98
Emma Stokes	Ph.D. student	St Andrews Univ.	Manipulation by injured chimps	Aug 96–Sept 98
Mikala Lauridsen	Ph.D. student	Copenhagen Univ.	Attitudes to forest among locals	Sept 96–Sept 97
Fred Babweteera	M.Sc. student	Makerere Univ.	Climber tangles in gaps	Oct 96–continuing
Jo Lee	Research visitor	Oxford Univ.	Mapping trail system in NR	Apr–May 97
Emily Bethell	Undergrad.	Univ. Coll. London	Vigilance in chimps + mapping	July–Aug 97
Jim Paterson	Senior scientist	Univ of Alberta	Baboons and Khaya bark-eating	July–Aug 97
Brad McVitie	Grad. student	Univ of Alberta	Baboons and Khaya bark-eating	July–Aug 97
Ryan Raaum	Grad. student	Univ of Alberta	Baboons and Khaya bark-eating	July–Aug 97
Ben Mines	Undergrad.	Oxford Univ.	Understorey birds	July–Sept 97
Rhiannon Meredith	Undergrad.	Oxford Univ.	Herbarium documentation	July–Sept 97
Andrew Whiten	Senior scientist	St Andrews Univ.	Supervisor of Clea Assersohn	Aug–Sept 97
Lucy Beresford-Stooke	Research officer	Edinburgh Univ.	Monkeys/Deputy Co-Director	Sept 97–Mar 98
Richard Byrne	Senior scientist	St Andrews Univ.	Supervisor of Emma Stokes	Sept–97
Polycarp Mwima	M.Sc. student	Makerere Univ.	Regeneration of Khaya	Sept 97–Mar 98
Eric Sande	Ph.D. student	Makerere Univ.	Nahan's francolin	Oct 97–Dec 99
Byamukama Biryahwaho	M.Sc. student	Makerere Univ.	Forest outliers	Oct 97–Apr 98
Tweheyo Mnason	M.Sc. student	Makerere Univ.	Figs and primates	Oct 97–Apr 98
Liz Rogers	Senior scientist	Edinburgh Univ.	Supervisor of Katie Fawcett	Dec 97–Jan 98
Paula Pebsworth	Ph.D. student	Colorado State Univ.	Self-medication by chimps	Feb–May 98
Mike Huffman	Senior scientist	Kyoto Univ.	Supervisor of Paula Pebsworth	Apr–Jun 98
Chris Perrins	Senior scientist	Oxford Univ.	Supervisor of Jeremy Lindsell	Mar–98
Kate Arnold	Ph.D. student	St Andrews Univ.	Reconciliation in male chimps	Oct 98–Sept 99
Donna Sheppard	M.Sc. student	Univ. of Alberta	Redtail monkey ecology	Jun 98–May 99
Charles Kahindo	M.Sc. student	Makerere Univ.	Bird communities in gaps	Dec 98–Jun 99
Bjornar Slembe	M.Sc. student	AUN	Bristlebill in logged vs. unlogged	Jan 99–Apr 99
Hugh Notman	Ph.D. student	Univ of Alberta	Chimp vocalizations	Jan 99–Apr 00
Lucilla Spini	Ph.D. student	Oxford Univ.	Juvenile chimpanzees	Jan 99–Mar 99
Adam Alberts Okecha	M.Sc. student	Makerere Univ.	Baboon diets	Sept 99–Feb 00
Clement Okia	M.Sc. student	Makerere Univ.	Regeneration of rattan	Sept 99–Feb 00
Mabe Akhos	M.Sc. student	Makerere Univ.	Attitudes to eco-tourism	Sept 99–Dec 99
Michael Mbogga	M.Sc. student	Makerere Univ.	Distribution of *Broussonettia*	Sept 99–Feb 00
David Harris	Undergrad.	Oxford Univ.	Chimp party size and food supply	Oct–Dec 99
Helen Jones	Undergrad.	Oxford Univ.	Chimp party size and food supply	Oct–Dec 99
Siddhartha Singh	M.Sc. student	Oxford Univ.	Nocturnal copulations by chimps	July–Aug 99
Ben Dempsey	Undergrad.	Oxford Univ.	Male–male behaviour in chimps	July–Sept 99

Table F.1: (continued)

Name	Status	From	Area of interest	Time at Budongo
Julie Munn	M.Sc. student	Aus. Nat. Univ.	Effects of snares on chimps	Oct 99–June 00
Joseph Bahati	Ph.D. student	Makerere Univ.	Regeneration of mahogany	Jan–Dec 00
Sally Seraphin	M.Sc. student	Oxford Univ.	Chimp faecal hormones	Apr–00
Janette Wallis	Senior scientist	Univ. of Oklahoma	Chimp female sexuality	Apr & Sept 00
Tweheyo Mnason	Ph.D. student	AUN	Chimpanzee diet	Apr 00–Sept 00
Melissa Emery	Ph.D. student	Harvard Univ.	Chimp female endocrines	June 00–Aug 00
Lori Oliver	M.Sc. student	Bucknell Univ.	Mate choice in chimpanzees	June 00–Dec 00
Jenny Greenham	Senior scientist	Reading Univ.	Phytochem. of chimp foods	Sept–01
Chris Fairgrieve	Rapporteur	Scotland	Report on DFID Workshop	Mar–01
Kate Hill	Senior scientist	Oxford Brookes Univ.	Workshop speaker	Mar–01
Graham Preece	M.Sc. student	Univ. Wales — Bangor	Black and white colobus monkeys	May–June 01
Cristy Watkins	M.Sc. student	Calif. State Univ.	Local community attitudes	June–Aug 01
Lucy Bates	M.Sc. student	Oxford Univ.	Female chimp social groupings	July–Aug 01
Elinor Croxall	Undergrad.	Oxford Univ.	Day and night nests of chimps	Aug–Sept 01
Melissa Emery	Ph.D. student	Harvard Univ.	Chimp female endocrines	Aug 01–June 04
Chie Hashimoto	Senior scientist	Kyoto Univ. & Kalinzu	Research visit	
Martha Mwesiga	M.Sc. student	Makerere Univ.	Uses of forest products	Sept–01
Caroline Kyomuhendo	M.Sc. student	Makerere Univ.	Land use change around Budongo	Aug 01–Mar 02
Simon Nampindo	M.Sc. student	Makerere Univ.	Elephant damage in Rabongo	Oct 01–Jan 02
Paul Mugabi	M.Sc. student	Makerere Univ.	Potential of less-used tree species	Aug–Nov 01
Robin May	Volunteer	Univ. of Birmingham	Help with lab. and museum	Oct–Dec 01
Wilma van Riel	Volunteer	Holland	Help with herbarium	Oct–Dec 01
Juliet Craig	Volunteer	Canada	Assist. to M. Emery and tree tagging	Dec 01–Feb 02
Michio Nakamura	Senior scientist	Univ. Kyoto	Chimp grooming behaviour	Feb–02
Lucy Bates	Ph.D. student	St Andrews Univ.	Female chimp decision-making	March 02–Sept 03
Sean O'Hara	Ph.D. student	Cambridge Univ.	Sexual behaviour in chimps	March 02–Sept 03
Catherine O'Hara	Ph.D. student	Cambridge Univ.	Genetics and mate choice in chimps	March 02–Sept 03
Richard Gregory	Volunteer	Oxford Univ.	Tree tagging, mapping	May–June 02
Michelle Higgins	Volunteer	Australia	Asst. to Caroline Kyomuhendo	March–April 02
Rory Lynch	Volunteer	Australia	Cata_ogue library	March–April 02
Dr Grant Joseph	Volunteer	S. Africa	Live trapping	Sept–02
Samantha Ralston	Volunteer	S. Africa	Live trapping	Sept–02
David Tumsiime	B.Sc. student	Makerere Univ.	Snaring	Sept–Oct 02
Allison Phillips	M.Sc. student	Oxford Brookes Univ.	Chimp habituation	Oct–Dec 02
Janette Wallis	Senior scientist	Oklahoma Univ.	Chimps at Sonso and Kasokwa	Oct 02–Mar 03
Svein Dale	Senior scientist	AUN, Norway	Tropical forest bird ecology	Oct 02
Are Pilskog	M.Sc. student	AUN, Norway	Tropical forest bird ecology	Oct 02–Mar 03
Siv Midtlein	Res. Asst.	AUN, Norway	Tropical forest bird ecology	Oct 02–Mar 03
John Bosco	M.Sc. student	Makerere Univ.	African grey parrots	Nov 02
Katie Slocombe	Ph.D. student	St Andrews Univ.	Chimpanzee vocalizations	Feb 03–Apr 04
Nicky Jenner	Volunteer	UK	Assist. to Lucy Bates	Feb–Apr 03
Ilona Johnston	Volunteer	UK	Assist. to BFP as required	Jan–Mar 03
Christophe Boesch	Senior scientist	MPIEA, Leipzig	Explore collaboration with BFP	July–03

Table F.1: (continued)

Name	Status	From	Area of interest	Time at Budongo
Nick Newton-Fisher	Senior scientist	UK	Chimp hunting	Sept 03–Aug 04
Emily Bethell	Ph.D. student	Portsmouth Univ.	Attention in chimps	Jan–Apr 04
Kerry Slater	Ph.D. student	Pretoria Univ.	Grooming in chimps	Nov 03
Tyler Weldon	Grad. student	Oklahoma Univ.	Kasokwa chimps	Feb–May 04
Donald Cole	Grad. student	Oklahoma Univ.	Kasokwa chimps	Feb–May 04
Lisa Riley	M.Sc. student	Univ. Roehampton	Primate niches	Mar–Jun 04
Leela Hazzah	Volunteer	Kenya	Mapping project	May 04
Mary Reuling	Volunteer	Kenya	Mapping project	May 04
Christina Connolly	M.Sc. student	London Univ. — Imperial Coll.	Private forests	May–Jun 04
Anita Stone	M.Sc. student	Univ Wales — Bangor	Seed dispersal	Jun–Aug 04
Zarin Machanda	Ph.D. student	Harvard Univ.	Chimps at Sonso	Jun–Aug 04
Simon Townsend	Undergrad.	Oxford Univ.	Male relationships in chimps	Jul–Sept 04
Crystal Davis	Undergrad.	Stanford Univ.	Mother–infant relations in chimpanzees	Jul–Sept 04
Zinta Zommers	M.Sc. student	Oxford Univ.	Ape–human disease transmission	Aug–Sept 04
Valerie Kosheleff	Grad. student	Univ. S. California	Arboreal vs. terrestrial habitat use	Sept–Dec 04

The setting

The BFP has its base at the site of the old Budongo Sawmills in the heart of the forest; in 1991, with help from USAID, we renovated two large wooden sawmill houses (see Fig. F.1), and built staff accommodation and ancillary facilities. Things did not always go smoothly; no sooner had we completed the building of the staff housing than a tree fell on one of the houses and demolished it. Today the Project employs 25 people, about half of whom live at camp, the other half coming in to work from nearby villages. Because the site of the Project is right in the forest, we are surrounded by wildlife and it takes less than a minute to get from the breakfast table into the forest. The sounds of the forest wildlife at night (tree hyraxes in particular) are amazing, and at dawn each day the black and white colobus monkeys make a wonderful deep roaring sound all around camp as the dominant males of different groups announce their territorial claims to each other. Chimpanzees can be heard most days as well as in the middle of the night, and sometimes come right up to our camp.

Fig. F.1:   One of the two main houses at Sonso camp (photo: V.R.).

Fig F.2:   Field assistants of the BFP (photo: V.R.).

The staff

We have three main categories of staff: field assistants who work in the forest collecting data (Fig. F.2), domestic staff who work around camp, and transect cutters who initially cut and then maintain our system of trails in the forest. In recent years we have also had snare removers and education officers.

Since everything depends on the staff, it is appropriate to list them; their work is very much appreciated by all who have ever come to Budongo to do research or take part in the Project in any way. Their names up to the time of writing, together with those who have retired or left, are as follows:

Field assistants
Zephyr Kiwede — senior field assistant
Geresomu Muhumuza — deputy senior field assistant
James Kakura
Raymond Ogen
George Otai
Gideon Monday
Joshua Nkosi
Joshua Odaga
(Former field assistants: Dissan Kugonza, Nabert Mutungire, John Tinka, Julius Kyamanywa, Alfred Tholith, Stephen Hatari, Godfrey Biroch, Alfred Afayo, Joseph Karamaji, Kennedy Andama, Richard Kyamanywa)

Domestic staff
Richard Odong-Too (camp maintenance, driver and mechanic)
Evace Karuguba (first cook)
Mary Ndobya (second cook)
Gino Ezati (night watchman)
Oliver Lomuru (firewood and water collection)
Moses Kiwede (slashing)
Fred Odong (road maintenance)
Wilson Obida (baboon control)
(Former domestic staff: Joy Tuhairwe, Charles Kahawa, Lilian Kyogabairwe, Grace Karamaji)

Snare removal
Ofen Anzima
Dominic Andii
Gideon Atayo
Moses Lemi
(Former snare remover: Pascal Muhindo)

Education
Zed Tumwiine (with Zephyr Kiwede)

Transect cutters
Claveri Kyombe (head cutter)
Augustino Okellowange
Tandima Operasio
Onyera Ascencio
Sebastiano Wele
Morris Athoithim
(Former transect cutters: Selestino Orach, Sebbi Okwonga, Gideon Ocokuru)

Project data

Two categories of data are collected by the BFP: Project data and students' data. The former are for general use by Project staff and students alike; the latter are for the use of individual students in connection with their personal research projects (but after five years can be used by anyone). The data end up in a variety of forms: as computer files, handouts, reports, theses, and publications in the scientific and popular press.

The lists of reports, dissertations and publications emanating from BFP give some idea of the wealth of data collected since 1992.[105] Data exist on the forest vegetation, tree growth, phenology, dragonflies, canopy arthropods, rodents, primate and bird censusing, diets of primates and bird species, bird identi-fication and ringing, behaviour of primates and birds, seed dispersal, physical aspects of the forest such as soil analysis and rainfall; there are maps of our trail system and other aspects of the forest and its recent history, data on the human population around Budongo, including its use of timber and non-timber forest products, and crop-raiding. Indeed, it is these data that have provided the material for this book, and I am deeply indebted to all those whose work has made this possible.[106]

The BFP has established a herbarium of Budongo tree and shrub specimens, a museum with skeletal material and much more, a laboratory for bench work, and a library that contains most of the Project's reports and publications, as well as many dissertations; all these are located at the Sonso site. It is a tribute to our staff, students and volunteers that these have been possible and I am very grateful for what they have done.

Students, senior scientists and volunteers

The BFP has become a much visited site over the years, mainly by students doing research projects for Masters and PhD theses, but also by senior scientists and volunteers who have in many cases greatly assisted the Project. I have listed them in Table F.1.[107]

Awards

In the year 2000 we received two awards, the President's Award from the American Society of Primatologists, and the Chairman's Award from the Committee for Research and Exploration of the National Geographic Society, in recognition of the Project's contributions to primatology and conservation over the preceding decade.

---

[105] These lists can be found on the Project's website, www. budongo org.

[106] A 45-page summary of all the work done for BFP from 1990 to 2000 has been compiled by Zoe Wales, entitled *What We Know About Budongo Forest* (BFP Report No. 81).

[107] I greatly apologise in advance if anyone has been omitted; please inform me if this is the case.

References

Some of the references listed below are unpublished theses and reports. In the case of theses, these can normally be obtained from the library of the relevant University. In the case of Budongo Forest Project reports, enquiries should be made to the author.

Akhos, M.W. 2000. Economic impact of ecotourism on local rural communities: a case study of Budongo Forest Ecotourism Project. M.Sc. thesis, Makerere University, Kampala.

Aldrich-Blake, P. 1970. The ecology and behaviour of the blue monkey. Ph.D. thesis, Bristol University, Bristol, UK.

Anderson, D.P., Nordheim, E.V., Boesch, C. and Moermond, T.C. 2002. Factors influencing fission–fusion grouping in chimpanzees in the Taï National Park, Côte d'Ivoire. In *Behavioural diversity in chimpanzees and bonobos* (ed. C. Boesch, G. Hohmann and L.F. Marchant), pp. 90–101. Cambridge University Press, Cambridge, UK.

Arnold, K. 2001. Affiliation, aggression and reconciliation in male chimpanzees of the Budongo Forest, Uganda. Ph.D. thesis, School of Psychology, University of St Andrews, St Andrews, UK.

Arnold, K. and Whiten, A. 2001. Post-conflict behaviour of wild chimpanzees (*Pan troglodytes schweinfurthii*) in the Budongo Forest, Uganda. *Behaviour*, **138**, 649–90.

Arnold, K. and Whiten, A. 2003. Grooming among the chimpanzees of the Budongo Forest, Uganda: tests of five explanatory models. *Behaviour*, **140**, 519–52.

Ashford, R., Reid, G. and Butynski, T. 1990. The intestinal faunas of man and mountain gorillas in a shared habitat. *Ann. Trop. Med. Parasitol.*, **84**, 337–40.

Asquith, P. 1984. The inevitability and utility of anthropomorphism in description of primate behaviour. In *The meaning of primate signals* (ed. R. Harré and V. Reynolds), pp. 138–74. Cambridge University Press, Cambridge, UK.

Assersohn, C. 2000. Development of foraging in wild chimpanzees. Ph.D. thesis, School of Pyschology, University of St Andrews, St Andrews, UK.

Assersohn, C. and Whiten, A. 1998. Development of foraging and food sharing in chimpanzees of the Budongo Forest, Uganda (Abstract), *XVII Congr. Int. Primatol. Soc.*, Antananarivo.

Assersohn, C., Whiten, A., Kiwede, Z.T., Tinka, J. and Karamaji, J. 2004. Use of leaves to inspect ectoparasites in wild chimpanzees: a third cultural variant? *Primates*, 45, 255–58.

Babweteera, F. 1997. Influence of gap size and age on the diversity and abundance of climbers in Budongo Forest Reserve, Uganda. Budongo Forest Project Report No. 47.

Babweteera, F. 1998. Influence of gap size and age on the diversity and abundance of climbers in Budongo Forest Reserve, Uganda. M.Sc. thesis, Makerere University, Kampala.

Babweteera, F., Plumptre, A. and Obua, J. 2000. Effect of gap size and age on climber abundance and diversity in Budongo Forest Reserve, Uganda. *Afr. J. Ecol.*, **38**, 230–7.

Bahati, J. 1995. Impact of arboricidal treatments on the natural regeneration and species composition of Budongo Forest, Uganda. M.Sc. thesis, Makerere University, Kampala.

Bakuneeta, C., Inagaki, H. and Reynolds, V. 1993. Identification of wild chimpanzee hair samples from feces by electron microscopy. *Primates*, **34**, 233–5.

Bakuneeta, C., Johnson, K., Plumptre, R. and Reynolds, V. 1995. Human uses of tree species whose seeds are dispersed by chimpanzees in the Budongo Forest, Uganda. *Afr. J. Ecol.*, **33**, 276–8.

Bannon, L. 1997. Understanding of national and local laws among villagers living to the south-west of the Budongo Forest Reserve, Uganda, with special reference to hunting. Budongo Forest Project Report No. 46.

Barrows, M. 1996. A survey of intestinal parasites of the primates in Budongo Forest, Uganda. B.Vet. Med. thesis, Glasgow University, Glasgow, UK.

Bates, L. 2001. The association patterns of female chimpanzees within the Budongo Forest Reserve, Uganda. M.Sc. thesis, Oxford University, Oxford.

Bates, L. 2005. Cognitive aspects of range use and food location by the chimpanzees of Budongo, Uganda. Ph.D. thesis, University of St Andrews, St Andrews, UK.

Bauer, H.R. and Philip, M. 1983. Facial and vocal individual recognition in the common chimpanzee. *Psych. Rec.*, **33**, 161–70.

Bearder, S.K. and Martin, R.D. 1980. *Acacia* gum and its use by bushbabies, *Galago senegalensis* (Primates: Lorisidae). *Int. J. Primatol.*, **1**, 103–28.

Bennett, H. 1996. Study of kinship, social relations and possible avoidance of incest among Budongo chimpanzees. M.A. thesis, Edinburgh University, Edinburgh, UK.

Beresford-Stooke, L. 1999. Primate population densities after pitsawing in Budongo Forest, Uganda. Budongo Forest Project Report No. 61.

Bethell, E. 1998. Vigilance in foraging chimpanzees. Dissertation, Department of Zoology, University College London.

Bethell, E. 2003. Rank effects on chimpanzee social monitoring. *Primate Eye*, **80**, 4 (Abstract).

Bethell, E., Whiten, A., Muhumuza, G. and Kakura, J. 2000. Active plant food division and sharing by wild chimpanzees. *Primate Report*, **56**, 67–71.

Boesch, C. 1991. The effects of leopard predation on grouping patterns in forest chimpanzees. *Behaviour*, **117**, 2220–41.

Boesch, C. 1996. Social partying in Taï chimpanzees. In *Great ape societies* (ed. W.C. McGrew *et al.*). Cambridge University Press, Cambridge, UK.

Boesch, C. 2003. Is culture a golden barrier between human and chimpanzee? *Evol. Anthropol.*, **12**, 85–91.

Boesch, C. and Boesch, H. 1989. Hunting behaviour of wild chimpanzees in the Taï National Park. *Amer. J. Phys. Anthropol.*, **78**, 547–73.

Boesch, C. and Boesch-Achermann, H. 2000. *The chimpanzees of the Taï Forest: behavioural ecology and evolution*. Oxford University Press, Oxford.

Boesch, C. and Tomasello, M. 1998. Chimpanzee and human cultures. *Curr. Anthropol.*, **39**, 591–614.

Bradley, B.J. and Vigilant, L. 2002. False alleles derived from microbial DNA pose a potential source of error in microsatellite genotyping of DNA from faeces. *Mol. Ecol. Notes*, **2**, 602–5.

Brasnett, N.V. 1946. The Budongo Forest, Uganda. *J. Oxford Univ. For. Soc.*, Third Series (1), 11–17.

Brownlow, A., Plumptre, A.J., Reynolds, V. and Ward, R. (2001). Sources of variation in the nesting behavior of chimpanzees (*Pan troglodytes schweinfurthii*) in the Budongo Forest, Uganda. *Amer. J. Primatol.*, **55**, 49–55.

Buechner, H.K., Buss, I.O., Longhurst, W.M. and Brooks, A.C. 1963. Numbers and migration of elephants. *J. Wildlife Management*, 27, 37.

Burch, D. 1994. Disabling injuries in the wild chimpanzees of the Budongo Forest, Uganda. M.Sc. thesis, Oxford University, Oxford.

Chadwick-Jones, J. 1998. *Developing a social psychology of monkeys and apes*. Psychology Press, Hove, UK.

Chapman, C.A. 1990. Association pattern of spider monkeys: the influence of ecology and sex on social organization. *Behav. Ecol. Sociobiol.*, **26**, 409–14.

Chapman, C.A. 1995. Primate seed dispersal: coevolution and conservation implications. *Evol. Anthropol.*, **4**, 74–82.

Chapman, C.A. and Peres, C.A. 2001. Primate conservation in the New Millennium: the role of scientists. *Evol. Anthropol.*, **10**, 16–33.

Chapman, C.A. and Wrangham, R.W. 1993. Range use of the forest chimpanzees of Kibale: implications for the understanding of chimpanzee social organization. *Amer. J. Primatol.*, **31**, 263–73.

Chapman, C.A., White, F.J. and Wrangham, R.W. 1993. Defining party size in fission-fusion social organizations. *Folia Primatol.*, **63**, 31–34.

Chapman, C.A., Wrangham, R.W. and Chapman L.J. 1994. Indices of habitat-wide fruit abundance in tropical forest. *Biotropica*, **26**, 160–71.

Chapman, C.A., Wrangham R.W. and Chapman L.J. 1995. Ecological constraints on group size: an analysis of spider monkey and chimpanzee subgroups. *Behav. Ecol. Sociobiol.*, **36**, 59–70.

Chase, E. 2002. Kitera. Attitudes towards chimpanzees in Kasokwa Forest, Western Uganda: an anthropological and ecological approach. B.A. thesis in Human Sciences, Oxford University, Oxford.

Clark, A. 1993. Rank differences in the vocal production of Kibale Forest chimpanzees as a function of social context. *Amer. J. Primatol.*, **31**, 159–79.

Clark, A.P. and Wrangham, R.W. 1994. Chimpanzee arrival pant-hoots: do they signify food or status? *Int. J. Primatol.*, **15**, 185–206.

Clark Arcadi, A. 1996. Phase structure of wild chimpanzee pant-hoots: patterns of production and interpopulation variability. *Amer. J. Primatol.*, **38**, 159–78.

Clark Arcadi, A. and Wrangham, R.W. 1999. Infanticide in chimpanzees: review of cases and a new within-group observation from the Kanyawara study group in Kibale National Park. *Primates*, **40**, 337–51.

Clutton-Brock, T.H. and Harvey, P.H. 1977. Primate ecology and social organization. *J. Zool.*, **183**, 1–39.

Conklin, N.L. and Wrangham, R.W. 1994. The value of figs to a hind-gut fermenting frugivore: a nutritional analysis. *Biochem. Anal. System.*, **22**, 137–51.

Crockford, C. and Boesch, C. 2003. Context-specific calls in wild chimpanzees, *Pan troglodytes verus*: analysis of barks. *Anim. Behav.*, 66, 115–25.

Croxall, E. 2002. The nature, context and significance of primate nest sites. B.A. thesis in Archaeology and Anthropology, Oxford University, Oxford.

Curran, B.K. and Tshombe, R.K. 2001. Integrating local communities into the management of protected areas. In *African rain forest ecology and conservation* (ed. W. Weber, L.J.T. White, A. Vedder and L. Naughton-Treves). Yale University Press, New Haven.

Darwin, C.R. 1871. *The descent of man, and selection in relation to sex.* John Murray, London.

Dawkins, H.C. 1954. Contact arboricides for rapid tree-weeding in Uganda. Proceedings of 4th World Forestry Conference, Dehra Dun, India.

Dawkins, H.C. 1955. The refining of mixed forest: a new objective for tropical silviculture. *Empire For. Rev.*, 34, 188–91.

Dempsey, B. 2000. A study of social dominance hierarchies and competition in relation to male chimpanzees. B.A. thesis, Oxford University, Oxford.

de Waal, F.B.M. 1982. *Chimpanzee politics.* Harper & Row, New York.

de Waal, F.B.M. and Roosmalen, A. von 1979. Reconciliation and consolation among chimpanzees. *Behav. Ecol. Sociobiol.*, **5**, 55–66.

de Waal, F.B.M. and Seres, M. 1997. Propagation of hand-clasp grooming among captive chimpanzees. *Am. J. Primatol.*, **43**, 339–46.

Dixson, A.F. 1997. Evolutionary perspectives on primate mating systems and behavior. *Ann. New York Acad. Sci.*, **807**, 42–61.

Driciru, F. 2001. Collaborative forest management initiatives by the Forest Department. In The 2nd Budongo Conference. Development through conservation (ed. J. Wallis). Budongo Forest Project Report No. 76.

Edroma, E.L., Rosen, N. and Miller, P.S. (eds.) 1997. *Conserving the chimpanzees of Uganda: a population and habitat viability assessment for* Pan troglodytes schweinfurthii. IUCN/SSC Conservation Breeding Specialist Group, Apple Valley, MN.

Eggeling, W.J. 1940a. Budongo: an East African mahogany forest. *Empire For. J.*, **19**, 179–97.

Eggeling, W.J. 1940b. *The indigenous trees of Uganda.* Government Printer, Entebbe, Uganda.

Eggeling, W.J. 1947. Observations on the ecology of the Budongo rain forest, Uganda. *J. Ecol.*, **34**, 20–87.

Emery Thompson, M., Wrangham, R.W. and Reynolds, V. [In press]. Urinary estrone conjugates and reproductive parameters in Kibale (Kanyawara) and Budongo (Sonso) chimpanzees. In *Primates of Western Uganda* (ed. N.E. Newton-Fisher, H. Notman, J.D. Paterson and V. Reynolds). Kluwer, New York.

Emery Thompson, M. and Wrangham, R.W. [In press]. Comparison of sex differences in gregariousness in fission–fusion species: reducing bias by standardizing for party size. In *Primates of Western Uganda* (ed. N.E. Newton-Fisher, H. Notman, J.D. Paterson and V. Reynolds). Kluwer, New York.

Fairgrieve, C. 1993. Report on blue monkeys. Budongo Forest Project Report No. 14.

Fairgrieve, C. 1995*a*. The comparative ecology of blue monkeys (*Cercopithecus mitis stuhlmanni*) in logged and unlogged forest, Budongo Forest Reserve, Uganda: the effects of logging on habitat and population density. Ph.D. thesis, University of Edinburgh, Edinburgh, UK.

Fairgrieve, C. 1995*b*. Infanticide and infant eating in the blue monkey (*Cercopithecus mitis stuhlmanni*) in the Budongo Forest Reserve, Uganda. *Folia Primatol.*, **64**, 69–72.

Fairgrieve, C. 1995*c*. Blue monkey infanticide. Budongo Forest Project Report No. 25.

Fairgrieve, C. 1997. Meat eating by blue monkeys (*Cercopithecus mitis stuhlmanni*): predation of a flying squirrel (*Anomalurus derbianus jacksonii*). *Folia Primatol.*, **68**, 354–6.

Fairgrieve, C. 2001. Training course in development through conservation. 18–24 March 2001. Budongo Forest Project Report No. 77.

Fairgrieve, C. 2003. Feeding ecology and dietary differences between blue monkey (*Cercopithecus mitis stuhlmanni* Matschie) groups in logged and unlogged forest, Budongo Forest Reserve, Uganda. *Afr. J. Ecol.*, **41**, 141–9.

Fawcett, K.A. 2000. Female relationships and food availability in a forest community of chimpanzees. Ph.D. thesis, University of Edinburgh, Edinburgh, UK.

Fawcett, K. and Muhumuza, G. 2000. Death of a wild chimpanzee community member: possible outcome of intense sexual competition. *Amer. J. Primatol.*, **51**, 243–7.

Friedmann, H. and Williams, J.G. 1973. The birds of Budongo Forest, Bunyoro Province, Uganda. *J. E. Africa Nat. Hist. Soc. Nat. Mus.*, **141**, 1–18.

Ghiglieri, M.P. 1984. *The chimpanzees of Kibale Forest: a field study of ecology and social structure*. Columbia University Press, New York.

Goldberg, T.L. 1997. Inferring the geographic origins of 'refugee' chimpanzees in Uganda from mitochondrial DNA sequences. *Conserv. Biol.*, **11**, 1441–6.

Goldberg, T.L. and Ruvolo, M. 1997. The geographic apportionment of mitochondrial genetic diversity in East African chimpanzees, *Pan troglodytes schweinfurthii*. *Mol. Biol. Evol.*, **14**, 976–84.

Goodall, J. 1968. The behaviour of free-living chimpanzees in the Gombe Stream Reserve. *Anim. Behav. Monog.*, **1** (3), 161–311.

Goodall, J. 1977. Infant killing and cannibalism in free-living chimpanzees. *Folia Primatol.*, **28**, 259–82.

Goodall, J. 1986. *The chimpanzees of Gombe: patterns of behaviour*. Belknap Press, Harvard, Cambridge, Mass.

Goodall, J. 1992. Unusual violence in the overthrow of an alpha male chimpanzee at Gombe. In *Topics in primatology, Vol. 1: Human origins* (ed. T. Nishida, W.C. McGrew, P. Marler, M. Pickford, and F.B.M. de Waal), pp. 131–42. Tokyo University Press, Tokyo.

Grieser-Johns, B. 1996. Responses of chimpanzees to habituation and tourism in the Kibale Forest, Uganda. *Biol. Cons.*, **78**, 257–62.

Grove, S. 1995. *A nature conservation source-book for forestry professionals*. Commonwealth Secretariat, London.

Hamai, M., Nishida, T., Takasaki, H. and Turner, L. 1992. New records of within-group infanticide and cannibalism in wild chimpanzees. *Primates*, **33**, 151–62.

Hamilton, A.C. 1984. *Deforestation in Uganda*. Oxford University Press, Nairobi, Kenya.

Harcourt, A.H. 1989. Social influences on competitive ability: alliances and their consequences. In *Comparative socioecology* (ed. V. Standen and R.A. Foley), pp. 223–42. Blackwell Science Publications, Oxford.

Harré, R. 1984. Vocabularies and theories. In *The Meaning of Primate Signals* (eds. R. Harré and V. Reynolds) pp. 90–106. Cambridge Univ. Press, Cambridge, UK.

Harris, C.M. 1934. *Provisional Working Plan Report for the Bunyoro Forests, Uganda*. Forest Department, Govt. of Uganda, Kampala.

Harris, D., Jones, H. and Reynolds, V. [In progress]. The effects of patch food quality on chimpanzee (*Pan troglodytes schweinfurthii*) group size, dynamics and party duration. Unpublished typescript.

Hasegawa, T. and Hiraiwa-Hasegawa, M. 1983. Opportunistic and restrictive matings among wild chimpanzees in the Mahale Mountains, Tanzania. *J. Ethol.*, 1, 75–85.

Hasegawa, T. 1990. Sex differences in ranging patterns. In *The Chimpanzees of the Mahale Mountains: sexual and life history strategies* (ed. T. Nishida), pp. 99–114. Tokyo University Press, Tokyo.

Hasegawa, T. and Hiraiwa-Hasegawa, M. 1990. Sperm competition and mating behavior. In *The Chimpanzees of the Mahale Mountains: sexual and life history strategies* (ed. T. Nishida), pp. 115–32. Tokyo University Press, Tokyo.

Hashimoto, C. 1999. Snare injuries of chimpanzees in the Kalinzu Forest, Uganda. *Pan Africa News*, 6 (2), 20–2.

Hashimoto, C. and Furuichi, T. [In press]. Frequent promiscuous copulation by female chimpanzees in the Kalinzu Forest, Uganda. In *Primates of Western Uganda* (ed. J.D. Paterson, N.E. Newton-Fisher, H. Notman and V. Reynolds). Kluwer, New York.

Hashimoto, C., Furuichi, T. and Tashiro, Y. 2001. What factors affect the size of chimpanzee parties in the Kalinzu Forest, Uganda? Examination of fruit abundance and number of estrous females. *Int. J. Primatol.*, 22, 947–59.

Hayaki, H. 1990. Social context of pant-grunting in young chimpanzees. In *The chimpanzees of the Mahale Mountains: sexual and life history strategies* (ed. T. Nishida). University of Tokyo Press, Tokyo.

Hill, C.M. 1997. Crop-raiding by wild vertebrates: the farmer's perspective in an agricultural community in western Uganda. *Int. J. Pest Manage.*, 43, 77–84.

Hill, C.M. 1998. Conflicting attitudes towards elephants around the Budongo Forest Reserve, Uganda. *Env. Conserv.*, 25, 244–50.

Hill, C.M. 2000. Conflict of interest between people and baboons: crop raiding in Uganda. *Int. J. Primatol.*, 21, 77–84.

Hill, C.M. 2002. People, crops and wildlife, a conflict of interests. In *Human–wildlife conflict: identifying the problem and possible solutions* (ed. C.M. Hill, F.V. Osborn and A.J. Plumptre), pp. 60–7. Albertine Rift Technical Report Series, Vol. 1, Wildlife Conservation Society.

Hinde, R.A. 1976. Interactions, relationships and social structure. *Man*, 11, 1–17.

Hiraiwa-Hasegawa, M. 1990. Role of food sharing between mother and infant in the ontogeny of feeding behavior. In *The chimpanzees of the Mahale Mountains: sexual and life history strategies* (ed. T. Nishida). University of Tokyo Press, Tokyo.

Hirata, S., Morimura, N. and Matsuzawa, T. 1998. Green passage plan (tree-planting project) and environmental education using documentary videos at Bossou: a progress report. *Pan Africa News*, **5** (2), 18–20.

Hladik, C.M. and Simmen, B. 1996. Taste perception and feeding behavior in non-human primates and human populations. *Evol. Anthropol.*, **5**, 58–71.

Homsy, J. 1999. *Ape tourism and human diseases: how close should we get?* International Gorilla Conservation Programme, Nairobi.

Hosaka, K. 1995. Epidemics and wild chimpanzee study groups. *Pan Africa News*, **2**, 1–4.

Hosaka, K., Nishida, T., Hamai, M., Matsumoto-Oda, A. and Uehara, S. 2001. Predation of mammals by the chimpanzees of the Mahale Mountains, Tanzania. In *All creatures great and small, Vol. 1: African apes* (ed. B.M.F. Galdikas, N.E. Briggs, L.K. Sheeran, G.L. Shapiro and J. Goodall), pp. 107–30. Kluwer Academic, New York.

Howard, P. 1991. *Nature conservation in Uganda's tropical forest reserves*. IUCN, Gland and Cambridge.

Howard, P., Davenport, T. and Matthews, R. 1996. Budongo Forest Reserve: Biodiversity Report. Forest Department, Kampala.

Huffman, M. 1990. Some socio-behavioral manifestations of old age. In *The chimpanzees of the Mahale Mountains: sexual and life history strategies* (ed. T. Nishida). University of Tokyo Press, Tokyo.

Huffman, M. 1997. Current evidence for self-medication in primates: a multidisciplinary perspective. *Yearbook Phys. Anthropol.*, **40**, 171–200.

Huffman, M. 2001. Self-medicative behavior in the African great apes: an evolutionary perspective into the origins of human traditional medicine. *BioScience*, **51**, 651–61.

Huffman, M.A., Gotoh, S., Izutsu, D., Koshimizu, K., and Kalunde, M.S. 1993. Further observations on the use of the medicinal plant, *Vernonia amygdalina* (Del), by a wild chimpanzee, its possible effect on parasite load, and its phytochemistry. *Afr. Study. Mon.*, 14, 227–40.

Huffman, M.A., Page, J.E., Sukhdeo, M.V.K., Gotoh, S., Kalunde, M.S., Chandrasiri, T. and Neil Towers, G.H. 1996. Leaf-swallowing by chimpanzees, a behavioral adaptation for the control of strongyle nematode infections. *Int. J. Primatol.*, **17**, 475–503.

Humle, T. and Matsuzawa, T. 2002. Ant-dipping among the chimpanzees of Bossou, Guinea, and some comparisons with other sites. *Amer. J. Primatol.*, 58, 133–48.

Inagaki, H. and Tsukahara, T. 1993. A method of identifying chimpanzee hairs in lion feces. *Primates*, **34**, 107–10.

Isabirye-Basuta, G. 1988. Food competition among individuals in a free-ranging chimpanzee community in Kibale Forest, Uganda. *Behaviour*, **105**, 135–47.

Johnson, K.R. 1993. Local use of Budongo's forest products. M.Sc. thesis, Oxford University, Oxford.

Johnson, K.R. 1996a. Hunting in the Budongo Forest, Uganda. *SWARA*, **19**, 24–6.

Johnson, K.R. 1996b. Local attitudes towards the Budongo Forest, Western Uganda. *Indig. Knowl. Develop. Mon.*, **4**, 31.

Jolly, D., Taylor, D., Marchant, R., Hamilton, A., Bonnefille, R., Buchet, G. and Riollet, G. 1997. Vegetation dynamics in central Africa since 18,000 yr BP: pollen records from the interlacustrine highlands of Burundi, Rwanda and western Uganda. *J. Biogeog.*, **24**, 495–512.

Kalema, G. 1992. Report on chimpanzee parasites. Budongo Forest Project Report No. 9.

Kalema, G. 1995. Epidemiology of the intestinal parasite burden of mountain gorillas in Bwindi Impenetrable National Park, W.W. Uganda. *Zebra Found., Brit. Vet. Zool. Soc. Newsletter, Autumn*, 18–34.

Kalema, G. 1997. A survey of intestinal helminth parasites of a community of wild chimpanzees (*Pan troglodytes*) in the Budongo Forest, Uganda. In *Conserving the chimpanzees of Uganda: population and habitat viability assessment for* Pan troglodytes schweinfurthii (ed. E.L. Edroma, N. Rosen and P.S. Miller), pp. 189–94. IUCN/SSC Conservation Breeding Specialist Group, Apple Valley, MN.

Kalina, J. 1988. Ecology and behavior of the black-and-white casqued hornbill (*Bycanistes subcylindricus subquadratus*) in Kibale Forest, Uganda. Ph.D. dissertation, Michigan State University.

Kamugisha, J.R. 1993. *Management of natural resources and environment in Uganda: policy and legislation landmarks, 1890–1990*. Regional Soil Conservation Unit, SIDA, Nairobi, Kenya.

Kamugisha, J. and Nsita, S. 1997. *Forest management plan for Budongo Forest Reserve*, 4th edn. Forest Department, Ministry of Natural Resources, Kampala.

Kano, T. 1984. Observations of physical abnormalities among the wild bonobos (*Pan paniscus*) of Wamba, Zaire. *Amer. J. Phys. Anthropol.*, 63, 1–11.

Katende, A.B., Birnie, A. and Tengnas, B. 1995. *Useful trees and shrubs for Uganda*. RSCU, Nairobi, Kenya.

Kawai, M. 1958. On the system of social ranks in a natural troop of Japanese monkeys. *Primates*, **1–2**, 111–30. English translation in *Japanese monkeys: a collection of translations* (ed. S. Altmann), 1965, University of Alberta.

Kawanaka, K. 1990. Alpha males' interactions and social skills. In *The Chimpanzees of the Mahale Mountains* (ed. T. Nishida) pp. 171–188. University of Tokyo Press, Tokyo.

Kityo, P.W. and Plumptre, R.A. 1997. *The Uganda timber users' handbook*. Commonwealth Secretariat, London.

Kiwede, Z.T. 2000. A live birth by a primiparous female chimpanzee at the Budongo Forest. *Pan Africa News*, **7** (2), 23–5.

Kiwede, Z.T. 2001. Attitudes of local people towards crop raiding with particular reference to chimpanzees: a case study for Nyabyeya Parish. Certificate of forestry Report, Nyabyeya Forestry College, Nyabyeya, Uganda.

Kortlandt, A. 1962. Chimpanzees in the wild. *Sci. Amer.*, **206**, 128–38.

Kummer, H. 1968. *Social organization of Hamadryas baboons*. Chicago University Press, Chicago.

Kyamanywa, R. 2001. Kasokwa forest chimpanzees: behavioural patterns, ranging patterns, population size, ecological status and local community attitudes. Unpublished report, Budongo Forest Project.

Kyamanywa, R. 2002*a*. Research, monitoring and conservation of Kasokwa and Masindi chimpanzees. Unpublished report, Budongo Forest Project.

Kyamanywa, R. 2002*b*. Report on an investigation of trapped chimpanzee in Kasongoire Forest, 4 June 2002. Unpublished report, Budongo Forest Project.

Langlands, B.W. 1967. Burning in Eastern Africa with particular reference to Uganda. *E. Af. Geog. Rev.*, 5, 22.

Lauridsen, M. 1999. Workers in a forest: understanding the complexity of incorporating local people in modern management. M.Sc. thesis, University of Copenhagen.

Leendertz, F.H., Ellerbrok, H., Boesch, C., Couacy-Hymann, E., Mätz-Rensing, K., Hakenbeck, R., Bergmann, C., Abaza, P., Junglen, S., Moebius, Y., Vigilant, L., Formenty, P. and Pauli, G. 2004. Anthrax kills wild chimpanzees in a tropical rainforest. *Nature*, **430**, 451–2.

Lloyd, J. and Mugume, S. 2000. Investigation into the survival of a small chimpanzee community under immediate threat. Unpublished report, Jane Goodall Institute, Uganda Wildlife Education Centre, Uganda.

Lucht, W. 1998. Eine neue Eucnemide aus der Kronenregion afrikanischer Laubbaume. *Mitt. Int. Ent. Ver.*, **23**, 67–72.

Lukas, D. [In press]. To what extent does living in a group mean living with kin? *Molec. Ecol.*

Manson, J.H. and Wrangham, R.W. 1991. Intergroup aggression in chimpanzees and humans. *Curr. Anthropol.*, **32**, 369–90.

Marler, P. 1969. Vocalizations of wild chimpanzees: an introduction. In *Recent advances in primatology, Vol 1: behaviour* (ed. C.R. Carpenter), pp. 94–100. Karger, Basle.

Marler, P. 1990. Song learning: the interface between behaviour and neuroethology. *Proc. Roy. Soc. Lond. Ser. B*, **329**, 109–14.

Marriott, H. 1999. Micro-demography of a farming population of Western Uganda: a preliminary analysis investigating population pressure on the Budongo Forest Reserve. *Soc. Biol. Hum. Affairs*, **64**, 1–11.

Marsh, L. (ed.) 2002. *Primates in fragments: ecology and conservation.* Kluwer Academic, Hingham, MA and Dordrecht, Netherlands.

Marshall, A.J., Wrangham, R.W. and Clark Arcadi, A. 1999. Does learning affect the structure of vocalizations in chimpanzees? *Anim. Behav.*, **58**, 825–30.

Martin, D., Graham, C.E. and Gould, K.G. 1978. Successful artificial insemination in the chimpanzee. *Symp. Zool. Soc. Lond.*, **43**, 249–60.

Matsumoto-Oda, A. 1999. Mahale chimpanzees: grouping patterns and cycling females. *Amer. J. Primatol.*, **47**, 197–207.

Matsumoto-Oda, A., Hosaka, K., Huffman, M.A. and Kawanaka, K. 1998. Factors affecting party size in chimpanzees of the Mahale Mountains. *Int. J. Primatol.*, **19**, 999–1012.

Mbogga, M. 2000. The population dynamics and invasive potential of *Broussonetia papyrifera* in Budongo Forest Reserve, Uganda. M.Sc. thesis, Makerere University, Kampala.

McGrew, W.C. 1975. Patterns of plant food sharing by wild chimpanzees. In *Contemporary primatology* (ed. S. Kondo, M. Kawai and A. Ehara), pp. 304–9. Karger, Basle.

McGrew, W.C. 1992. *Chimpanzee material culture: implications for human evolution.* Cambridge University Press, Cambridge, UK.

McGrew, W.C. 1998. Culture in non-human primates? *Ann. Rev. Anthropol.*, **27**, 301–28.

McGrew, W.C. and Tutin, C.E.G. 1978. Evidence for a social custom in wild chimpanzees. *Man*, 13, 234–51.

McVittie, B. 1998. A report on the collection of nest hairs and nesting behaviours from chimpanzees (*Pan troglodytes schweinfurthii*). Budongo Forest Project Report No. 57.

Mitani, J.C. and Brandt, K.L. 1994. Social factors influence the acoustic variability in the long-distance calls of male chimpanzees. *Ethology*, **96**, 233–52.

Mitani, J., Hunley, K.L. and Murdoch, M.E. 1999. Geographic variation in the calls of wild chimpanzees: a reassessment. *Amer. J. Primatol.*, **47**, 133–51.

Mitani, J.C., Merriwether, D.A. and Zhang, C. 2000. Male affiliation, cooperation and kinship in wild chimpanzees. *Anim. Behav.*, **59**, 885–93.

Mitani, J.C., Watts, D.P. and Lwanga, J.S. 2002. Ecological and social correlates of chimpanzee party size and composition. In *Behavioural diversity in chimpanzees and bonobos* (ed. C. Boesch, G. Hohmann and L.F. Marchant). Cambridge University Press, Cambridge, UK.

Mitani, J.C., Watts, D.P., Pepper, J.W. and Merriwether, A. 2002. Demographic and social constraints on male chimpanzee behaviour. *Anim. Behav.*, **64**, 727–37.

Mitani, J., Hasegawa, T., Gros-Louis, J., Marler, P. and Byrne, R. 1992. Dialects in wild chimpanzees? *Amer. J. Primatol.*, **27**, 233–43.

Mittermeier, R.A. and Cheney, D.L. 1987. Conservation of primates and their habitats. In *Primate societies* (ed. B.B. Smuts, D.L. Cheney, R.M. Seyfarth, R.W. Wrangham and T.T. Struhsaker), pp. 477–90. University of Chicago Press, Chicago.

Morin, P.A., Chambers, K.E., Boesch, C. and Vigilant, L. 2001. Quantitative polymerase chain reaction analysis of DNA from noninvasive samples for accurate microsatellite genotyping of wild chimpanzees (*Pan troglodytes verus*). *Molec. Ecol.*, **10**, 1835–44.

Muhumuza, G. 2002. The impact of community conservation education on conservation of Budongo Forest Reserve. Certificate of Forestry Report, Nyabyeya Forestry College, Nyabyeya, Uganda.

Muller, M.N. 2002. Agonistic relations among Kanyawara chimpanzees. In *Behavioural diversity in chimpanzees and bonobos* (ed. C. Boesch, G. Hohmann and L.F. Marchant), pp. 112–23. Cambridge University Press, Cambridge, UK.

Muller, M.N. and Wrangham, R.W. 2004. Dominance, cortisol and stress in wild chimpanzees (*Pan troglodytes schweinfurthii*). *Behav. Ecol. Sociobiol.*, 55, 332–40.

Munn, J. 2003. The impact of injuries on free-living chimpanzees. M. Phil. thesis, Australian National University.

Munn, J. [In press]. The effects of injury on the locomotion of free-living chimpanzees in the Budongo Forest Reserve, Uganda. In *Primates of Western Uganda* (ed. N.E. Newton-Fisher, H. Notman, J.D. Paterson and V. Reynolds). Kluwer, New York.

Munn, J. and Kalema, G. 2000. Death of a chimpanzee (*Pan troglodytes schweinfurthii*) in a trap in Kasokwa Forest Reserve, Uganda. *Afr. Primates*, **4**, 58–61.

Mwima, P.M., Obua, J. and Oryem-Origa, H. 2001. Effect of logging on the natural regeneration of Khaya anthotheca in Budongo Forest Reserve, Uganda. *Int. For. Rev.* **3**, 131–5.

Nakamura, M. and Itoh, N. 2001. Sharing of wild fruits among male chimpanzees: two cases from Mahale, Tanzania. *Pan Africa News*, **8**, 28–31.

Naughton-Treves, L. 1997. Farming the forest edge: vulnerable places and people around Kibale NP, Uganda. *Geog. Rev.*, **87**, 27–46.

Newton-Fisher, N.E. 1997. Tactical behaviour and decision making in wild chimpanzees. Ph.D. thesis, Cambridge University, Cambridge, UK.

Newton-Fisher, N.E. 1999*a*. The diet of chimpanzees in the Budongo Forest Reserve, Uganda. *Afr. J. Ecol.*, **37**, 344–54.

Newton-Fisher, N.E. 1999*b*. Termite eating and food sharing by male chimpanzees in the Budongo Forest, Uganda. *Afr. J. Ecol.*, **37**, 369–71.

Newton-Fisher, N.E. 1999*c*. Infant killers of Budongo. *Folia Primatol.*, **70**, 167–9.

Newton-Fisher, N.E. 1999*d*. Association by male chimpanzees: a social tactic? *Behaviour*, **136**, 705–30.

Newton-Fisher, N.E. 2000. Male core areas: ranging by Budongo Forest chimpanzees. *Pan Africa News*, **7**, 10–12.

Newton-Fisher, N.E. 2002*a*. Relationships of male chimpanzees in the Budongo Forest, Uganda. In *Behavioural diversity in chimpanzees and bonobos* (ed. C. Boesch, G. Hohmann and L.F. Marchant). Cambridge University Press, Cambridge, UK.

Newton-Fisher, N.E. 2002*b*. Ranging patterns of male chimpanzees in the Budongo Forest, Uganda: range structure and individual differences. In *New perspectives in primate evolution and behaviour* (ed. C. Harcourt and B. Sherwood). Westbury Academic and Scientific Publishing, Otley, UK.

Newton-Fisher, N.E. 2003. The home range of the Sonso community of chimpanzees from the Budongo Forest, Uganda. *Afr. J. Ecol.*, **41**, 150–6.

Newton-Fisher, N.E. 2004. Hierarchy and social status in Budongo chimpanzees. *Primates*, **45**, 81–7.

Newton-Fisher, N.E., Notman, H. and Reynolds, V. 2002. Hunting of mammalian prey by Budongo Forest chimpanzees. *Folia Primatol.*, **73**, 281–3.

Newton-Fisher, N.E., Reynolds, V. and Plumptre, A.J. 2000. Food supply and chimpanzee (*Pan troglodytes schweinfurthii*) party size in the Budongo Forest, Uganda. *Int. J. Primatol.*, **21**, 613–28.

Newton-Fisher, N.E., Notman, H., Paterson, J.D., and Reynolds, V. (Eds.). [In press]. *Primates of Western Uganda*. Kluwer, New York.

Nishida, T. 1968. The social party of wild chimpanzees in the Mahale Mountains. *Primates*, 9, 167–224.

Nishida, T. 1980. The leaf-clipping display: a newly discovered expressive gesture in wild chimpanzees. *J. Hum. Evol.*, 2, 357–70.

Nishida, T. 1983. Alpha status and agonistic alliance in wild chimpanzees. *Primates*, **24**, 318–36.

Nishida, T. 1989. Social interactions between resident and immigrant female chimpanzees. In *Understanding chimpanzees* (ed. P.G. Heltne and L.A. Marquardt), pp. 68–89. Harvard University Press, Cambridge, Mass.

Nishida, T. (ed.) 1990. *The chimpanzees of the Mahale Mountains: sexual and life history strategies*. University of Tokyo Press, Tokyo.

Nishida, T. 1996. The death of Ntologi, the unparalleled leader of M group. *Pan Africa News*, **3**, 4.

Nishida, T. and Turner, L.A. 1996. Food transfer between mother and infant chimpanzees of the Mahale Mountains National Park, Tanzania. *Int. J. Primatol.*, **17**, 947–68.

Nishida, T., Haraiwa-Hasegawa, M. and Takahata, Y. 1985. Group extinction and female transfer in wild chimpanzees in the Mahale National Park, Tanzania. *Z. fur Tierpsychol.*, **67**, 284–301.

Nishida, T., Ohigashi, H. and Koshimizu, K. 2000. Tastes of chimpanzee plant foods. *Curr. Anthropol.*, **41**, 431–8.

Nishida, T., Takasaki, H. and Takahata, Y. 1990. Demography and reproductive profiles. In *The chimpanzees of the Mahale Mountains: sexual and life history strategies* (ed. T. Nishida). University of Tokyo Press, Tokyo.

Nishida, T., Hosaka, K., Nakamura, M. and Hamia, M. 1995. A within-group gang attack on a young adult male chimpanzee: ostracism of an ill-mannered member? *Primates*, **36**, 207–11.

Nishida, T., Kano, T., Goodall, J., McGrew, W.C. and Nakamura, M. 1999. Ethogram and ethnography of Mahale chimpanzees. *Anthropol. Sci.*, 107, 141–88.

Nishida, T., Corp, N., Hamai, M., Hasegawa, T., Hiraiwa-Hasegawa, M., Hosaka, K., Hunt, K., Itoh, N., Kawanaka, K., Matsumoto-Oda, A., Mitani, J.C., Nakamura, M., Norikoshi, K., Sakamaki, T., Turner, L., Uehara, S. and Zamma, K. 2003. Demography, female life history, and reproductive profiles among the chimpanzees of Mahale. *Amer. J. Primatol.*, **59**, 99–121.

Nissen, H.W. 1931. A field study of the chimpanzee. *Comp. Psych. Mon.*, **8**, 1–121.

Notman, H. 1996. The ontogenetic development and inter-population variability of pant-hoot vocalizations in wild chimpanzees. M.Sc. thesis, Oxford University, Oxford.

Notman, H. 2003. The meaning, structure and function of chimpanzee pant hoots from the Budongo Forest, Uganda. Ph.D. thesis, University of Calgary.

Notman, H. [In press]. Structural variation in pant hoots from the Budongo Forest: a new perspective on proximate mechanisms for variation in chimpanzee vocalizations. In *Primates of Western Uganda* (ed. N.E. Newton-Fisher, H. Notman, J.D. Paterson and V. Reynolds). Kluwer, New York.

Notman, H. and Munn, J. 2003. A case of infant carrying by an adult male chimpanzee in the Budongo Forest. *Pan Africa News*, **10**, 7–9.

Nsubuga, A.M., Robbins, M.M., Roeder, A.D., Morin, A., Boesch, C. and Vigilant, L. 2004. Factors affecting the amount of genomic DNA extracted from ape faeces and the identification of an improved sample storage method. *Molec. Ecol.*, **13**, 2089–94.

Obua, J., Banana, A.Y. and Turyahabwe, N. 1998. Attitudes of local communities towards forest management practices in Uganda: the case of Budongo Forest Reserve. *Comm. For. Rev.*, **77**, 113–18.

O'Hara, S. J. 2003. Anthropogenic killing of a study community chimpanzee in Budongo Forest, Uganda. *Newsletter of the Albertine Rift Conservation Society (ARCOS)*, **10**, 11.

O'Hara, S. J. [In progress]. Sexual behaviour and male violence in wild chimpanzees. Ph.D. thesis, Cambridge University, Cambridge, UK.

O'Hara, S. J. and O'Hara, C. 2001. Chimpanzee nest site preference in Budongo and Bugoma Forests, Uganda. Poster presented at meeting of Primate Society of Great Britain entitled 'Primates of the Western Forests of Uganda', London, December 2001.

O'Hara, S.J. and Lee, P.C. [In progress]. Post-coital penile cleaning in wild chimpanzees.

Okecha, A.A. 2000. Feeding ecology of olive baboons (*Papio anubis*) in Budongo Forest Reserve, Uganda. M.Sc. thesis, Makerere University, Kampala.

Okecha, A.A. [In press]. Diet of olive baboons (*Papio anubis*) in Budongo Forest Reserve, Uganda. In *Primates of Western Uganda* (ed. N.E. Newton-Fisher, H. Notman, J.D. Paterson and V. Reynolds). Kluwer, New York.

Okia, C. 2001. Regeneration of rattan palm (*Calamus deëratus*) in Budongo Forest Reserve, Uganda. M.Sc. thesis, Makerere University, Kampala.

Oliver, L. 2002. Female mating patterns in the Budongo community of chimpanzees (*Pan troglodytes schweinfurthii*). M.Sc. thesis, Bucknell University, Lewisburg.

Parkin, D. 1979. The categorization of work: cases from coastal Kenya. In *Social anthropology of work* (ed. S. Wallman), pp. 317–35. Academic Press, London.

Paterson, J.D. 1991. The ecology and history of Uganda's Budongo Forest. *For. Conserv. Hist.*, **35**, 179–87.

Paterson, J.D. 1996. Preliminary report on the Sonso baboon troop, Budongo Forest. Budongo Forest Project Report No. 36.

Paterson, J.D. 1997a. Report of field project: *Khaya anthotheca*: growth, bark and primates. Budongo Forest Project Report No. 52.

Paterson, J.D. 1997b. Pith-ing and bark-ing: baboons on the edge of the forest. *Amer. J. Primatol.*, **42** (Abstract), 140.

Paterson, J.D. 1997c. Seed predation patterns in the diet of the Sonso forest baboon troop. *Amer. J. Primatol.*, **42** (Abstract), 140.

Paterson, J.D. 2001. Pest primate status along the southern edge of the Budongo Forest (Abstract). *XVIII Congr. Int. Primatol. Soc.*, Adelaide.

Paterson, J.D. [In press]. Aspects of diet, foraging, and seed predation in Ugandan Forest Baboons. In *Primates of Western Uganda* (ed. N.E. Newton-Fisher, H. Notman, J.D. Paterson and V. Reynolds). Kluwer, New York.

Paterson, J.D. and Teichroeb, J.A. 2001. Phytochemicals and energy: *Khaya* bark consumption by Budongo Forest primates. Budongo Forest Project Report No. 70.

Paterson, J.D., Teichroeb, J.A., Rabatach, L., Fournel, D. and Faccini, P. 2001. Phytochemicals and energy yield for *Khaya* bark consumed by Budongo Forest baboons. Budongo Forest Project Report No. 71.

Pepper, J.W., Mitani, J.C. and Watts, D.P. 1999. General gregariousness and specific social preferences among wild chimpanzees. *Int. J. Primatol.*, **20**, 613–32.

Peterson, R.B. 2001. Conservation—For Whom? A study of immigration onto DR Congo's Ituri Forest Frontier. In *African rain forest ecology and conservation* (ed. W. Weber, L.J.T. White, A. Vedder and L. Naughton-Treves). Yale University Press, New Haven.

Phillips, A. 2003. Factors impacting on the habituation of chimpanzees (*Pan troglodytes schweinfurthii*) in the Budongo Forest, Uganda. M.Sc. thesis, Oxford Brookes University.

Plumptre, A.J. 1995. The importance of 'seed trees' for the natural regeneration of selectively logged tropical forest. *Comm. For. Rev.*, **74**, 253–8.

Plumptre, A.J. 1996. Changes following 60 years of selective timber harvesting in the Budongo Forest Reserve, Uganda. *For. Ecol. Man.*, **89**, 101–13.

Plumptre, A.J. [In press]. The diets, preferences and overlap of the primate community in the Budongo Forest Reserve, Uganda: effects of logging on primate diets. In *Primates of Western Uganda* (ed. N.E. Newton-Fisher, H. Notman, J.D. Paterson and V. Reynolds). Kluwer, New York.

Plumptre, A.J. and Grieser-Johns, A.D. 2001. Changes in primate communities following logging disturbance. In *The cutting edge: conserving wildlife in logged tropical forests* (ed. R.A. Fimbel, A. Grajal and J.G. Robinson), pp. 71–92. Columbia University Press, New York.

Plumptre, A.J. and Reynolds, V. 1994. The effect of selective logging on the primate populations in the Budongo Forest Reserve, Uganda. *J. Appl. Ecol.*, **31**: 631–41.

Plumptre, A.J. and Reynolds, V. 1996. Censusing chimpanzees in the Budongo Forest, Uganda. *Int. J. Primatol.*, **17**, 85–99.

Plumptre, A.J. and Reynolds, V. 1997. Nesting behaviour of chimpanzees: implications for censuses. *Int. J. Primatol.*, **18**, 475–85.

Plumptre, A.J., Reynolds, V. and Bakuneeta, C. 1994. The contribution of fruit eating primates to seed dispersal and natural regeneration after selective logging. Final Report, ODA–FRP Project No. R4738.

Plumptre, A.J., Reynolds, V. and Bakuneeta, C. 1997. The effects of selective logging in monodominant tropical forests on biodiversity. Final Report, ODA–FRP Project No. R6057.

Plumptre, A.J., Reynolds, V. and Bakuneeta, C. 1997. The effects of selective logging in monodominant tropical forests on biodiversity. Final Report, ODA Project R6057, Oxford and London.

Plumptre, A.J., Mugume, S., Cox, D. and Montgomery, C. 2001. Chimpanzee and large mammal survey of Budongo Forest Reserve and the Kibale National Park. Wildlife Conservation Society Report (unpublished).

Plumptre, A.J., Cox, D. and Mugume, S. 2003. The status of chimpanzees in Uganda. Albertine Rift Technical Report Series No. 2, Wildlife Conservation Society.

Power, M. 1991. *The egalitarians, human and chimpanzee: an anthropological view of social organization.* Cambridge University Press, Cambridge, UK.

Preece, G.A. 2001. Factors influencing variation in the population densities of black and white colobus monkeys (*Colobus guereza occidentalis*) within selectively logged forest at the Budongo Reserve, Uganda. M.Sc. thesis, University College of North Wales, Bangor.

Preece, G.A. [In press]. Factors influencing variation in the population densities of *Colobus guereza* within selectively logged forest at the Budongo Reserve: the importance of lianas during a subsistence diet. In *Primates of Western Uganda* (ed. N.E. Newton-Fisher, H. Notman, J.D. Paterson and V. Reynolds). Kluwer, New York.

Pusey, A., Williams, J. and Goodall, J. 1997. The influence of dominance rank on the reproductive success of female chimpanzees. *Science*, **277**, 828–31.

Quiatt, D. 1984. Devious intentions of monkeys and apes? In *The meaning of primate signals* (ed. R. Harré and V. Reynolds), pp. 9–42. Cambridge University Press, Cambridge, UK.

Quiatt, D. 1994. Leaf sponge drinking by a Budongo Forest chimpanzee. *Amer. J. Primatol.*, **33**, 236.

Quiatt, D. [In press]. Instrumental leaf-use by chimpanzees of the Budongo Forest (Sonso community). In *Primates of Western Uganda* (ed. N.E. Newton-Fisher, H. Notman, J.D. Paterson and V. Reynolds). Kluwer, New York.

Quiatt, D. and Reynolds, V. 1993. *Primate behaviour: information, social knowledge, and the evolution of culture.* Cambridge University Press, Cambridge, UK.

Quiatt, D., Reynolds, V. and Stokes, E.J. 2002. Snare injuries to chimpanzees (*Pan troglodytes schweinfurthii*) at 10 study sites in east and west Africa. *Afr. J. Ecol.*, **40**, 303–5.

Quiatt, D., Rutan, B. and Stone, T. 1994. Budongo Forest chimpanzees: composition of feeding groups during the rainy season, with attention to the social integration of disabled individuals (Abstract). *XV Congr. Int. Primatol. Soc.*

Reynolds, V. 1965. *Budongo: a forest and its chimpanzees.* Methuen, London.

Reynolds, V. 1967. *The apes: the gorilla, chimpanzee, orangutan and gibbon—their history and their world*. Cassell, London and E.P. Dutton, New York.

Reynolds, V. 1986. Primate social thinking. In *Primate ontogeny, cognition and social behaviour* (ed. J.G. Else and P.C. Lee). Cambridge University Press, Cambridge, UK.

Reynolds, V. 1992. Chimpanzees in the Budongo Forest, 1962–1992. *J. Zool. Lond.*, **228**, 695–9.

Reynolds, V. 1993a. Sustainable forestry: the case of the Budongo Forest, Uganda. *SWARA*, **16**, 13–17.

Reynolds, V. 1993b. Conservation of chimpanzees in the Budongo Forest Reserve. *Primate Cons.*, **11**, 41–3.

Reynolds, V. 1997/8. Demography of chimpanzees *Pan troglodytes schweinfurthii* in Budongo Forest, Uganda. *Afr. Primates*, **3**, 25–8.

Reynolds, V. [In press]. Threats to chimpanzees of the Budongo Forest. In *Primates of Western Uganda* (ed. N.E. Newton-Fisher, H. Notman, J.D. Paterson and V. Reynolds). Kluwer, New York.

Reynolds, V. and Reynolds, F. 1965. Chimpanzees of the Budongo Forest. In *Primate Behavior: field studies of monkeys and apes* (ed. I. De Vore). Holt, Rinehart and Winston, New York.

Reynolds, V., Plumptre, A.J., Greenham, J. and Harborne, J. 1998. Condensed tannins and sugars in the diet of chimpanzees (*Pan troglodytes schweinfurthii*) in the Budongo Forest, Uganda. *Oecologia*, **115**, 331–6.

Reynolds, V., Wallis, J. and Kyamanywa, R. 2003. Fragments, sugar, and chimpanzees in Masindi District, Western Uganda. In *Primates in fragments* (ed. L. Marsh). Kluwer, New York.

Richard, A.F. and Dewar, R.E. 2001. Politics, negotiation and conservation: a view from Madagascar. In *African rain forest ecology and conservation* (ed. W. Weber, L.J.T. White, A. Vedder and L. Naughton-Treves). Yale University Press, New Haven.

Rukondo, T. 1997. The long term effect of canopy treatment on tree diversity in Budongo Forest: an evaluation of a 25 hectare plot study. M.Sc. thesis, Makerere University, Kampala.

Seabo, G.M., Abraham, Y.B. and Nyombi, K. 2002. The impact of Budongo Forest Project (specifically life trap project) on the rural livelihoods and its contribution to chimpanzee conservation. Field Study Report, Makerere University, Kampala.

Seraphin, S.B. 2000. The reproductive ecology and stress physiology of free-ranging male chimpanzees in Budongo Forest, Uganda. M.Sc. thesis, Oxford University, Oxford.

Seraphin, S.B., Whitten, P. and Reynolds, V. 2001. Reproductive challenge and competition stress in the life of male chimpanzees (*Pan troglodytes schweinfurthii*) free-ranging in Budongo Forest, Uganda. *Amer. J. Primatol.*, **54**, 70.

Seyfarth, R.M., Cheney, D.L. and Marler, P. 1980. Vervet monkey alarm calls: semantic communication in a free-ranging primate. *Anim. Behav.*, **28**, 1070–94.

Sheppard, D.J. 2000. Ecology of the Budongo Forest redtail: patterns of habitat use and population density in primary and regenerating forest sites. M.Sc. thesis, University of Calgary.

Sheppard, D.J. and Paterson, J.D. 2001a. *Cercopithecus ascanius*: logged and unlogged diet and ranging comparisons in the Budongo Forest, Uganda (Abstract). *XVIII Congr. Int. Primatol Soc.*, Adelaide.

Sheppard, D.J. and Paterson, J.D. 2001*b*. *Cercopithecus ascanius*: logged and unlogged diet and ranging comparisons in the Budongo Forest, Uganda. Typescript of paper to the International Primatology Society, Adelaide. Budongo Forest Project Report No. 68.

Shiel, D. 1996. The ecology of long term change in a Ugandan rain forest. D.Phil. thesis, University of Oxford, Oxford.

Shiel, D. 1999a. Developing tests of successional hypotheses with size-structured populations, and an assessment using long-term data from a Ugandan rain forest. *Plant Ecol.*, **140**, 117–127.

Shiel, D. 1999b. Tropical forest diversity, environmental change and species augmentation: after the intermediate disturbance hypothesis. *J. Veg. Sci.*, **10**, 851–60.

Shiel, D. 2001. Long-term observations of rain forest succession, tree diversity and responses to disturbance. *Plant Ecol.*, **155**, 183–99.

Shiel, D. 2003. Observations of long-term change in an African rain forest. In *Long-term changes in tropical tree diversity as a result of natural and man-made disturbances* (ed. H. ter Steege). Tropenbos Series 22, Wageningen.

Shiel, D. and Burslem, D.F.R.P. 2003. Disturbing hypothesis in tropical forests. *Trends Ecol. Evol.*, **18**, 18–26.

Shimizu, K., Douke, C., Fujita, S., Matsuzawa, T., Tomonaga, M., Tanaka, M., Matsubayashi, K. and Hayashi, M. 2003. Urinary steroids, FSH and CG measurements for monitoring the ovarian cycle and pregnancy in the chimpanzee. *J. Med. Primatol.*, **32**, 15–22.

Silk, J.B. 1978. Patterns of plant food sharing among mother and infant chimpanzees at Gombe National Park, Tanzania. *Folia Primatol.*, **29**, 1129–41.

Singer, B. 2002. The road to sustainable mahogany trade in Uganda. M.Sc. thesis, University College, London University.

Singh, S. 1999. An investigation of nocturnal mating tactics in the chimpanzee using good gene sexual selection theory and sperm competition theory. M.Sc. thesis, Oxford University, Oxford.

Slocombe, K.E. and Zuberbühler, K. 2005. Agonistic screams in wild chimpanzees vary as a function of social role. *J. Comp. Psychol.*, **119**, 67–77.

Smith, R.J. 1995. Some effects of limb injuries on the chimpanzees of the Budongo Forest. Budongo Forest Project Report No. 29.

Spini, L. 1998. The frequency and significance of social grooming in the Sonso community of chimpanzees at the Budongo Forest, Uganda. M.Sc. thesis, Oxford University, Oxford.

Ssuna, R. 2002. Report on the investigation of a trapped chimpanzee in Kasongoire Forest. Unpublished Report.

Stanford, C.B. 1998. *Chimpanzee and red colobus: the ecology of predator and prey*. Harvard University Press, Cambridge, Mass.

Stanford, C.B., Wallis, J., Mpongo, E. and Goodall, J. 1994. Hunting decisions in wild chimpanzees. *Behaviour*, **131**, 1–18.

Stokes, E.J. 1999. *Feeding skills and the effect of injury on wild chimpanzees*. Ph.D. thesis, University of St Andrews, St Andrews, UK.

Stokes, E.J. and Byrne, R.W. 2001. Cognitive capacities for behavioural flexibility in wild chimpanzees (*Pan troglodytes*): the effect of snare injury on complex manual food processing. *Anim. Cogn.*, **4**, 11–28.

Sugiyama, Y. 1968. Social organization of chimpanzees in the Budongo Forest, Uganda. *Primates*, **9**, 225–58.

Sutton, R., Bowes-Lyon, D., Prince, R. and Fergusson, J. 1996. A study of health care in Masindi District, Western Uganda. Budongo Forest Project Report No. 31.

Suzuki, A. 1971. Carnivority and cannibalism observed among forest-living chimpanzees. *J. Anthropol. Soc. Nippon*, **79**, 30–48.

Suzuki, A. 1972. On the problems of conservation of the chimpanzees in East Africa and of the preservation of their environment. *Primates*, **12**, 415–18.

Synnott, T.J. 1972. Requiem for a chimp. *Africana*, **4** (10), 10–11.

Synnott, T.J. 1975. Factors affecting the regeneration and growth of seedlings of *Entandrophragma utile* (Dawe and Sprague) Sprague. Ph.D. thesis, Makerere University, Kampala.

Synnott, T.J. 1985. A checklist of the flora of Budongo Forest Reserve, Uganda, with notes on ecology and phenology. *CFI Occasional Papers* No. 27.

Takahata, Y. 1990*a*. Adult males' social relations with adult females. In *The chimpanzees of the Mahale Mountains: sexual and life history strategies* (ed. T. Nishida). University of Tokyo Press, Tokyo.

Takahata, Y. 1990*b*. Social relationships among adult males. In *The chimpanzees of the Mahale Mountains: sexual and life history strategies* (ed. T. Nishida). University of Tokyo Press, Tokyo.

Teleki, G. 1973. *The predatory behavior of wild chimpanzees*. Bucknell University Press, Lewisburg.

Thalmann, O., Hebler, H., Poinar, S., Pääbo S. and Vigilant L. 2004. Unreliable mtDNA data due to nuclear insertions: a cautionary tale from analysis of humans and other great apes. *Molec. Ecol.*, **13**, 321–35.

Tinka, J. and Reynolds, V. 1997. Budongo Forest chimpanzee grooms a redtail monkey. *Pan Africa News*, **4**, 6.

Tumusiime, D. 2002. An assessment of snaring in the Budongo Forest. B.Sc. thesis, Makerere University Faculty of Forestry and Nature Conservation.

Tutin, C.E.G. 1979. Mating patterns and reproductive strategies in a community of wild chimpanzees (*Pan troglodytes schweinfurthii*). *Behav. Ecol. Sociobiol.*, **6**, 39–48.

Tutin, C.E.G. and Fernandez, M. 1991. Responses of wild chimpanzees and gorillas to the arrival of primatologists: behaviour observed during habituation. In *Primate responses to environmental change* (ed. H. Box), pp. 187–97. Chapman and Hall, New York.

Tweheyo, M. 2000. The importance of figs to primates in the Budongo Forest Reserve, Uganda. M.Sc. thesis, Makerere University, Kampala.

Tweheyo, M. 2003. Abundance, distribution and phenology of chimpanzee food in the Budongo Forest Reserve, Uganda. Ph.D. thesis, Agricultural University of Norway.

Tweheyo, M. and Lye, A.K. 2003. Phenology of figs in Budongo Forest Uganda and its importance for the chimpanzee diet. *Afr. J. Ecol.*, **31**, 306–16.

Tweheyo, M. and Obua, J. 2001. Feeding habits of chimpanzees (*Pan troglodytes*), red-tail monkeys (*Cercopithecus ascanius schmidti*) and blue monkeys (*C. mitis stuhlmanni*) on figs in Budongo Forest Reserve, Uganda. *Afr. J. Ecol.*, **39**, 133–9.

Tweheyo, M., Hill, C.M. and Obua, J. [In press]. Patterns of crop raiding by primates around Budongo Forest Reserve, Uganda. *Wildlife Biol.*

Tweheyo, M., Lye, A.K. and Weladji, B.R. 2004. Chimpanzee diet and habitat selection in the Budongo Forest Reserve, Uganda. *For. Ecol. Manage.*, **188**, 267–78.

Tweheyo, M., Reynolds, V., Huffman, M.A., Pebsworth, P., Goto, S., Mahaney, W.C., Milner, M., Waddell, A., Dirszowsky, R. and Hancock, R.G.V. [In press]. Geophagy in chimpanzees (*Pan troglodytes schweinfurthii*) of the Budongo Forest Reserve, Uganda—a multidisciplinary study. In *Primates of Western Uganda* (ed. N.E. Newton-Fisher, H. Notman, J.D. Paterson and V. Reynolds). Kluwer, New York.

Uganda Wildlife Authority, 2003. *Conservation for Uganda's chimpanzees 2003–2008*. Wildlife Conservation Society and Jane Goodall Institute.

Van Elsacker, L. and Verheyen, R.F. 1995. A review of terminology on aggregation patterns in bonobos (*Pan paniscus*). *Int. J. Primatol.*, **16**, 37–52.

Vigilant, L. 2002. Technical challenges in the microsatellite genotyping of a wild chimpanzee population using feces. *Evol. Anthropol.* (Suppl. 1), 162–5.

Vigilant, L., Hofreiter, M., Siedel, H. and Boesch, C. 2001. Paternity and relatedness in wild chimpanzee communities. *Proc. Nat. Acad. Sci.*, **98**, 12890–5.

Vigilant, L. Lukas, D., Immel, U.-D., Reynolds, V. and Boesch, C. 2003. Dispersal and relatedness in wild chimpanzees (*Pan troglodytes*). *Folia Primatol.*, **74**, 227.

Vigilant, L., Lukas, D., Immel, U.-D., Reynolds, V. and Boesch, C. [In press]. Relatedness in a male-philopatric primate.

Walaga, C. 1993. Species distribution and the development of the climax vegetation of the Budongo Forest Reserve, Uganda. M.Sc. thesis, Makerere University, Kampala.

Wales, Z. 2000. *What we know about Budongo Forest*. Budongo Forest Project Report No. 81.

Waller, J.C. and Reynolds, V. 2001. Limb injuries resulting from snares and traps in chimpanzees (*Pan troglodytes schweinfurthii*) of the Budongo Forest, Uganda. *Primates*, **42**, 135–9.

Wallis, J. 1985. Synchrony of estrous swelling in captive group-living chimpanzees (*Pan troglodytes*). *Int. J. Primatol.*, **6** (4), 335–50.

Wallis, J. 1992*a*. Chimpanzee genital swelling and its role in the pattern of sociosexual behavior. *Amer. J. Primatol.*, **28**, 101–13.

Wallis, J. 1992*b*. Socioenvironmental effects on timing of first postpartum cycles in chimpanzees. In *Topics in primatology, Vol 1: Human origins* (ed. T. Nishida, W.C. McGrew, P. Marler, M. Pickford and F.B.M. de Waal), pp. 119–30. University of Tokyo Press, Tokyo.

Wallis, J. 1994. Socioenvironmental effects on first full anogenital swellings in adolescent female chimpanzees. In *Current Primatology, Vol 2: Social Development, Learning and Behavior* (ed. J.J. Roeder, B. Thierry, J.R. Anderson and N. Herrenschmidt), pp. 25–32. University of Louis Pasteur, Strasbourg.

Wallis, J. 2002*a*. Seasonal aspects of reproduction and sexual behavior in two chimpanzee populations: a comparison of Gombe (Tanzania) and Budongo (Uganda). In *Behavioural diversity in chimpanzees and bonobos* (ed. C. Boesch, G. Hohmann and L.F. Marchant), pp. 181–91. Cambridge, University Press, Cambridge, UK.

Wallis, J. 2002*b*. Monitoring the behavioral ecology and viability of forest fragment chimpanzees, Masindi District, Uganda. *Amer. Soc. Primatol. Bull.*, **26** (1), 9.

Wallis, J. and Goodall, J. 1993. Genital swelling patterns of pregnant chimpanzees in Gombe National Park. *Amer. J. Primatol.*, **31**, 89–98.

Wallis, J. and Lee, D.R. 1999. Primate conservation: the prevention of disease transmission. *Int. J. Primatol.*, **20**, 803–26.

Wallis, J. and Reynolds, V. 1999. Seasonal aspects of sociosexual behavior in two chimpanzee populations: a comparison of Gombe (Tanzania) and Budongo (Uganda) (Abstract). *Amer. J. Primatol.*, **49**, 111.

Wathyso, M. 2000. Economic impact of ecotourism on local rural communities: a case study of Budongo Forest Ecotourism Project. M.Sc. thesis, Makerere University, Kampala.

Watkins, C. 2001. The implications of human behavioral ecology for chimpanzee conservation in Budongo Forest, Uganda. Masters thesis, California State University, Fullerton, CA.

Watkins, C. [In press]. The implications of perception, value, and an evolved psychology of selfishness for chimpanzee conservation in Budongo Forest, Uganda. In *Primates of Western Uganda* (ed. N.E. Newton-Fisher, H. Notman, J.D. Paterson and V. Reynolds). Kluwer, New York.

Webb, H.R. 1948. Mahogany regeneration in the Budongo Forest, Uganda. *E. Afr. Agric. J.*, January.

White, L.J.T. and Tutin, C.E.G. 2001. Why chimpanzees and gorillas respond differently to logging. In *African rain forest: ecology and conservation* (ed. W. Weber, L.J.T. White, A. Vedder and L. Naughton-Treves). Chapter 26. Yale University Press, New Haven.

Whiten, A. and Byrne, R.W. 1988. Tactical deception in primates. *Behav. Brain Sci.*, **11**, 233–73.

Whiten, A., Horner, V. and Marshall-Pescini, S. 2003. Cultural panthropology. *Evol. Anthropol.*, **12**, 92–105.

Whiten, A., Goodall, J., McGrew, W.C., Nishida, T., Reynolds, V., Sugiyama, Y., Tutin, C.E.G., Wrangham, R.W. and Boesch, C. 1999. Cultures in chimpanzees. *Nature*, **399**, 682–5.

Whiten, A., Goodall, J., McGrew, W.C., Nishida, T., Reynolds, V., Sugiyama, Y., Tutin, C.E.G., Wrangham, R.W. and Boesch, C. 2001. Charting cultural variation in chimpanzees. *Behaviour*, **138**, 1481–516.

Willan, R. 1989. Management of tropical moist forests in Africa. FAO Forestry Paper 88, p. 89. FAO, Rome.

Wilson, M.L. and Wrangham, R.W. 2003. Intergroup relations in chimpanzees. *Ann. Rev. Anthropol.*, **32**, 363–92.

Wong, J. 1995. An acoustic analysis of the inter-population variability in the pant-hoot vocalizations of male chimpanzees. M.Sc. thesis, Oxford University, Oxford.

Woodford, M.H., Butynski, T.M. and Karesh, W.B. 2002. Habituating the great apes: the disease risks. *Oryx*, **36** (2), 153–60.

Wrangham, R.W. 1977. Feeding behaviour of chimpanzees in Gombe National Park, Tanzania. In *Primate ecology* (ed. T. H. Clutton-Brock), pp. 504–38. Academic Press, London.

Wrangham, R.W. 1979. Sex differences in chimpanzee dispersion. In *The great apes* (ed. D.A. Hamburg and E.R. McCown), pp. 481–90. Benjamin Cummings, Menlo Park, CA.

Wrangham, R.W. 2000*a*. Why are male chimpanzees more gregarious than mothers? A scramble competition hypothesis. In *Male primates* (ed. P. Kappeler), pp. 248–58. Cambridge University Press, Cambridge, UK.

Wrangham, R.W. 2000*b*. Kibale snare removal program 1997–2000: Interim Report. Unpublished report to the Jane Goodall Institute, Uganda Wildlife Education Centre, Uganda.

Wrangham, R.W. and Goldberg, T. 1997. An overview of chimpanzee conservation and management strategies. In *Conserving the chimpanzees of Uganda: a population and habitat viability assessment for* Pan troglodytes schweinfurthii (ed. E. Edroma, N. Rosen and P.S. Miller). IUCN/SSC, Apple Valley, MN.

Wrangham, R.W., Clark, A.P. and Isabirye-Basuta, G. 1992. Female social relationships and social organization of Kibale Forest chimpanzees. In *Topics in primatology, Vol. 1: Human origins* (ed. T. Nishida, W.C. McGrew, P. Marler, M. Pickford and F.B.M. de Waal), pp. 81–98. University of Tokyo Press, Tokyo.

Wrangham, R.W., Conklin, N.L., Etot, G., Obua, J., Hunt, K.D., Hauser, M.D. and Clark, A.P. 1993. The value of figs to chimpanzees. *Int. J. Primatol.*, **14**, 243–56.

Zihlmann, A., Bolter, D. and Boesch, C. 2004. Wild chimpanzee dentition and its implications for assessing life history in immature hominin fossils. *Proc. Nat. Acad. Sci.*, **101**, 10541–3.

Index

DATE DUE